Up and Running with DAX for Power BI

A Concise Guide for Non-Technical Users

Alison Box

Apress®

Up and Running with DAX for Power BI: A Concise Guide for Non-Technical Users

Alison Box
Billingshurst, West Sussex, UK

ISBN-13 (pbk): 978-1-4842-8187-1 ISBN-13 (electronic): 978-1-4842-8188-8
https://doi.org/10.1007/978-1-4842-8188-8

Copyright © 2022 by Alison Box

This work is subject to copyright. All rights are reserved by the Publisher, whether the whole or part of the material is concerned, specifically the rights of translation, reprinting, reuse of illustrations, recitation, broadcasting, reproduction on microfilms or in any other physical way, and transmission or information storage and retrieval, electronic adaptation, computer software, or by similar or dissimilar methodology now known or hereafter developed.

Trademarked names, logos, and images may appear in this book. Rather than use a trademark symbol with every occurrence of a trademarked name, logo, or image we use the names, logos, and images only in an editorial fashion and to the benefit of the trademark owner, with no intention of infringement of the trademark.

The use in this publication of trade names, trademarks, service marks, and similar terms, even if they are not identified as such, is not to be taken as an expression of opinion as to whether or not they are subject to proprietary rights.

While the advice and information in this book are believed to be true and accurate at the date of publication, neither the authors nor the editors nor the publisher can accept any legal responsibility for any errors or omissions that may be made. The publisher makes no warranty, express or implied, with respect to the material contained herein.

Managing Director, Apress Media LLC: Welmoed Spahr
Acquisitions Editor: Joan Murray
Development Editor: Laura Berendson
Coordinating Editor: Jill Balzano

Cover image designed by Freepik (www.freepik.com)

Distributed to the book trade worldwide by Springer Science+Business Media LLC, 1 New York Plaza, Suite 4600, New York, NY 10004. Phone 1-800-SPRINGER, fax (201) 348-4505, e-mail orders-ny@springer-sbm.com, or visit www.springeronline.com. Apress Media, LLC is a California LLC and the sole member (owner) is Springer Science + Business Media Finance Inc (SSBM Finance Inc). SSBM Finance Inc is a **Delaware** corporation.

For information on translations, please e-mail booktranslations@springernature.com; for reprint, paperback, or audio rights, please e-mail bookpermissions@springernature.com.

Apress titles may be purchased in bulk for academic, corporate, or promotional use. eBook versions and licenses are also available for most titles. For more information, reference our Print and eBook Bulk Sales web page at http://www.apress.com/bulk-sales.

Any source code or other supplementary material referenced by the author in this book is available to readers on GitHub.

Printed on acid-free paper

To Madeleine, John, and Alan

Table of Contents

About the Author xi

About the Technical Reviewer xiii

Acknowledgments xv

Introduction xvii

Chapter 1: Show Me the Data 1

Stars and Snowflakes 4

Fact Tables 4

Dimensions 5

The Star Schema 5

Finding Nonmatching Values 7

Chapter 2: DAX Objects, Syntax, and Formatting 15

DAX Syntax 16

DAX Formatting 19

Chapter 3: Calculated Columns and Measures 23

Calculated Columns 23

Creating Simple Calculated Columns 24

Looking at the RELATED Function 26

DAX Measures 32

Implicit Measures 32

Explicit Measures 34

Creating a Measures Table 35

Creating Simple DAX Measures 35

What Exactly Is a Measure? 41

Chapter 4: Evaluation Context 47
The Filter Context 47
Evaluations Using a Single Filter 49
Calculation in the Total Row 52
Evaluations Using Multiple Filters 53
The Row Context 56
Chapter 5: Iterators 59
The SUMX Function (and Other "X" Functions) 61
Total Row Grief 66
Chapter 6: The CALCULATE Function 71
Why You Need CALCULATE 71
Using Single Filters 77
Using Multiple Filters 78
AND and OR Filters 79
Complex Filters 81
Chapter 7: DAX Table Functions 85
Types of DAX Functions 85
Table Functions 87
Examples of Table Expressions 88
Why Do We Need Table Expressions? 89
The FILTER Function 89
FILTER Used to Reduce Rows 90
FILTER as the Filter Argument of CALCULATE 91
Column Filters vs. Table Filters 99
Table Filters Are Less Efficient 100
Table Filters Return Different Results 104
Using the KEEPFILTERS Function 108

Chapter 8: The ALL Function and All Its Variations 109
The ALL Function........ 110
Applied to the Fact Table........ 111
Using ALL on Dimension Tables........ 117
Using ALL on a Column........ 120
The ALLEXCEPT Function........ 127
The ALLSELECTED Function........ 129
ALL as a Modifier to CALCULATE........ 131
Chapter 9: Calculations on Dates: Using DAX Time Intelligence........ 143
Power BI Date Hierarchies........ 144
Creating a Date Table........ 146
Using Time Intelligence Functions........ 149
Previous Month/Year – PREVIOUSMONTH/YEAR........ 153
Same Period Last Year – SAMEPERIODLASTYEAR........ 153
Values for Any Time Ago – DATEADD........ 154
Year to Date – DATESYTD........ 154
Total to Date or Cumulative Totals........ 155
Rolling Annual Totals and Averages........ 156
Calculating the Last Transaction Date and the Last Transaction Value........ 158
Finding the Difference Between Two Dates........ 162
Chapter 10: Empty Values vs. Zero........ 165
The BLANK() Function........ 165
The ISBLANK Function........ 168
Testing for Zero........ 168
Using Measures to Find Blanks and Zero........ 169
Using the COALESCE Function........ 171
Chapter 11: Using Variables: Making Our Code More Readable........ 173
Improved Performance........ 174
Improved Readability........ 176

Reduced Complexity 177
Variables As Constants 178
Chapter 12: Returning Values in the Current Filter 183
The SELECTEDVALUE Function 184
The CONCATENATEX Function 189
Using Parameter Tables 195
The Values Function 199
A Table or a Scalar Function? 200
Replacing "Lost Filters" 205
Converting Columns to Tables 207
Chapter 13: Controlling the Direction of Filter Propagation 209
Programming Bidirectional Filters 210
Why You Should Never Use Bidirectional Relationships 213
Chapter 14: Working with Multiple Relationships Between Tables 217
Activating Inactive Relationships 219
Comparing Values in the Same Column 221
Chapter 15: Understanding Context Transition 227
Overview of DAX Evaluations Contexts 228
Row Context Revisited 228
Filter Context Revisited 229
How Row Context Becomes Filter Context 229
How Context Transition Can Return "Surprising Results" 237
Filters Using AVERAGE 238
Filters Using MAX 242
Filters Using Measures 247
Aggregating Totals Using Context Transition 251
Aggregating in Dimensions 252
Aggregating in Virtual Tables 259

Chapter 16: Leveraging Context Transition.......... 271

Ranking Data: Looking at RANKX.......... 272

Binning Measures into Numeric Ranges.......... 275

Calculating TopN Percent.......... 279

Create the Slicers.......... 280

Create the Measure to Find the Top or Bottom Percent Selected in Slicers.......... 281

Calculating "Like for Like" Yearly Sales Using SUMMARIZE.......... 285

Using Context Transition in Calculated Columns.......... 293

Calculating Running Totals.......... 293

Calculating the Difference from the Value in the Previous Row.......... 294

Chapter 17: Virtual Relationships: The LOOKUPVALUE and TREATAS Functions... 297

LOOKUPVALUE Function.......... 298

The TREATAS Function.......... 302

Chapter 18: Table Expansion.......... 311

Revisiting Filters.......... 313

Column Filters Revisited.......... 313

The ALL Function Revisited.......... 317

Expanded Tables Explained.......... 318

Leveraging Expanded Tables.......... 323

"Reaching" Dimensions.......... 324

Table Expansion vs. CROSSFILTER.......... 333

Using Snowflake Schemas.......... 337

Chapter 19: The CALCULATETABLE Function.......... 343

CALCULATETABLE vs. FILTER.......... 344

CALCULATETABLE and Table Expansion.......... 349

Calculating "New" Entities.......... 350

Calculating "Returning" Entities.......... 356

Index.......... 361

About the Author

Alison Box is a Director of Burningsuit Ltd (`www.burningsuit.co.uk`) and an IT trainer and consultant with over 30 years experience of delivering computer applications training to all skill levels, from basic users to advanced technical experts. Currently, she specializes in delivering training in Microsoft Power BI Service and Desktop, Data Modeling, DAX (Data Analysis Expressions), and Excel. Alison also works with organizations as a DAX and Data Analysis consultant. She was one of the first Excel trainers to move into delivering courses in Power Pivot and DAX, from where Power BI was born. Part of her job entails promoting Burningsuit as a knowledge base for Power BI and includes writing regular blog posts on all aspects of Power BI that are published on her website.

About the Technical Reviewer

A native to Northern Indiana, **Jake Halsey** has over a decade of experience working with various products, services, and development tools in the IT industry. Working in the Fort Wayne and Chicago areas as a senior-level software developer and application administrator, he regularly performs complex data analysis and prepares professional reports. He's particularly excited about his work on this book as it has enabled him to add Power BI and DAX to his own list of tools to prepare effective data visualizations and has personally found the examples created by Alison Box to be realistic, practical, and accessible to readers getting started in their journey with Power BI.

Acknowledgments

Writing a book can often be a lonely experience, but a book can only come to fruition with help from outside. I would like to acknowledge and thank those people around me, both professional and personal, that have been instrumental in assisting me in writing this book. Firstly, many thanks to Jake Halsey, my technical reviewer, for his invaluable and encouraging comments and his thorough review of the many DAX examples and listings. I would also like to thank Joan Murray, the Acquisitions Editor at Apress, who, on receiving the original manuscript of the entire book, agreed on the benefit of publishing a book on DAX that had a non-technical focus. I'm also grateful to my Coordinating Editor, Jill Balzano, for her professional approach that makes working with Apress a pleasurable experience. Last but not least, my heartfelt and enduring thanks to my family for their consistent support and encouragement, without whom I would have found it hard to see this book through.

Introduction

Up and Running with DAX for Power BI is a condensed self-teaching resource for learning DAX inside Power BI Desktop. DAX (Data Analysis Expressions) is the formula language of Microsoft Power BI and was first introduced in 2009 as the programming language of the Excel add-in, Power Pivot, from which Power BI was born. With the ever-increasing adoption of Power BI as the preferred data analytics platform, the ability to use DAX is fast becoming a necessary requirement to find and share the important insights into your data. This book is a concise guide for non-technical users that focuses on the core concepts that underpin this language, taking you from zero knowledge to being able to use DAX to write the challenging calculations that are often necessary for reporting on your data.

If you need to use DAX, there is quite a lot of help out there: books, videos, and experts with a lot of opinions and copious examples of mind-boggling DAX code that, to use, you can simply copy and paste without ever understanding how it works. Yet even with the help of these resources, the DAX mantra continues: "DAX is difficult"! But this is a misconception, and it's the first barrier to learning DAX that you will encounter. Although there is no doubt that DAX can often be challenging to understand, labelling it "difficult" might appear as an excuse for those people who haven't made the effort to understand what goes on under the hood.

When you have shaken off the misconception that DAX is difficult and decided you want to understand how DAX works, currently, there are two hurdles you will face, both of which this book tackles. Firstly, many resources have been written specifically with the DAX developer or other highly skilled technicians in mind. However, the intended audience for this book is either Excel users or people with no technical or coding background. In fact, it's aimed at someone probably just like you who simply wants to get on with their day job while still becoming a competent user of DAX. In this book, little technical knowledge is assumed. Difficult concepts are explained with easy-to-follow examples that everyone can understand, and the content is structured to gradually build up confidence in working with DAX. The second obstacle you will encounter is that most books on DAX can be considered as "reference works." For example,

The Definitive Guide to DAX[1] comprises over 700 pages covering every aspect of DAX in meticulous detail. You may feel that using such works as "teach yourself" resources is a daunting prospect because the abundance of information fast becomes overwhelming. To get up and running with DAX, it's not necessary to wade through copious pages on rarefied DAX functions and the technical aspects of the language. There are just a few mandatory concepts that must be fully understood before DAX can be mastered, and it's on these concepts that this book focuses. You will also probably want to learn DAX from something more easily consumable and less of an investment in your time. This is why I felt there was a need for a more concise approach to explaining the DAX language.

To get the most from the information contained in this book, being a competent user of Power BI Desktop will be an advantage. This includes the ability to create data models and generate reports using Power BI's data visualizations. However, where specific knowledge of these areas is required, I have provided links to the relevant information for you to self-explore. You will find that within each successive chapter, the book builds on the knowledge gained and the skills learned, and by the final chapters, you will have acquired the necessary understanding of DAX to author complex calculations.

In Chapters 1 to 3, we cover the precursor knowledge that's required before you can begin to author DAX expressions, such as understanding the structure of your data model and using DAX syntax. You will then be able to create some basic calculated columns and measures. You will find that up to this point, DAX is definitely easy! It's then in Chapter 4 that we broach the first major DAX concept, which is the evaluation context. Here, we look at the distinct ways in which calculated columns and measures are calculated. We then move you on in Chapter 5 to the second important concept, understanding iterators, where calculations are performed on each row of a table, just as you would copy down on Excel formulas.

You will take a big leap forward in your understanding of DAX in Chapter 6, where you meet the most important of all DAX functions, CALCULATE. It's at this juncture that you will start to use DAX as a programming language, where the outcomes of your expressions happen in memory. At this point too, DAX veers well away from Excel conceptionally, and you will begin to author more powerful calculations than the simple sums and averages of basic measures.

In Chapter 7, we explore the idea of table expressions that are used to generate in-memory virtual tables. As you move into more advanced areas of DAX, you will start

[1] Marco Russo and Alberto Ferrari (2020), *The Definitive Guide to DAX*, 2nd ed, [Microsoft Press]

to appreciate that most DAX expressions involve generating these virtual tables. These are typically subsets of real tables and are used to programmatically filter the data in preparation for aggregations via the DAX measure. At this point, you may feel that DAX is definitely getting a little more challenging. This is because you can't see virtual tables, you just have to imagine them, and the inner workings of expressions are mostly hidden from us. Once you have completed Chapter 8 where we take a detailed look at the ALL function that, along with CALCULATE, comprises most DAX expressions, you are now ready to use DAX to solve a wide variety of data analysis scenarios. For example, in Chapter 9, you will learn to compare data over time periods, and in Chapter 12, you are taken through the creation of user-driven calculations using parameter tables. In Chapter 14, you will discover how to make dynamic comparisons across categories of data, such as finding which customers who bought product "X" also bought product "Y".

Chapter 15 will bring you to the most challenging of all DAX concepts to understand. This is the concept of context transition where you will learn to perform aggregations at higher granularities. Once you have mastered the use of this concept, the list of data insights you can now uncover greatly increases. You will be able to rank customers or products by sales, bin totals into numeric ranges, dynamically find top or bottom percent by value, and find the average total sales over years, quarters, and months. In fact, most DAX calculations you author will use context transition in some way.

It may seem odd that it's not until you are almost at the end of your journey through DAX that we tell you at last how DAX really works and how it all fits together. The reason for this is that it's not until you reach Chapter 18 that you will have the skills to understand the last DAX concept, that of table expansion. Although this concept is mostly theoretical, once you know how the data model functions behind the scenes when your expressions are evaluated, the knowledge you have gained throughout this book will now all fall into place. In Chapter 18, finally, all the pieces of the DAX jigsaw fit together, and you are now a fully fledged DAX expert.

Finally, on a personal note, I've written the book that I wish had been around when I was first learning DAX, which was back in the days when Power Pivot was first launched. There was very little to help me, and I've never forgotten the many hours of deciphering DAX code that it took me to get to the position of thinking "yes, I can do this!" I'm hoping that, with the help of this book, it will be an easier journey for you and that this book will be a useful resource as DAX becomes as mainstream as Excel formulas.

Let's not lose sight either of the objective of learning DAX, which is not an end in its own right. It's not so you can impress your colleagues by showing off your skills in writing copious lines of DAX code. No, the objective of learning DAX is as a means to an end. It is to enable you to analyze your data in ways that give you those insights that up to now you've been struggling to find.

"The goal is to turn data into information, and information into insight."[2]

If you want to follow along with the examples we've used in this book, these are the files you will need:

Chapters 1 to 9	-	1 DAX Sample Data.pbix
Chapter 10	-	2 DAX Blanks & Zeros.pbix
Chapters 11 to 13	-	1 DAX Sample Data.pbix
Chapter 14	-	3 DAX USERELATIONSHIP.pbix
Chapters 15 and 16	-	1 DAX Sample Data.pbix
Chapter 17	-	4 DAX LOOKUPVALUE.pbix
		5 DAX TREATAS.pbix
Chapters 18 and 19	-	6 Expanded Tables.pbix

[2] Carly Fiorina, former president and chair of Hewlett-Packard Co, "Information: the currency of the digital age," Oracle OpenWorld, San Francisco, December 6, 2004

CHAPTER 1

Show Me the Data

The key to understanding DAX is getting to grips with the challenging concepts that underpin the expressions. Most DAX expressions you'll write will amount to only a few lines of code, but it's what goes on under the hood that is the secret to understanding their evaluation. For example, take this DAX expression:

=MAXX (Customers, [Total Sales])

It comprises a function, a table name, and a measure name. It should be simple to understand. However, to unravel the calculation behind this expression, you would need to have a firm grasp of the following concepts: row context, filter context, iterators, and context transition. With DAX, the devil is definitely in the detail. This is why you can't just copy and paste other people's expressions, hack them around, and hope for the best that they'll work. You'll find it difficult to learn DAX using this approach. You must concentrate on the core principles of the function language. You'll find that DAX becomes less difficult to understand if you simply pay attention to the detail.

However, before we can start writing code, we must begin our journey into the language of DAX with the mandatory preparatory work.

Note If you would like a detailed explanation of the DAX language and when it first appeared, its history is here: *https://en.wikipedia.org/wiki/Data_analysis_expressions*.

It would, for instance, be impossible to create the correct DAX expressions without understanding the structure and shape of the data that lies beneath. This is because the construct of your expressions will depend directly on the arrangement of the tables in your data model. This is why any DAX expert will say to you "show me the data" before they attempt to write the relevant DAX code.

© Alison Box 2022
A. Box, *Up and Running with DAX for Power BI*, https://doi.org/10.1007/978-1-4842-8188-8_1

Therefore, in this chapter, you will familiarize yourself with the data we will be using in our DAX examples throughout this book, and we will pay particular attention to the structure of this data. You will learn the various terms that are used to describe the constituent parts and the major precepts that underpin the structure. Only when you understand these principles can you move on to author DAX code.

Our sample data[1] comprises a fictitious sales scenario and what better product to sell than wine (perhaps a more attractive prospect than selling cycles or electrical equipment).[2] In everything that follows in this book, you must imagine that you're engaged in selling this product, and by using DAX to analyze your sales through the metrics that matter to you, you'll gain insights into your data that can help drive successful business results and profitability.

Note I appreciate that your data may not be sales related. However, our wine sales data is generic data. It comprises the names of entities, numbers, and dates, and your data will be no different from this.

We've imported six tables into Power BI Desktop as follows:

Winesales - Records our sales transactions.

Wines - Records the names and details of the wines we sell.

Customers - Who we sell our wines to.

SalesPeople - The people making the sales.

Regions - Our customers are grouped into these regions.

DateTable - Records every date, starting from the first day of the month when sales start and ending with the last date in the current financial year, categorizing these dates into year, quarter, and month.

Note As we'll discover later, it's simpler to have single-word table names, and that's why we've named the tables "Winesales", "SalesPeople", and "DateTable".

---

[1] To follow along with the examples, use the Power BI Desktop file "1 DAX Sample Data.pbix".

[2] This is a reference to the ubiquitous sample data, "AdventureWorks" and "Contoso Corporation" used by many books on DAX

Our tables are related in many-to-one relationships as shown in Figure 1-1.

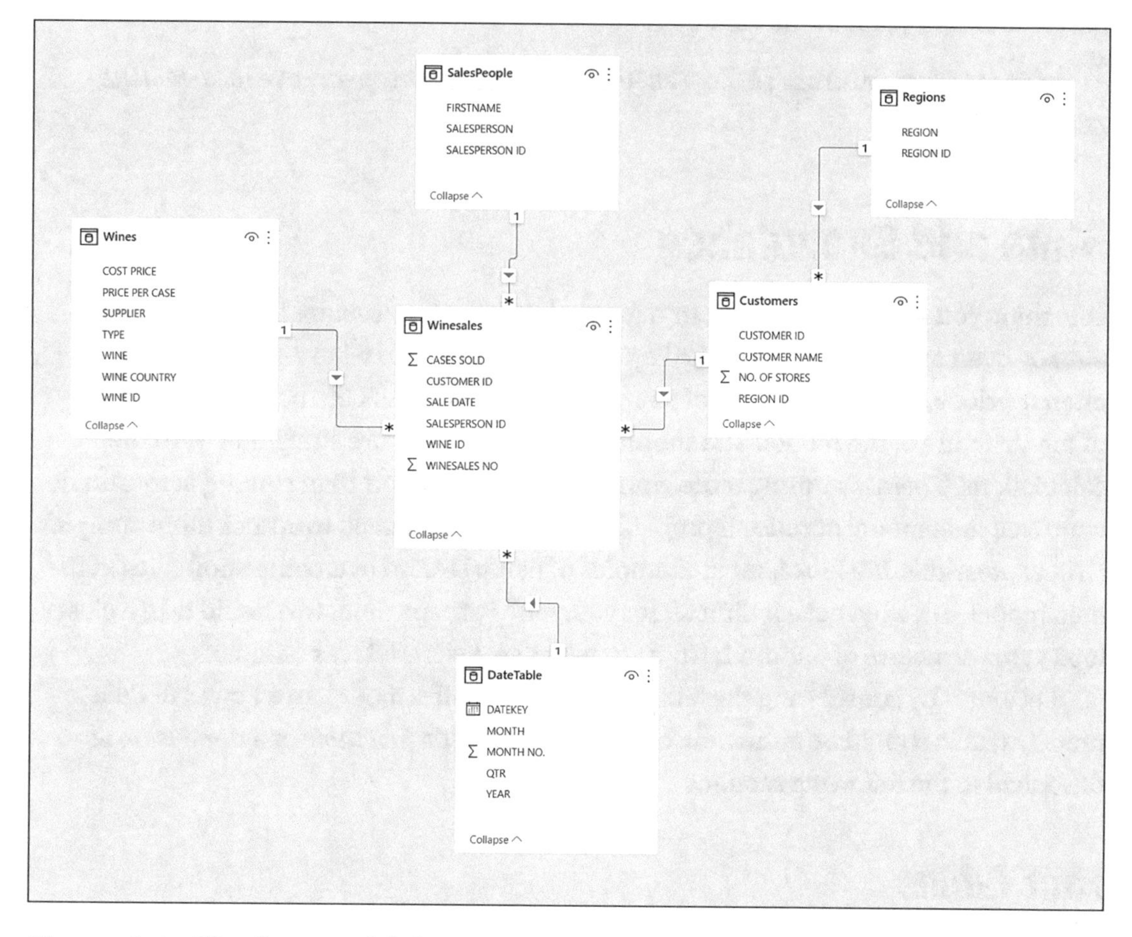

***Figure 1-1.** The data model that is used throughout this book*

To view these relationships, click on the **Model** button on the top left of the report canvas. This view shows the relationships between the tables, and this structure is known as the *data model.*

You will observe that the DateTable, SalesPeople, Customers, and Wines tables are all related to the Winesales table. Notice the "1" and "*" to denote the one side and the many side, respectively. The columns used to create the relationships have the same column names in both tables; for example, WINE ID in the Wines table is related to WINE ID in the Winesales table. The Regions table is the odd one out in that it's not directly related to the Winesales table but *indirectly* via the Customers table.

If you would like more information on creating relationships between tables in Power BI Desktop, follow this link:

https://docs.microsoft.com/en-us/power-bi/desktop-create-and-manage-relationships

Stars and Snowflakes

One thing you may notice about our data model is that its structure is simple. As has already been mentioned, one of the key aspects of DAX, and what newbies to DAX often overlook, is that the details of your DAX expressions will be inextricably tied to the structure of the model. The simpler the model, the more straightforward the calculations. There is nothing more worrying to a DAX expert than coming across an ill-contrived data model because it probably means they will need to author more complex DAX expressions. We look later at examples of using DAX to overcome anomalies in the data model, but why make it difficult for yourself? Perhaps then, we should take a closer look at the structure of our model and see why I've described it as "simple."

Let's start by considering the tables that comprise the model. In a Power BI data model, a table should be either one of two types, either a *fact table* or a *dimension* as described in the following sections.

Fact Tables

This type of table stores "events." The term "event" is used loosely here to describe activities such as sales, orders, or survey results. Fact tables answer the question *what*? That is, what are you analyzing in your report? You can identify the fact table by asking yourself these three questions:

1. Which table holds the data that you want to analyze in your report?
2. If you delete this table, will the remaining tables still be related to other tables in the data model?
3. Which table sits on the many side of all the other relationships?

Let's answer these questions using our data. We want to report on our sales that are recorded in the Winesales table. If we delete the Winesales table, we'll just have unrelated tables floating around in Model view. The Winesales table sits on the many side of all the other relationships. Clearly, the Winesales table is our fact table. By definition, fact tables sit on the many side of a many-to-one relationship. Another attribute of the fact table is that its data will change frequently and it'll probably have many more rows than a dimension.

Dimensions

These tables store the descriptions of the entities in your model. Dimensions answer the question *how*? That is, how do you want to analyze your data? In our data model, we can analyze our sales by wines, salespeople, customers, regions, and dates using the data in the columns within these tables. The data in dimensions does not necessarily change regularly, and dimensions tend to have fewer rows than fact tables.

There's no table property that you set to configure the table type as a dimension or a fact table. It's determined by which side of the relationship the table sits on. Tables that sit on the "one" side are always dimension-type tables, while tables that are *only* related on the "many" side are fact tables.

The reason it's so important to distinguish between these two different types of tables is that they support two different types of behavior in the data model, as follows:

- Dimension tables support *grouping* and *filtering*.
- Fact tables support *summarization.*

As we'll learn later, DAX *measures* are usually designed to summarize data from the fact table that's been grouped and filtered by a dimension table.

The Star Schema

You'll notice in Model view that we've placed the fact table in the middle of the view and arranged our dimensions around the fact table. This arrangement can be described as a star shape, giving a name to the structure, *star schema*. In a perfect star schema, all dimensions are directly related to the fact table. There is an imperfection in our data model because the Regions table is a dimension related to another dimension. Dimensions that are not directly related to a fact table but are indirectly related via dimension tables are described as *snowflake* dimensions. You can imagine that if we had a number of dimensions related to other dimensions in chains outward from the fact table, the schema would more resemble a snowflake.

Because data is infinitely variable, the tables in your data model may not be arranged obediently in a perfect star schema. Having multiple fact tables, for instance, isn't necessarily a problem. The thing to bear in mind, however, is that the more your model diverges from a star schema, the more you will need DAX to manage it. We will be resolving problems inherent in the structure of the data model later in this book when we explore the CROSSFILTER and TREATAS functions where we will create "virtual" relationships.

As we'll discover when we learn to control filters and more specifically calculate distinct counts, it can be difficult to work with dimensions that are not related directly to the fact table. Therefore, it sometimes makes sense to integrate a snowflake dimension into its parent table and therefore tidy up the schema back to a star, a process known as denormalization. You can find more information on this and star schemas generally here: `https://docs.microsoft.com/en-us/power-bi/guidance/star-schema`.

One thing that Power BI prevents is *ambiguity* in the data model, where there are multiple paths through which filters can propagate. Therefore, if you attempt to relate a dimension to two or more other dimensions, this will result in an *inactive* relationship being created, indicated by a dotted line. For example, in Figure 1-2, we've related the SalesPeople dimension to both the Customers dimension *and* the Regions dimension, and this results in an inactive relationship between SalesPeople and Regions.

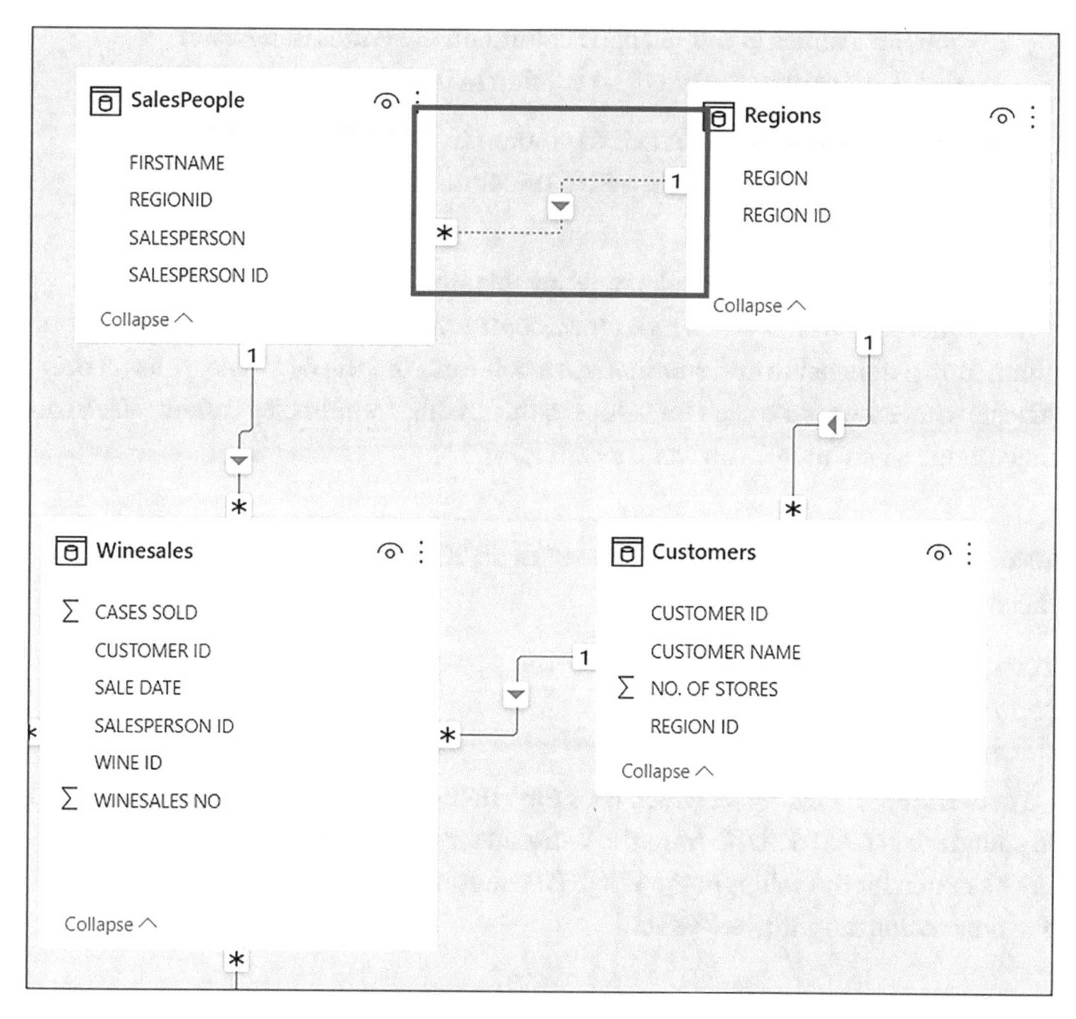

Figure 1-2. *An inactive relationship is created to avoid ambiguity*

We look at the concept of ambiguity and working with inactive relationships in later chapters, but for the moment, let's just be thankful that we aren't allowed to do anything that impedes the normal mechanism of the model.

Finding Nonmatching Values

A question that is often asked is what happens when there are missing values in the linking columns used to create relationships. There are two different scenarios here, taking the Wines dimension as our example:

1. You have values in the WINE ID column in the Wines dimension that don't exist in the WINE ID column in the Winesales fact table.

2. You have values in the WINE ID column in the Winesales fact table that don't exist in the WINE ID column of the Wines dimension.

Let's take scenario #1 first. Understanding this situation allows us to answer the following question: Which wines haven't we sold? When you build a visual that takes a column from a dimension and summarizes a column from the fact table, you will only see items where there's a match for values in the linking columns. By default, all visuals remove items where there is no value to show.

Note For information on building Power BI visuals, including the Table visual shown in Figure 1-3, visit

`https://docs.microsoft.com/en-us/power-bi/visuals/power-bi-report-add-visualizations-i`

For example, in Figure 1-3, which uses the WINE column from the Wines dimension and summarizes CASES SOLD from the Winesales table, we only see the wines where there's a match for the values in the WINE ID column in both tables. In other words, we're only seeing the wines we've sold.

WINE ▲	CASES SOLD
Bordeaux	54,070
Champagne	49,158
Chardonnay	41,883
Chenin Blanc	24,739
Chianti	27,323
Grenache	35,965
Malbec	34,290
Merlot	23,084
Piesporter	10,253
Pinot Grigio	23,449
Rioja	33,951
Sauvignon Blanc	47,415
Shiraz	17,644
Total	**423,224**

Figure 1-3. *By default, you only see items where there are matching values in the linking columns*

How can we see the wines we haven't sold in this visual? To do this is straightforward and requires no DAX. In the Visualisations pane, in the Values bucket, click on the drop-down of the column from the dimension, for example, the WINE column, and select **Show items with no data**. You'll now see a blank value beside items that have no match in the fact table, in our case, "Lambrusco" wine. This tells us that we haven't sold this wine; see Figure 1-4.

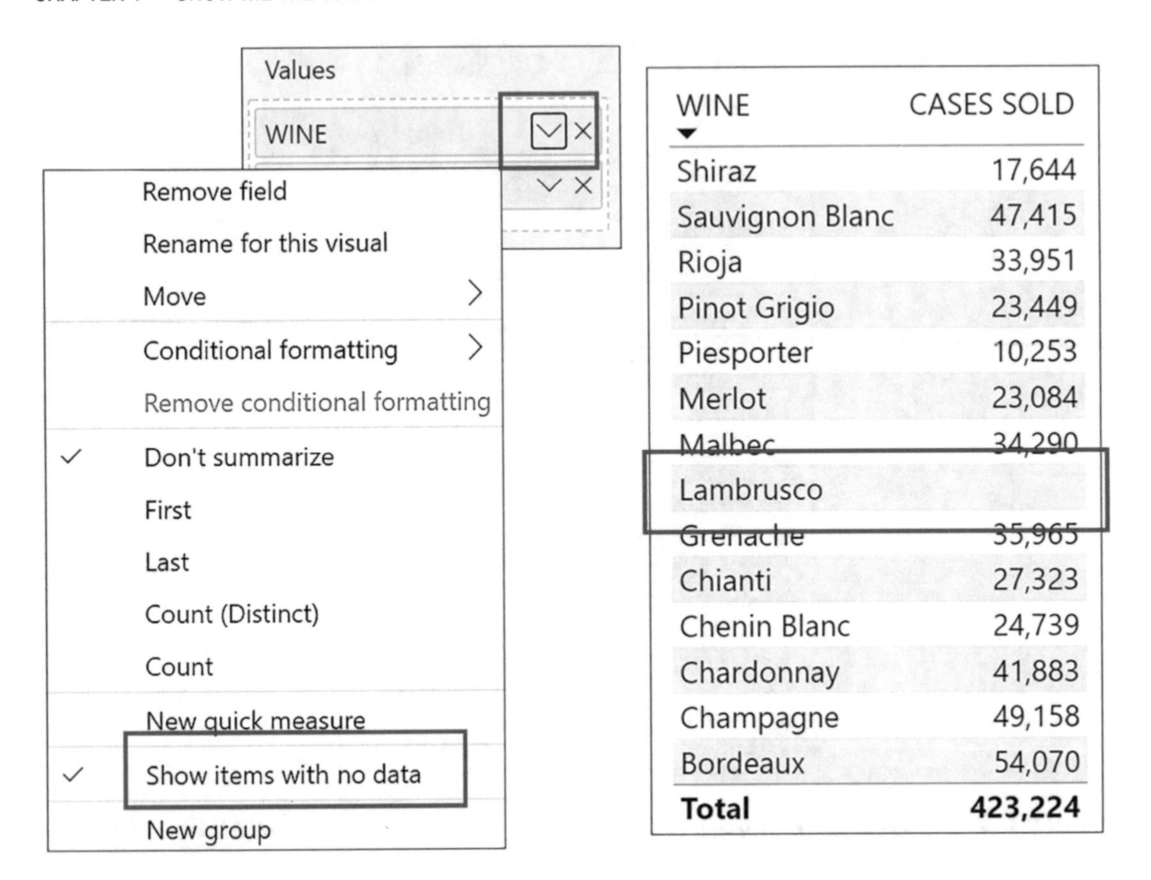

***Figure 1-4.** Finding the items for which there is no data*

If we look at the Wines dimension in Data view (click the button above Model on the left of the report canvas), we will see that "Lambrusco" has a WINE ID of 14. Examining the values in the WINE ID column of the Winesales table using the filter shows there is no WINE ID 14 in this column; see Figure 1-5.

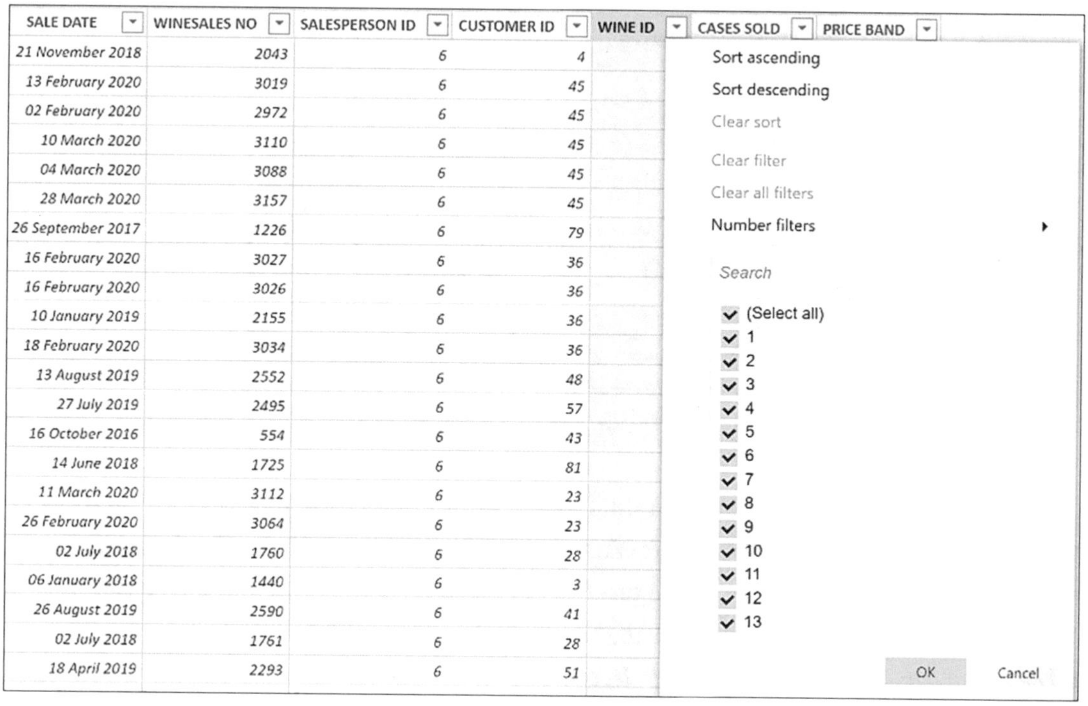

SALE DATE	WINESALES NO	SALESPERSON ID	CUSTOMER ID	WINE ID	CASES SOLD	PRICE BAND
21 November 2018	2043	6	4			
13 February 2020	3019	6	45			
02 February 2020	2972	6	45			
10 March 2020	3110	6	45			
04 March 2020	3088	6	45			
28 March 2020	3157	6	45			
26 September 2017	1226	6	79			
16 February 2020	3027	6	36			
16 February 2020	3026	6	36			
10 January 2019	2155	6	36			
18 February 2020	3034	6	36			
13 August 2019	2552	6	48			
27 July 2019	2495	6	57			
16 October 2016	554	6	43			
14 June 2018	1725	6	81			
11 March 2020	3112	6	23			
26 February 2020	3064	6	23			
02 July 2018	1760	6	28			
06 January 2018	1440	6	3			
26 August 2019	2590	6	41			
02 July 2018	1761	6	28			
18 April 2019	2293	6	51			

Sort ascending
Sort descending
Clear sort
Clear filter
Clear all filters
Number filters
Search
(Select all)
1
2
3
4
5
6
7
8
9
10
11
12
13
OK
Cancel

Figure 1-5. *The fact table does not contain the value from the dimension in the linking column*

As the name of the "Show items with no data" option implies, it can be used whenever you want to see items where there is no calculation to show, for example, where a measure doesn't return a value for an item. It doesn't mean there is never a value to show, as in the case of "Lambrusco" wine; rather, it means that the current filters on the model result in there being no value to show.

Let's now move on to scenario #2 where there are values in the WINE ID column of the Winesales fact table that don't exist in the WINE ID column of the Wines dimension.

Note The sample file does not contain the data described in scenario #2. However, Figure 1-6 shows you what this data would look like.

You can see in Figure 1-6, we have just this scenario. The wine ID's shown have no match in the Wines dimension.

ERSON ID	CUSTOMER ID	WINE ID	CASES SOL
6	35	100	
3	16	444	
4	20	555	
1	12	100	
2	17	133	
3	45	11	
6	11	7	
2	75	13	

Figure 1-6. *Values in the fact table that are not in the dimension*

When such values occur in your data, you'll see the outcome in any visual as soon as you take a column from the dimension and analyze a column from the fact table, as shown in Figure 1-7, where we have put the data into a Table visual and also a slicer. Here, we have a "blank" wine name that represents all the WINE ID values in the fact table for which there are no matches in the dimension. You'll also see the same outcome in a slicer even though it doesn't use the relationship and only shows values from the dimension.

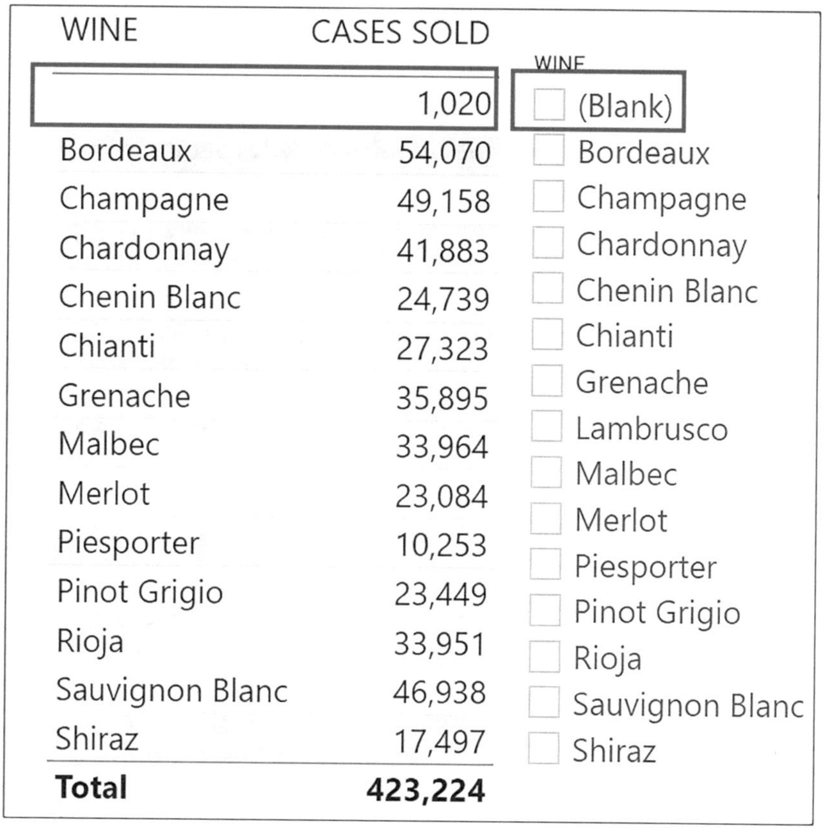

WINE	CASES SOLD
	1,020
Bordeaux	54,070
Champagne	49,158
Chardonnay	41,883
Chenin Blanc	24,739
Chianti	27,323
Grenache	35,895
Malbec	33,964
Merlot	23,084
Piesporter	10,253
Pinot Grigio	23,449
Rioja	33,951
Sauvignon Blanc	46,938
Shiraz	17,497
Total	**423,224**

WINE

- (Blank)
- Bordeaux
- Champagne
- Chardonnay
- Chenin Blanc
- Chianti
- Grenache
- Lambrusco
- Malbec
- Merlot
- Piesporter
- Pinot Grigio
- Rioja
- Sauvignon Blanc
- Shiraz

Figure 1-7. *The "Blank" entry shows there are values in the fact table that don't match to values in the dimension*

The "Blank" entry is a result of what we sometimes refer to as "dirty data." Why are there values in the fact table for which there is no match in the dimension? How are you going to resolve this scenario? This is a question that only the data modeler can answer, and ultimately the solution lies in correcting the data at its source.

We hope you appreciate how important it is to identify nonmatching values in your data and to understand that you don't need DAX to do this. Finding out where there's no data can be equally as valuable as knowing where there is, and the star schema allows us to do this.

In this chapter, you have familiarized yourself with the data we will be using henceforth. You also now understand concepts that underpin the data model and how it comprises fact tables and dimensions related to many-to-one relationships. The simplest structured data model is the star schema where dimensions are related directly to the fact table. However, it is possible to have dimensions indirectly related to the fact table via other dimensions creating snowflake dimensions. This is mandatory precursor knowledge to understanding DAX because what you will learn as we progress through this book is that many DAX calculations will involve manipulating the tables in the data model, and in doing so, the way the tables are structured is paramount.

CHAPTER 2

DAX Objects, Syntax, and Formatting

Now that you understand the structure of the data we'll be using throughout this book, the next step is learning how to construct DAX expressions. In this chapter, we will compare and contrast DAX expressions to Excel formulas as this will provide context for your knowledge. You will learn to reference the objects used in DAX expressions, the syntax of the expressions, and how you can format your DAX code, making it easier to read and debug.

To follow along with the examples in this chapter, in the Data view of Power BI desktop, select the Winesales table in the Fields list, and on the **Table Tools** tab, click the **New Column** button. This will display the DAX "formula bar" as shown in Figure 2-1.

© Alison Box 2022
A. Box, *Up and Running with DAX for Power BI*, https://doi.org/10.1007/978-1-4842-8188-8_2

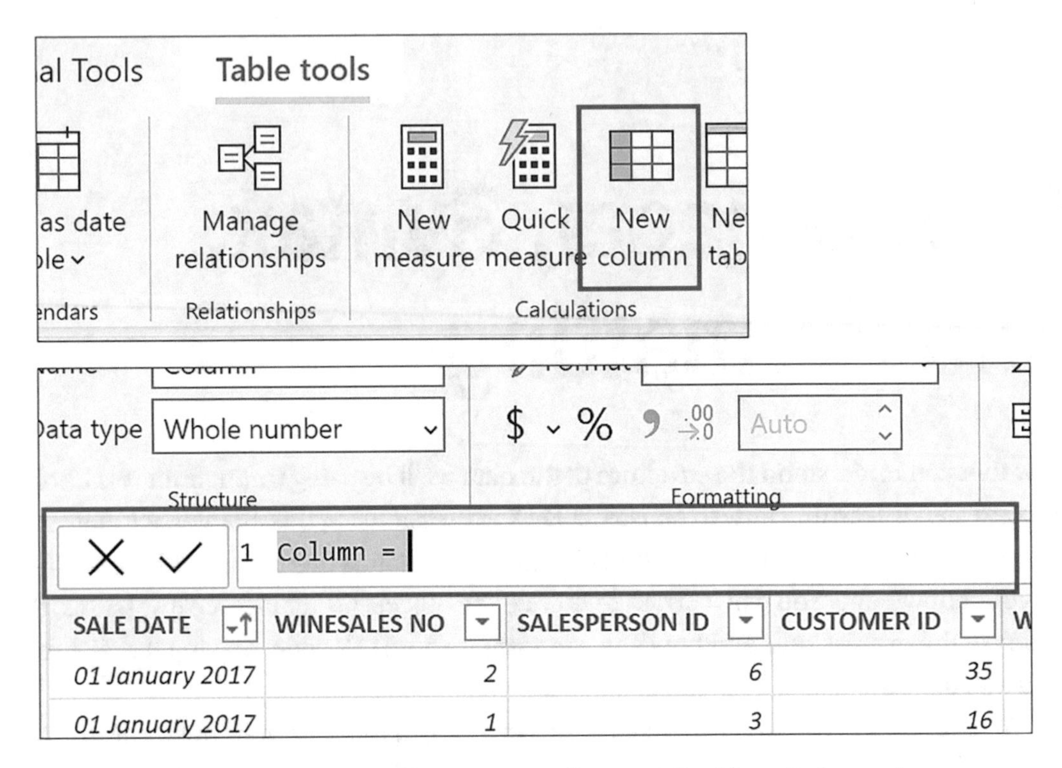

Figure 2-1. *To follow along with the examples, use the New Column button*

The first similarity to Excel is the "formula bar" that pops up when you create new columns or measures. We will start to type some expressions into the formula bar presently. However, at this stage, you don't need to know what the expressions are calculating. You are just learning how to type the correct syntax.

DAX Syntax

Notice on the left, just like Excel, the formula bar has a Cancel button (the cross) and an Enter button (the tick). However, the formula bar is in effect a code editor and can extend to many lines if the SHIFT + ENTER key combination is pressed (see the section below on Formatting). This is why, unlike the Excel formula bar, each line of the DAX code editor is numbered. You will, for instance, notice that in Figure 2-1, we are on line 1.

The next parallel with Excel is that DAX expressions are constructed in the same way as Excel formulas. For example, all DAX expressions begin with an equals sign, and commas separate the arguments of functions. Also, just like Excel, DAX expressions are *case insensitive*; it makes no difference in what case you type your DAX code.

However, one of the major differences between DAX and Excel is that in DAX, you can't reference "cells." The only two objects that are referenced in DAX expressions are tables and columns.

You reference a table by just naming it. For example, to count the rows in the Winesales table, this would be this expression:

```
= COUNTROWS ( Winesales )
```

Notice that when you start to type this expression into the DAX editor, just like Excel, the DAX editor matches what you're typing in a list of suggestions. This list is referred to as the DAX **IntelliSense**; see Figure 2-2. Just click on a suggestion in the IntelliSense list to place it into your code. You can't put anything into your expression that isn't on the list.

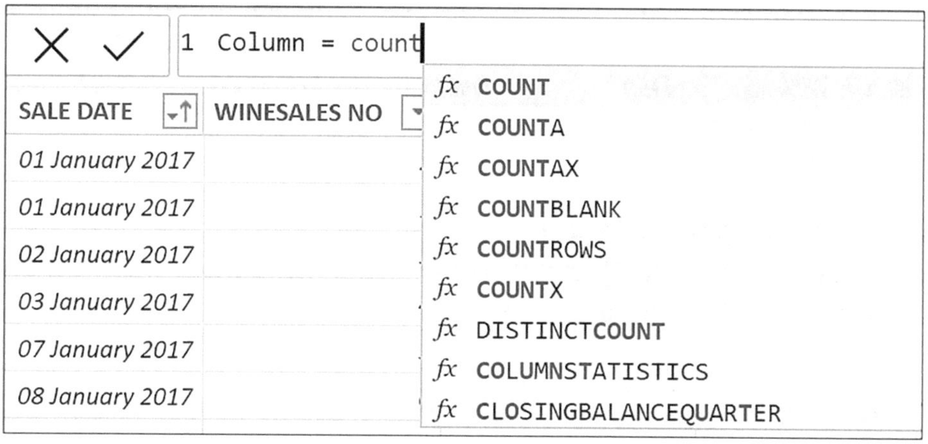

Figure 2-2. *The DAX IntelliSense list*

Also notice in the COUNTROWS expression that spaces have been used before and after the brackets. Typing spaces is arbitrary as they will be ignored by the DAX editor and can be used wherever you feel they improve the clarity of the expression (see the section below on Formatting).

If the table name contains a space, the table name must be surrounded with single quotes:

```
= COUNTROWS (  'Wine Sales' )
```

To reference a column, you surround the column name with square brackets ([ ]) and *always precede the column name with the table name*. For example, to sum the CASES SOLD column in the Winesales table, this would be the expression:

```
= SUM ( Winesales[CASES SOLD] )
```

As mentioned before, in DAX, there is no such thing as a cell, only tables and columns.

Table 2-1 shows a comparison of equivalent Excel formulas and DAX expressions, and you can see how similar the syntax is between the two.

Table 2-1. *Comparing Excel formulas and DAX expressions*

Excel	**DAX**
=IF (B2 > 50 , "Yes" , "No") *When this formula is copied down, the "B2" will change relatively to "B3, B4, B5 etc."*	=IF (Winesales[CASES SOLD] > 50 , "Yes" , "No") *Used in a calculated column, this expression is automatically applied to the entire column.*
= SUM (Winesales[CASES SOLD]) *This uses Excel Table syntax where the table is named "Winesales" and the column is named "CASES SOLD".*	= SUM (Winesales[CASES SOLD]) *Used in a measure or in a calculated column to find total cases in the CASES SOLD column in the Winesales table.*

Another contrast between Excel and DAX is the way you reference "AND" and "OR". In Excel, you use the AND() and OR() functions. In DAX, you typically use these ***operators*** instead; AND is **&&** (double ampersand) and OR is || (double pipe).

Note You'll find the pipe symbol "|" on your keyboard at the bottom left, above the backslash and to the right of SHIFT.

Table 2-2 shows a comparison of using "AND" and "OR" in Excel formulas and DAX expressions.

Table 2-2. *Contrasting AND and OR in Excel and DAX*

<table>
<tr><th>Excel</th><th>DAX</th></tr>
<tr><td>AND</td><td>AND</td></tr>
<tr><td>= IF (AND (Winesales[@CASES SOLD] > 50,
Winesales[@CASES SOLD] < 100), “Yes” , “No”)
Using Excel Table syntax where the table is named “Winesales” and the column is named “CASES SOLD”
Note the use of the “@” to denote “the current row.”</td><td>= IF (Winesales[CASES SOLD] > 50
&&
Winesales[CASES SOLD] < 100 ,
“Yes” , “No”)
Used in a calculated column.
Using the value in the current row is implicit in calculated columns.</td></tr>
<tr><td>OR</td><td>OR</td></tr>
<tr><td>= IF (OR (Winesales[@SALESPERSON ID] = 2 ,
Winesales[@SALESPERSON ID] = 6),
“Yes” , “No”)
Using Excel Table syntax where the table is named “Winesales” and the column is named “SALESPERSON ID”</td><td>= IF (Winesales[SALESPERSON ID] = 2
||
Winesales[SALESPERSON ID] = 6 , “Yes” , “No”)
Used in a calculated column.</td></tr>
</table>

Note DAX does have an AND function and an OR function, but in DAX, these functions only accept two arguments, so it’s usually better to use the operators.

A single ampersand (&) is used in DAX as the concatenation operator, just as it is in Excel.

DAX Formatting

Before we start authoring DAX expressions in earnest, let’s get into some good habits concerning the formatting of our DAX code. Consider the two expressions in Figure 2-3. They are the same expression but with two different layouts.

```
1 Average Cases for Black Ltd in 2019 = CALCULATE(AVERAGE(Winesales[CASES SOLD]),DateTable[Year]
  =2019,Customers[CUSTOMER NAME]="black ltd")
2
```

```
1 Average Cases for Black Ltd in 2019 =
2 CALCULATE (
3     AVERAGE ( Winesales[CASES SOLD] ),
4     DateTable[Year] = 2019,
5     Customers[CUSTOMER NAME] = "black ltd"
6 )
7
```

Figure 2-3. *Comparing unformatted and formatted expressions*

Question: Which layout makes the DAX code easier to understand? I think you'll agree that it's the second layout where we have separated the code onto different lines. In the DAX editor, you can use the keyboard combination SHIFT + ENTER to move onto a new line and use the TAB key to indent lines. Spaces can be used for clarity. It's also recommended that you start nested functions on a new line and close brackets at the same indent of the function it closes.

To add comments to your code, use the following:

-- - Single line comment (double dash)

// - Single line comment (double forward slash)

/* - Start a multiline comment (forward slash and asterisk)

*/ - End a multiline comment (asterisk and forward slash)

However, there are no hard and fast rules about how to format your DAX code. Whatever works for you.

If you want to quickly format your untidy DAX code, use the DAX formatter here:

`https://www.daxformatter.com/`

You can also find more information and guidelines on best practices here:

`https://www.sqlbi.com/articles/rules-for-dax-code-formatting/`

You should now be able to type your DAX code correctly. Use square brackets to reference columns and always precede your column references with the table name where the column resides. You understand that in DAX, we often use "AND" and "OR" operators rather than the equivalent functions used in Excel. Using separate lines in the code editor will greatly improve the clarity of the expression. However, DAX doesn't care how your code is formatted. It will execute your code however dire the layout of the expression looks!

This chapter concludes our preparatory work before we can move on to author DAX expressions and generate calculations. The next step is to understand that in DAX, we work with different types of expressions, and this will be the focus of the next chapter.

CHAPTER 3

Calculated Columns and Measures

In the previous chapter, you learned the syntax used by the DAX language, and now you're ready to write your first DAX expressions. In DAX, there are three types of expression: ***calculated columns***, ***measures***, and ***calculated tables***. However, in this chapter, we will only be addressing the first two types (we look briefly at calculated tables in Chapter 15).

Note You already know that DAX is the acronym for "Data Analysis Expressions." However, we often refer to "DAX ***expressions***" because it seems clearer to do so.

Firstly, you will learn how to write calculations using the calculated column. This will be the part of DAX that will be intuitive to you, particularly if you are an Excel user. Calculated columns will seem no different to you than using Excel formulas. When we move forward to learn how and why we need DAX measures, however, things may become a little more challenging. One of the biggest hurdles when learning DAX is understanding the difference between the calculated column and the measure, and this is something that we will also be exploring in this chapter. For instance, the same DAX expression that's used in a calculated column can't typically be used in a DAX measure, but perversely, most DAX expressions used in measures can be put into a calculated column.

Calculated Columns

When learning DAX, most people understand expressions that are entered into calculated columns because they are very similar to creating Excel formulas, particularly if you use formulas in Excel tables. In DAX, you will find many of your favorite Excel functions, such as IF, TODAY, ROUNDUP, and SUM, that can be used in a calculated column.

© Alison Box 2022
A. Box, *Up and Running with DAX for Power BI*, https://doi.org/10.1007/978-1-4842-8188-8_3

This is why newbies to DAX mistakenly think that DAX is just like Excel and create a plethora of calculated columns when they really should be creating measures, which are more efficient in every way. The thing to understand about the calculated column is that, just like copying down on an Excel formula, the calculated column is evaluated for ***every row*** in the table and therefore can be process heavy. We will see that this is very different from how measures are evaluated.

Creating Simple Calculated Columns

To create our first DAX expression in a calculated column, let's take a very simple calculation and multiply the CASES SOLD values in the Winesales table by 10 percent. In Chapter 2, you learned how to create a new column. Ensuring that the Winesales table is selected in Data view, you click on the **New Column** button on the **Table Tools** tab. In the DAX editor, enter the following expression:

```
10 PC of Cases = Winesales[CASES SOLD] * 0.1
```

When you've finished typing, you can press the enter key, or you can click on the tick to the left of the DAX editor. Your calculated column called "10 PC of Cases" is created and joins the Fields list; see Figure 3-1.

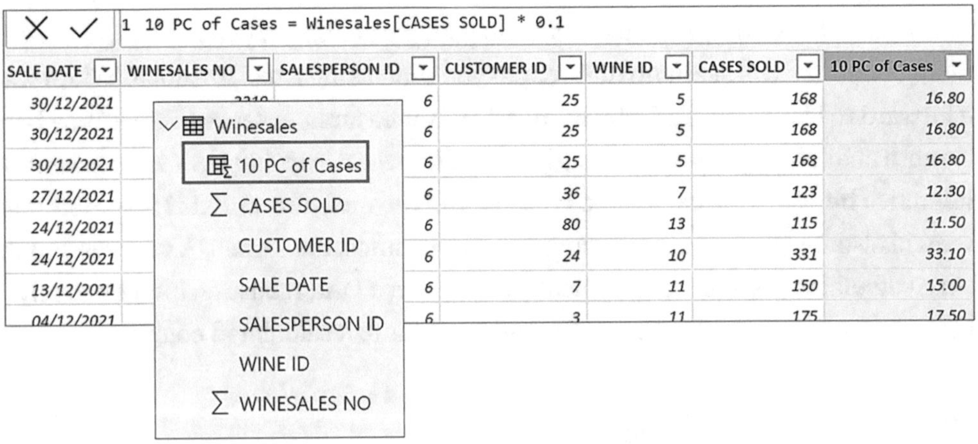

Figure 3-1. *The calculated column joins the Fields list*

Let's now see how we can use the IF function in DAX in a calculated column. We could, for instance, add a column in the Winesales table that's populated with either "Team A" or "Team B." This column will group our salespeople as follows: Salespeople with IDs 1, 3, and 6 are in Team A, and other salespeople are in Team B. We'll call this new column "Team".

In the DAX editor, enter this code noting the use of the double pipe for "OR":

```
Team =
IF (
      Winesales[SALESPERSON ID] = 1
       || Winesales[SALESPERSON ID] = 3
       || Winesales[SALESPERSON ID] = 6,
    "Team A",
    "Team B"
)
```

Similarly, you could group the values in the CASES SOLD column into "High" and "Low" volume where high volume is any sales where CASES SOLD is between 50 and 400 by using this DAX expression, noting the use of the double ampersand for "AND":

```
Volume =
IF (
    Winesales[CASES SOLD] >= 50
        && Winesales[CASES SOLD] <= 400,
    "High",
    "Low"
)
```

Creating these calculated columns has been an easy introduction to DAX because, as we've seen, the expressions are very similar to formulas in Excel. The reason we've included these calculated columns here is because they're simple examples that teach you DAX syntax and that every Excel user can do.

However, we wouldn't recommend you do such calculations here.[1] There's a more efficient way to create these columns, and that's to generate them in Power Query using Power Query's conditional column.

Looking at the RELATED Function

So we've established that there are common functions to both Excel and DAX such as the IF function. However, if using calculated columns isn't always the most efficient way to generate data, why would we need to use them? There are some functions that are specific to DAX and give us reasons to author our DAX expressions in the context of a calculated column. One of these functions is the RELATED function.

This function returns a value from a related table and is similar in purpose to the VLOOKUP function in Excel. However, RELATED will only return values from the *one side* of the relationship to the many side. For example, if you want to show the customer names related to the CUSTOMER ID's in the Winesales table, you could use this DAX expression in a calculated column in the Winesales table:

```
Customer Name from Customers Table =
                                RELATED ( Customers[CUSTOMER NAME] )
```

You will now see the names associated with each CUSTOMER ID in the calculated column; see Figure 3-2.

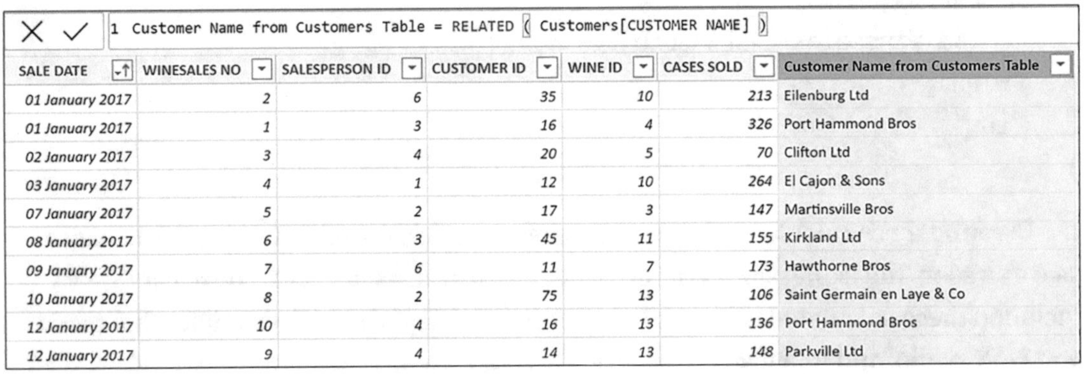

```
1 Customer Name from Customers Table = RELATED ( Customers[CUSTOMER NAME] )
```

SALE DATE	WINESALES NO	SALESPERSON ID	CUSTOMER ID	WINE ID	CASES SOLD	Customer Name from Customers Table
01 January 2017	2	6	35	10	213	Eilenburg Ltd
01 January 2017	1	3	16	4	326	Port Hammond Bros
02 January 2017	3	4	20	5	70	Clifton Ltd
03 January 2017	4	1	12	10	264	El Cajon & Sons
07 January 2017	5	2	17	3	147	Martinsville Bros
08 January 2017	6	3	45	11	155	Kirkland Ltd
09 January 2017	7	6	11	7	173	Hawthorne Bros
10 January 2017	8	2	75	13	106	Saint Germain en Laye & Co
12 January 2017	10	4	16	13	136	Port Hammond Bros
12 January 2017	9	4	14	13	148	Parkville Ltd

Figure 3-2. *The RELATED function returns values from related tables*

[1] The reason for this is that calculated columns have to be recalculated whenever the data is refreshed. This can have a big impact on the efficiency and performance of the report. You can find more information on this topic here: `https://docs.microsoft.com/en-us/power-bi/guidance/import-modeling-data-reduction`

Often, the generation of the calculated column using RELATED where you populate values from related tables is used solely for ad hoc reasons. Once you have the customer names alongside their transactions, you'll find it's often easier to cross-check your data analysis. Once the column has served its purpose, it can be removed.

Note If these were Excel tables and we wanted to populate the Winesales Excel table with the customer names in the Customers Excel table, we would use the VLOOKUP function in the Winesales Excel table like this:

=VLOOKUP ([@CUSTOMER ID] , Customers, 2, 0)

The "@" symbol means "use the value in the current row of the Excel table." Using the value from the current row is implicit in DAX calculated columns.

You can also use RELATED to pull through values from *indirectly* related tables into the fact table. For example, the Regions table is related to the Customers table that is in turn related to the Winesales table as shown in Figure 3-3.

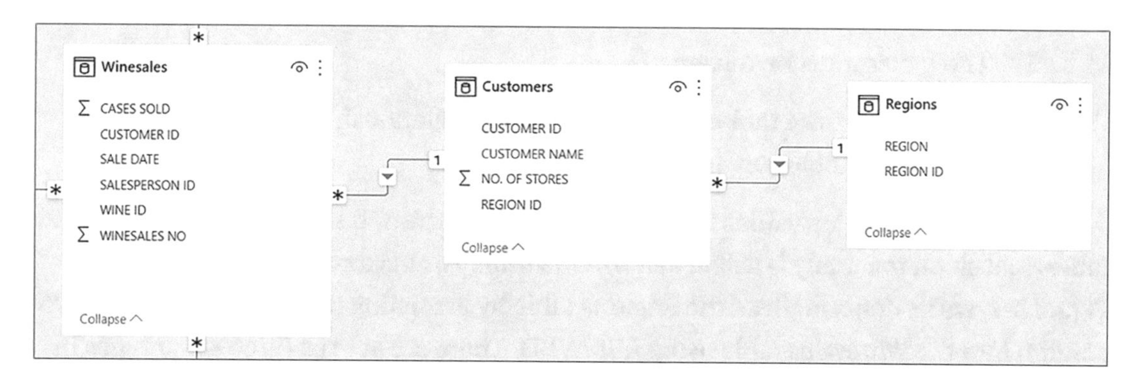

Figure 3-3. *The Regions table has an indirect relationship to the Winesales table*

Therefore, we could populate each REGION name alongside each sales transaction in the Winesales table by using this code (see Figure 3-4):

```
REGION NAME = RELATED ( Regions[REGION] )
```

```
1 REGION NAME = RELATED ( Regions[REGION] )
```

SALE DATE	WINESALES NO	SALESPERSON ID	CUSTOMER ID	WINE ID	CASES SOLD	REGION NAME
30 December 2021	2219	6	25	5	168	India
30 December 2021	2219	6	25	5	168	India
30 December 2021	2219	6	25	5	168	India
27 December 2021	2216	6	36	7	123	Wales
24 December 2021	2209	6	80	13	115	Argentina
24 December 2021	2208	6	24	10	331	United States
13 December 2021	2182	6	7	11	150	United States
27 November 2021	2153	6	5	10	313	Wales
26 November 2021	2151	6	11	13	98	Wales
26 November 2021	2150	6	29	4	236	China
25 November 2021	2148	6	51	12	109	Germany

Figure 3-4. *Using RELATED to return the Region names*

Notice that we've named this column REGION NAME to distinguish it from the REGION column in the Regions table.

Let's look more closely at the RELATED function. You should understand that you can only use this function in the following two circumstances:

1. The tables must be related.

2. Only values from tables on the *one* side of a relationship can be returned to tables on the *many* side.

The act of populating values from tables that sit on the one side of a relationship into tables that sit on the many is called *denormalization*. For instance, in the example in Figure 3-4, we've denormalized the Regions table by extracting the values in the REGION column into the Winesales table using RELATED. There are at least three advantages in doing this:

1. You now know in which Region each sales transaction was made.

2. If you need to use the region names in a visual, you can use the calculated column in the Winesales table. Therefore, you no longer need to see the Regions table in Report view. If this is the case, you can hide the Regions table. To hide a table in Report view, right-click the table name in the Fields list in either **Data view** or **Model view**, and select **Hide in report view**; see Figure 3-5.

3. You can perform a distinct count on the REGION NAME column in the Winesales table to calculate how many *different* Regions we've sold our wines in. We'll do this calculation later, but because the sales transactions must be directly associated with the regions in which they were made, this would be a difficult expression if we left the REGION values in the Regions table.

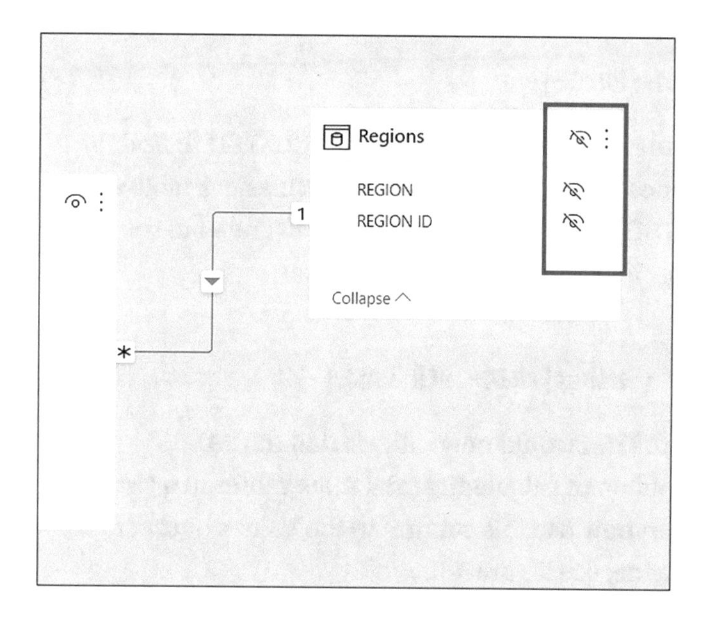

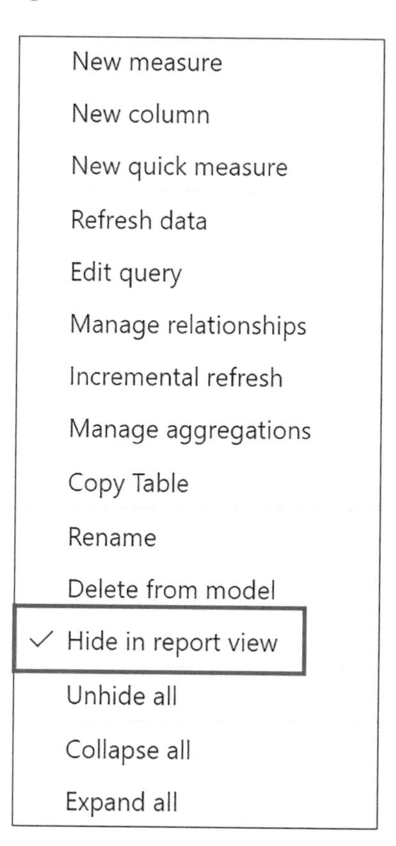

***Figure 3-5.** Hiding tables in Report view*

Understanding the RELATED function allows us to do another mandatory calculation in our data model. Perhaps you've noticed that although we have a Winesales table, we have no sales values. However, we can now calculate them. We can multiply the CASES SOLD column in the Winesales table with the PRICE PER CASE column in the Wines table, and because the Winesales table is related to the Wines table in a many-to-one relationship, we can use RELATED to do this.

We're going to look at two different methods of using RELATED to calculate the Sales revenue values.

For method #1, we could create two calculated columns. The first column, called "PRICE", uses RELATED to populate the PRICE PER CASE values into the Winesales table. The second column multiplies the "PRICE" column by the CASES SOLD column and is called "Sales":

```
PRICE =
RELATED ( Wines[PRICE PER CASE] )

SALES =
Winesales[CASES SOLD] * Winesales[PRICE]
```

Method #2 requires just one calculated column. You can use RELATED to find the PRICE PER CASE values from the Wines table for each row in the Winesales table *in memory* and then multiply by CASES SOLD. In other words, you don't need to see the price of each wine before you multiply it by the CASES SOLD values:

```
SALES =
Winesales[CASES SOLD] * RELATED ( Wines[PRICE PER CASE] )
```

What many people who are new to DAX would now think is that the SALES calculated column has solved the problem of calculating total sales values in a visual on the report canvas. For instance, we can now use this column in the Values bucket of a visual to find the total sales for each wine; see Figure 3-6.

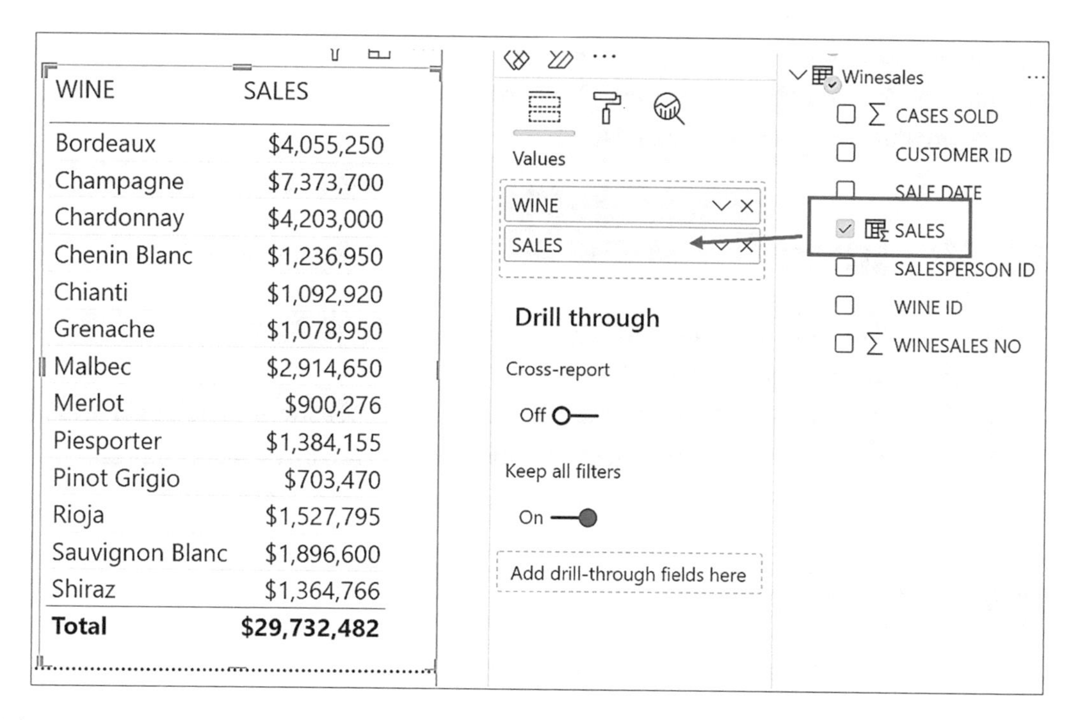

***Figure 3-6.** Using the SALES calculated column in the Values bucket to sum the sales*

However, this is probably not such a great idea. Think about it; firstly, the calculated column will be evaluated for *every row* in the Winesales fact table and *recalculated* whenever the data is refreshed. That's a lot of processing if you have millions of rows in your fact table.

Secondly, when you put this column into a visual containing items from dimensions, it performs *another* calculation to sum these values for each item from the dimensions. Does this sound a very efficient way of doing this calculation? Probably not. The upshot of inefficient data models is that reports built on the top of them become slow to refresh and render (refer to Footnote 1 where there is a link for more information on this topic).

Therefore, the question now is the following: If you shouldn't use a calculated column for the sales calculation, what should you use?

This is where *measures* can help us. We will revisit our sales calculation in Chapter 5, and rather than using a calculated column to perform the evaluation, we will be using a measure. But for now, we're going to leave calculated columns behind us (we will revisit the calculated column later in this book when we explore some complex expressions that require their use). If you're an Excel user, you'll feel quite at home creating calculated columns using the DAX functions that have a replica in Excel. Nevertheless, you probably won't have the same comfortable feeling when you come to writing DAX measures. This is where DAX becomes a little more challenging, so let's move forward and learn how to author DAX measures.

DAX Measures

We're now ready to look at the second type of DAX expression, the measure. There are two types of measures that you can use in visuals: *implicit* measures and *explicit* measures (however, we don't normally call them "explicit measures," just "measures" but implicit measures are always named accordingly). What's the difference between implicit and explicit measures? Well, let's start with the implicit measure first.

Implicit Measures

If you've created any Power BI visual, you've created an implicit measure. Have you ever wondered what the sigma symbol (∑) beside a numeric column in the Fields list means? It has a more precise purpose than signaling a column containing numbers. The sigma indicates that when you put this column into the Values bucket of a visual, the data in this column will automatically be aggregated. This is what we mean by an implicit measure.

The sigma normally indicates that the column will be summed, but you can perform other aggregations such as averages or find the maximum or minimum value by changing the function on an ad hoc basis. To do this, use the drop-down beside the column name in the Values bucket and, for example, change this to "Average" as shown in Figure 3-7 where the steps to generate an implicit measure have been numbered.

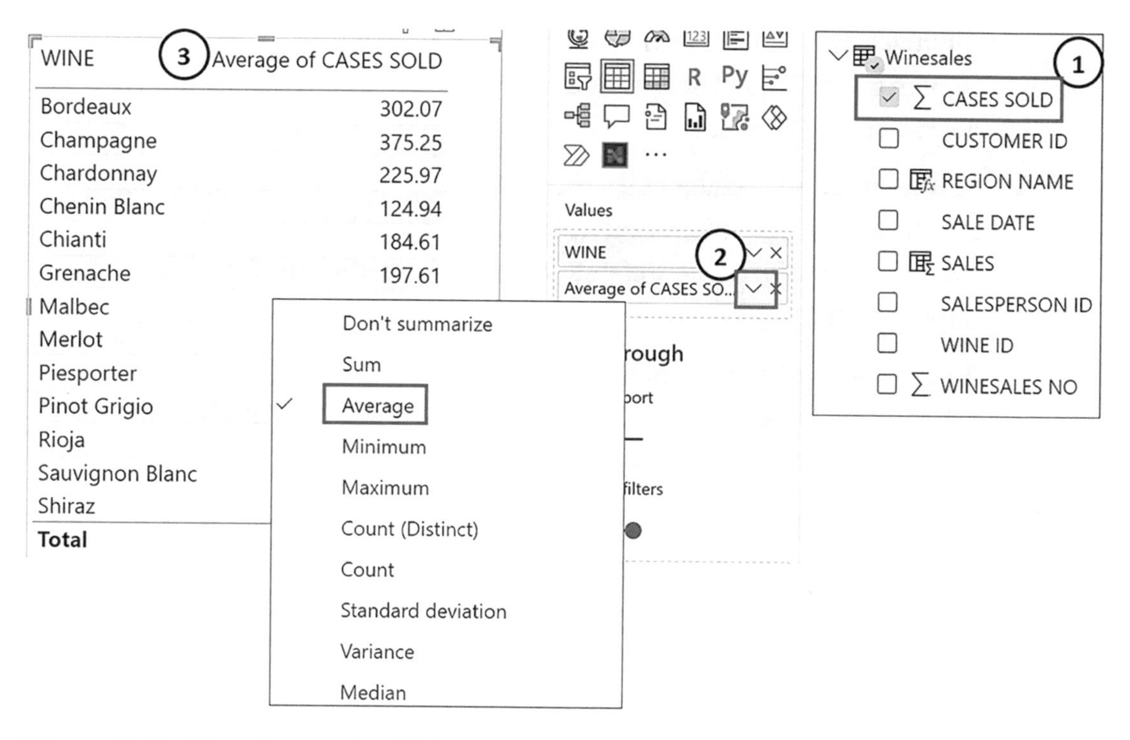

***Figure 3-7.** Creating an implicit measure*

1. The column CASES SOLD has a sigma beside it – ∑.
2. When this column is put into the Values bucket, it defaults to SUM, but you can change the function to AVERAGE by using the drop-down.
3. The implicit measure has calculated the average CASES SOLD for the items displayed in the visual, in this case, each wine.

However, there are several drawbacks to using implicit measures. Consider these scenarios:

- You may want to rename the implicit measure "Average of CASES SOLD" to something more concise. You can do this by double-clicking on the entry in the Values bucket, but you would have to repeat this every time you use an implicit measure and then want to rename it.
- If you rename the implicit measure, the name of a measure in the visual won't match the column name in the Fields list.

- Although you normally want to use the SUM function, you *often* want to use AVERAGE as well. You would have to keep changing the function to AVERAGE.
- In some visuals, you might like to format an implicit measure with two decimal places and sometimes with no decimal places. You would not be able to have different numeric formatting for the implicit measure in different visuals.
- What if you want to calculate 10 percent of the sum of the CASES SOLD values for each wine, or indeed, any calculation on the total values? You can't do this using an implicit measure.

This is the trouble with implicit measures; they just don't make the grade. So let's move the focus of this chapter to what we're really here for, and that's to learn how to create our own explicit measures using DAX.

Explicit Measures

If you create your own measures rather than relying on implicit measures, these are some of the benefits:

- You'll have more control over the aggregation performed by the measure and be able to name it accordingly.
- You'll be able to use different numeric formatting for different measures.
- Explicit measures will become a constituent part of the data model. Your measures will join the Fields list, and you, or people using your data, can use and reuse the measures whenever you need to visualize a particular calculation.
- By using DAX, you can go far beyond just simple aggregations of your data. You can perform complex calculations to get to the insights you really need.

So let's bite the bullet and create our first DAX measures. Once we've done this, we can then answer the pressing question that has yet to be answered, and that is *what exactly is a measure*?

Before we start, however, we need to find a place to store our measures. Explicit measures are table agnostic and can be stored in any table. However, it makes sense to create a table that will hold only measures.

Creating a Measures Table

To do this, on the Home tab, click on the **Enter Data** button. In the Create Table pane, give your table a name, for example, "Measures Table" (you can't name the table "Measures" because this is a reserved word), and load the table.

When you put a measure into this table and delete the column that's there, a "measures" icon will display beside the table in the Fields list, and the table will move to the top of the list; see Figure 3-8.

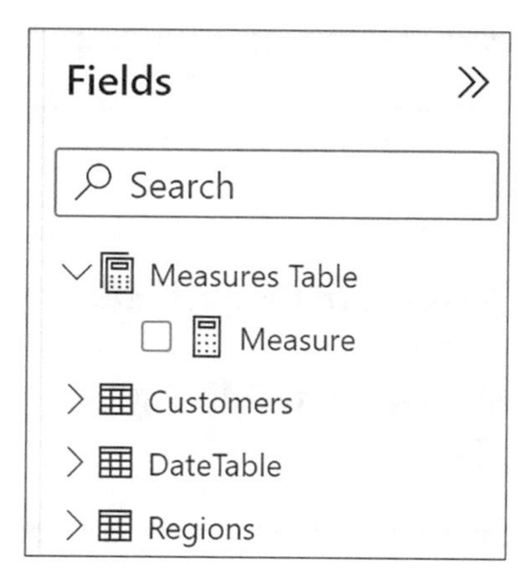

Figure 3-8. *The Measures table will sit at the top of the Fields list*

However, it's not mandatory to store your measures in a separate table. Some data modelers prefer to store measures in the fact table or in the table from where the data is being used by the measure.

Creating Simple DAX Measures

The first measure we're going to construct will replace the implicit measure that calculates the sum of the CASES SOLD. To create this measure, in **Report** view, right-click on your Measures table in the Fields list and select **New measure** from the shortcut menu. You could instead click on the **New Measure** button on the **Home** tab. However, if you use this method, ensure that the table you have selected in the Fields list is the table where you want to put your measure.

Tip If your measure is accidentally stored in the wrong table (or you just want to move it), use the Fields list in **Model** view where you can drag and drop measures between tables.

Once you have selected New measure, the DAX editor will appear at the top of the screen as it did when we created calculated columns. In the DAX editor, in front of the equals sign (=), name your measure, for example, "Total Cases", and type the following DAX expression:

```
Total Cases =
SUM ( Winesales[CASES SOLD] )
```

You can see this expression in the DAX editor in Figure 3-9.

```
1 Total Cases = sum(Winesales[CASES SOLD])
```

Figure 3-9. *Your first DAX measure in the DAX editor*

Press the Enter key and your measure will display in the Measures table. You can now delete "Column1" from this table.

Note DAX measure names are not case sensitive and can contain any characters. However, we would recommend restricting your measure names to containing just letters and/or numbers and spaces. We would also recommend that you keep the names of tables, columns, and measures simple and straightforward. I particularly like Chris Webb's blog on this topic: `https://blog.crossjoin.co.uk/2020/06/28/naming-tables-columns-and-measures-in-power-bi/`

DAX measures are only calculated when they are used, so you must put the measure into the Values bucket of a visual before you can see the calculation. For example, in Figure 3-10, in a Table visual, we've used the WINE column from the Wines dimension and then dragged the "Total Cases" measure into the Values bucket of the visual.

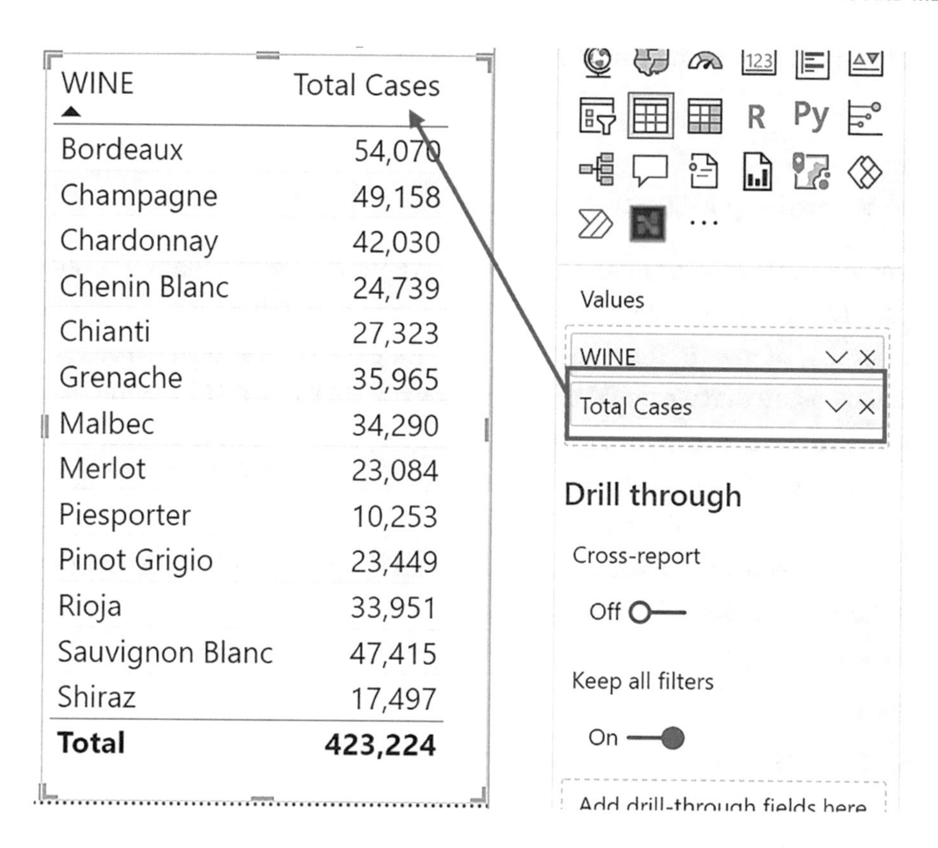

***Figure 3-10.** Measures are calculated when they are used*

One of the great advantages of using explicit measures is that the numeric formatting is stored with the measure. To format a measure, select the measure by clicking on it in the Fields list, and the measure expression shows in the DAX editor. Then, on the **Measures tools** tab, in the **Formatting** group of commands, you can select the numeric formatting you require, for example, a thousands separator; see Figure 3-11.

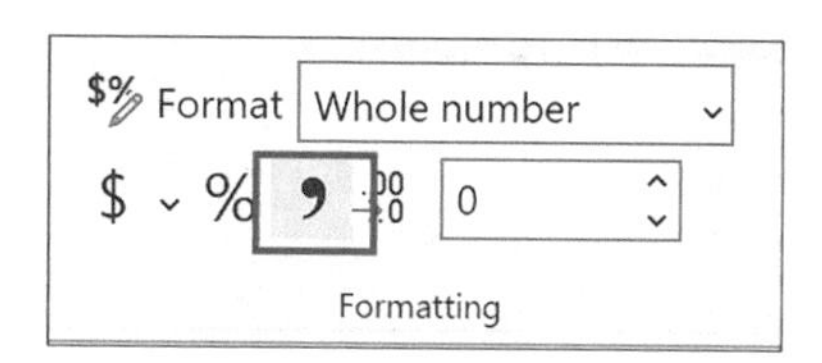

***Figure 3-11.** Use the Formatting group of commands on the Measures tools tab to format your measure*

Let's create our second measure, this time to calculate the average cases sold as follows:

```
Avg Cases =
AVERAGE ( Winesales[CASES SOLD] )
```

Another analysis you may need to perform on your data is calculating "how many," for example, the number of sales for each different wine. In other words, we need to count the number of rows in the Winesales table for each wine shown in the visual. The implicit measure that we could use here uses the DAX COUNT function that counts the *number of values* in the column you reference (for more information on the COUNT function, visit `https://docs.microsoft.com/en-us/dax/count-function-dax`). However, we want to count the *number of rows,* and therefore, only an explicit measure will do the job we want. The DAX function we need is the COUNTROWS function whose name describes its purpose. This function accepts a table as its only argument which is the table whose rows you want to count, so this would be the expression:

```
No. of Sales =
COUNTROWS ( Winesales )
```

One of the benefits of creating these simple measures is that you can use them to analyze any items from any dimension. As you generate visuals, taking items from different dimensions, the measures will consistently recalculate accordingly. For example, in Figure 3-12, we're using our measures in three Table visuals showing data from the following dimensions:

- WINE from the Wines dimension
- SALESPERSON from the SalesPeople dimension
- REGION from the Regions dimension

REGION	Total Cases	Avg Cases	No. of Sales
Argentina	18,377	211.23	87
Australia	15,794	183.65	86
Canada	6,317	180.49	35
China	27,389	195.64	140
Czech Republic	33,958	196.29	173
England	23,080	198.97	116
France	12,213	187.89	65
Germany	19,158	193.52	99
India	34,292	174.07	197
Ireland	1,160	290.00	4
Italy	35,374	194.36	182
Japan	22,153	203.24	109
New Zealand	23,813	177.71	134
Northern Ireland	3,489	183.63	19
Russia	1,043	260.75	4
Scotland	25,839	198.76	130
South Africa	26,002	192.61	135
United Arab Emirates	27,102	193.59	140
United States	14,210	205.94	69
Wales	52,461	185.37	283
Total	**423,224**	**191.76**	**2,207**

WINE	Total Cases	Avg Cases	No. of Sales
Bordeaux	54,070	300.39	180
Champagne	49,158	372.41	132
Chardonnay	42,030	224.76	187
Chenin Blanc	24,739	123.70	200
Chianti	27,323	184.61	148
Grenache	35,965	197.61	182
Malbec	34,290	201.71	170
Merlot	23,084	147.03	157
Piesporter	10,253	89.16	115
Pinot Grigio	23,449	139.58	168
Rioja	33,951	172.34	197
Total	**423,224**	**191.76**	**2,207**

SALESPERSON	Total Cases	Avg Cases	No. of Sales
Abel	69,871	185.83	376
Blanchet	65,581	191.20	343
Charron	68,137	196.36	347
Denis	84,018	193.14	435
Leblanc	69,304	195.22	355
Reyer	66,313	188.93	351
Total	**423,224**	**191.76**	**2,207**

Figure 3-12. *Measures are calculated according to the data comprising the visual*

Our final example of a simple DAX measure will accomplish an insightful calculation that would be difficult to repeat in Excel, that of the distinct count. In DAX, we have an aggregate function for this job. Its name is DISTINCTCOUNT, and we can simply reference the column required for the analysis. Let's discover how many *different* customers we sold our wines to by authoring this measure:

```
Distinct Customers =
DISTINCTCOUNT ( Winesales[CUSTOMER ID] )
```

While we're focusing on the DISTINCTCOUNT function, remember that we created this *calculated column* in the Winesales table:

```
REGION NAME =
RELATED ( Regions[REGION NAME] )
```

We can use this calculated column to create a measure to calculate in how many *different* regions we've sold our wines:

```
Distinct Regions =
DISTINCTCOUNT ( Winesales[REGION NAME] )
```

WINE	Distinct Customers	Distinct Regions
Bordeaux	57	18
Champagne	53	19
Chardonnay	58	19
Chenin Blanc	55	19
Chianti	50	18
Grenache	51	17
Malbec	53	20
Merlot	54	18
Piesporter	44	19
Pinot Grigio	51	17
Rioja	59	17
Sauvignon Blanc	58	20
Shiraz	50	18
Total	**84**	**20**

Figure 3-13. *Using the DISTINCTCOUNT function*

You will observe in Figure 3-13 that we've sold "Bordeaux" to 57 different customers and "Champagne" to 53 different customers. We've sold "Rioja" in 17 different regions.

What Exactly Is a Measure?

We've created a few simple explicit measures, but we still haven't answered the following question: What is a measure? The answer, like measures themselves, is not a straightforward one. *A measure is a DAX expression that is used in a Power BI visual to return a scalar value and is evaluated in a specific filter context.* In other words, DAX measures *filter the rows of tables* and typically perform an *aggregation* on the filtered data to return a scalar value (which is a single value) that is visualized in the report.

Note Not all DAX measures perform aggregations. As we will see later, some DAX measures can return text values. Nevertheless, they will be scalar in nature in that they will return a single value.

For example, a typical DAX measure might sum the values in a column containing quantities (e.g., the "Total Cases" measure) where the rows in the fact table are filtered for each year, and this analysis is visualized in a column chart where each year's totals (e.g., 2021) can be seen; see Figure 3-14.

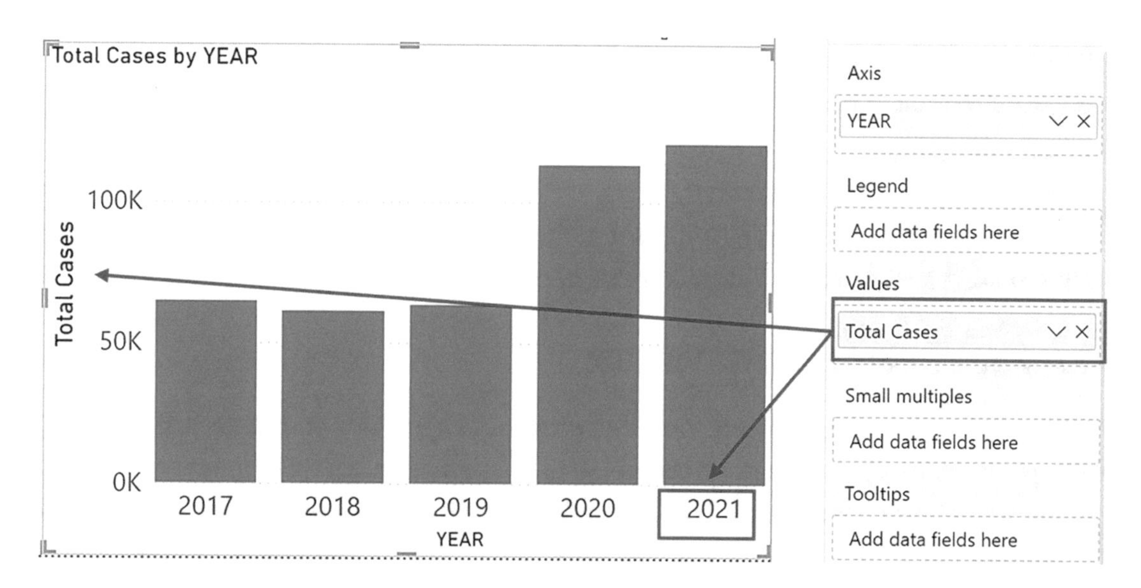

Figure 3-14. *Measures typically aggregate filtered data*

Let's take a closer look at these three aspects of the measure:

1. All visuals on the report canvas use measures.
2. Measures return scalar (single) values.
3. Measures are calculated where a filter has been placed on the data model. This is known as the *filter context* and is the subject of the next chapter.

All Report Visuals Use Measures

When we authored our calculated columns, these are seen in Data view and they returned a value for every row in the table. A measure, on the other hand, is used in Report view and is placed in the Values bucket of a visual. *All visuals use measures* in the Values bucket even if they are implicit measures (which, as already described, is a numeric column that you've dragged into the Values bucket).

Note There is an exception to this rule. The Key Influencers visual is best used with a non-aggregated column, rather than a measure. For more information on the Key Influencers visual, visit my blog: `www.burningsuit.co.uk/blog/2020/01/the-key-influencers-visual-versus-strictly-come-dancing/`

Another way to think of measures is that they are *report-level* calculations as opposed to the *row-level* calculations that you create in calculated columns.

Measures Return Scalar Values

All Power BI visuals are reporting tools that group and aggregate your data, just like an Excel pivot table or pivot chart. Therefore, to understand this aspect of the measure, let's put our Excel hats on and remind ourselves that in Power BI, Table and Matrix visuals are the equivalents of Excel pivot tables. For instance, consider the values in the Table visual in Figure 3-15.

WINE	Total Cases
Bordeaux	54,070
Champagne	49,158
Chardonnay	42,030
Chenin Blanc	24,739
Chianti	27,323
Grenache	35,965
Malbec	34,290
Merlot	23,084
Piesporter	10,253
Pinot Grigio	23,449
Rioja	33,951
Sauvignon Blanc	47,415
Shiraz	17,497
Total	**423,224**

Figure 3-15. *The value identified sits in the equivalent of the "Values" area of an Excel pivot table and would be in a "cell"*

If this were an Excel pivot table, the "Total Cases" values would be sitting in the "Values" area of the pivot table, and every value returned by the calculation would be sitting in a "cell." We've identified the "cell" for "Bordeaux" wine that holds the value of **54,070** being returned by this measure:

```
Total Cases =
SUM ( Winesales[Cases Sold] )
```

What does this value represent? It represents the sum of the values in the CASES SOLD column for all the rows in the Winesales table that equate to "Bordeaux" wines.

If the same data were sitting in an Excel pivot table, we could double-click on this value and drill through to display these rows on a separate sheet, as shown in Figure 3-16.

Row Labels	Sum of CASESSOLD
Bordeaux	54070
Champagne	49158
Chardonnay	42030
Chenin Blanc	24739
Chianti	27323
Grenache	35965
Malbec	34290
Merlot	23084
Piesporter	10253
Pinot Grigio	23449
Rioja	33951
Sauvignon Blanc	47415
Shiraz	17497
Grand Total	**423224**

SALEDATE	WINESALESNO	SALESPERSONID	CUSTOMERID	WINEID	CASESSOLD	TOTALSALES	WINE
23/12/2021	2207	3	12	1	290	21750	Bordeaux
14/12/2021	2184	1	34	1	190	14250	Bordeaux
13/12/2021	2181	4	3	1	330	24750	Bordeaux
06/12/2021	2169	5	11	1	188	14100	Bordeaux
20/11/2021	2145	4	44	1	149	11175	Bordeaux
14/11/2021	2134	3	37	1	329	24675	Bordeaux
15/10/2021	2083	3	16	1	197	14775	Bordeaux
07/10/2021	2065	5	39	1	451	33825	Bordeaux
05/10/2021	2060	3	18	1	304	22800	Bordeaux
21/09/2021	2037	3	25	1	240	18000	Bordeaux
19/09/2021	2033	1	17	1	382	28650	Bordeaux
13/09/2021	2022	4	10	1	328	24600	Bordeaux
11/09/2021	2019	5	28	1	495	37125	Bordeaux

Figure 3-16. *In an Excel pivot table, you can drill through*

We can't drill through on the value in the Power BI Table visual, but nevertheless, the measure in memory does the same. It filters a set of *specific rows* from a table. In our example, it filters the rows in the Winesales table for "Bordeaux" wines. However, the result of the measure must sit in the "cell" of the Table visual just as it sits in the cell of the pivot table. Therefore, the measure must return a *scalar value.* Typically, this would mean that the measure must aggregate the data; for example, sum the cases sold for "Bordeaux" wines.

Note Normally, a scalar value would be a single numeric value, but measures can return single text values as well, so here, the term "scalar" is used in a more general sense to mean a single value of any data type.

We've established that measures must return scalar or single values, a concept that we're sure you think is straightforward and easy to understand, but at some point, you'll attempt to create measures that return errors that look like that shown in Figure 3-17.

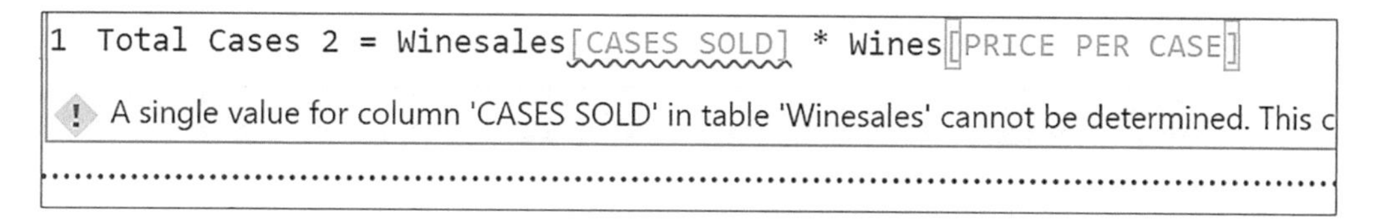

***Figure 3-17.** This error message displays when there is no aggregation*

The error message in Figure 3-17 reads:

"A single value for column 'CASES SOLD' in table 'Winesales' cannot be determined. This can happen when a measure formula refers to a column that contains many values without specifying an aggregation such as min, max, count, or sum to get a single result."

What is the reason for this error message? There is no *aggregation* in the measure; it's just multiplying two values.

Another example of where a measure does not return a scalar value is shown in Figure 3-18. Here, the VALUES function is being used in a Table visual (we look at the VALUES function in a later chapter). The measure should return a scalar value, which it does when evaluating individual rows in the Table visual, but when calculating the Total row of the visual, it returns multiple values, and so an error message is displayed when the measure is put into a Table visual that has the Total row turned on.

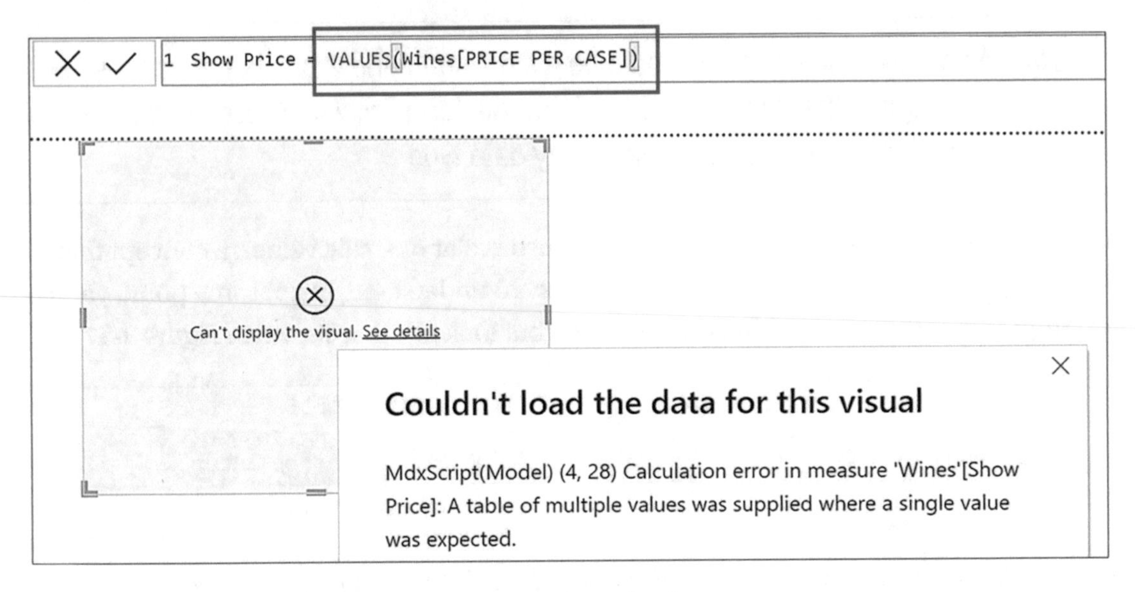

Figure 3-18. *Some measures return a table of values*

The error message reads:

"A table of multiple values was supplied where a single value was expected."

Even the most hardened DAX experts can be caught out by creating measures that don't return scalars!

In this chapter we have explored the difference between calculated columns and measures. You understand that calculated columns are row level calculations while measures are used in all visuals and are calculations that are performed at report level. However, we're still missing an explanation of the third and most important ingredient of the DAX measure, that all measures are evaluated in a *specific filter context.* To understand what is meant by this, you will need to move forward to the next chapter where we will focus on the context in which our expressions are evaluated and why this is so important in understanding DAX measures.

CHAPTER 4

Evaluation Context

You have learned to author simple calculated columns and measures, but one of the most fundamental questions for DAX users is how these two types of expression differ. At this stage, you understand that calculated columns are row-level calculations and that measures are calculations that are performed at the report level. However, we need to be more specific regarding this differentiation, and you need to understand that the definitive difference lies in the context in which the expressions are evaluated. In calculated columns, expressions are evaluated in the *row context*; in measures, they are evaluated in the *filter context*. It is the latter of these that will be the main focus in this chapter. Once you understand the implications of the filter context, the implications of the row context are more readily understood.

The Filter Context

In the last chapter, you learned that measures are report-level calculations and that they must return a scalar value. This brings us to the third and most important aspect of the measure, and that is that all DAX measures are *evaluated in a specific filter context*. To understand what is meant by a "specific filter context," let's compare these two different measures:

```
Total Cases =
SUM ( Winesales[CASES SOLD] )
```

```
Total Stores =
SUM ( Customers[NO. OF STORES] )
```

You can see the evaluation of these measures in Figure 4-1, but why does the first measure return different values for each wine but the second measure return the same value? The reason is the *filter context* that's active when both these measures are evaluated.

© Alison Box 2022
A. Box, *Up and Running with DAX for Power BI*, https://doi.org/10.1007/978-1-4842-8188-8_4

WINE	Total Cases	Total Stores
Bordeaux	54,070	1,181
Champagne	49,158	1,181
Chardonnay	42,030	1,181
Chenin Blanc	24,739	1,181
Chianti	27,323	1,181
Grenache	35,965	1,181
Lambrusco		1,181
Malbec	34,290	1,181
Merlot	23,084	1,181
Piesporter	10,253	1,181
Pinot Grigio	23,449	1,181
Rioja	33,951	1,181
Sauvignon Blanc	47,415	1,181
Shiraz	17,497	1,181
Total	**423,224**	**1,181**

Figure 4-1. *Measures will return different values or the same value because of the filter context that is active*

When a measure is placed into any visual, before the measure is evaluated, the DAX engine *in memory* places filters on tables in the data model depending on three factors:

1. The column or columns placed *in the visual* that group and categorize the data

2. The columns *in slicers* that are filtering the data in the visual

3. Any columns placed in *the Filters pane* that are filtering the data in the visual

These three factors come together to generate the filter context for the evaluation of the measure. We can't see these filters on the data model. We just have to imagine them.

Note The filtering of the data model happens in memory and is hidden from us. Therefore, in the Figures below and throughout this book, where we're simulating what happens in memory, the in-memory tables have a dashed border to distinguish them from the tables you can see in Data view.

In our Table visual in Figure 4-1, only factor #1 is relevant (there are no slicers or other filters).

Evaluations Using a Single Filter

The column in the visual that's grouping the data is the WINE column from the Wines dimension. The first value in this column to be calculated is the total cases for "Bordeaux" wine.

Before the "Total Cases" measure calculates the value for "Bordeaux," a filter is placed in memory on the Wines dimension to filter "Bordeaux" wines. If we could see the filter on this table, it might look something like Figure 4-2.

WINE ID	WINE	SUPPLIER	TYPE	WINE COUNTRY	PRICE PER CASE	COST PRICE
1	Bordeaux	Laithwaites	Red	France	$75.00	$25.00

Figure 4-2. *The in-memory Wines dimension that has been filtered to one row*

If we examine the data model (Figure 4-3), we can see that the Wines dimension is related to the Winesales fact table in a many-to-one relationship. The arrow tells us that if the Wines dimension is filtered, this filter is *propagated* onward to the Winesales fact table.

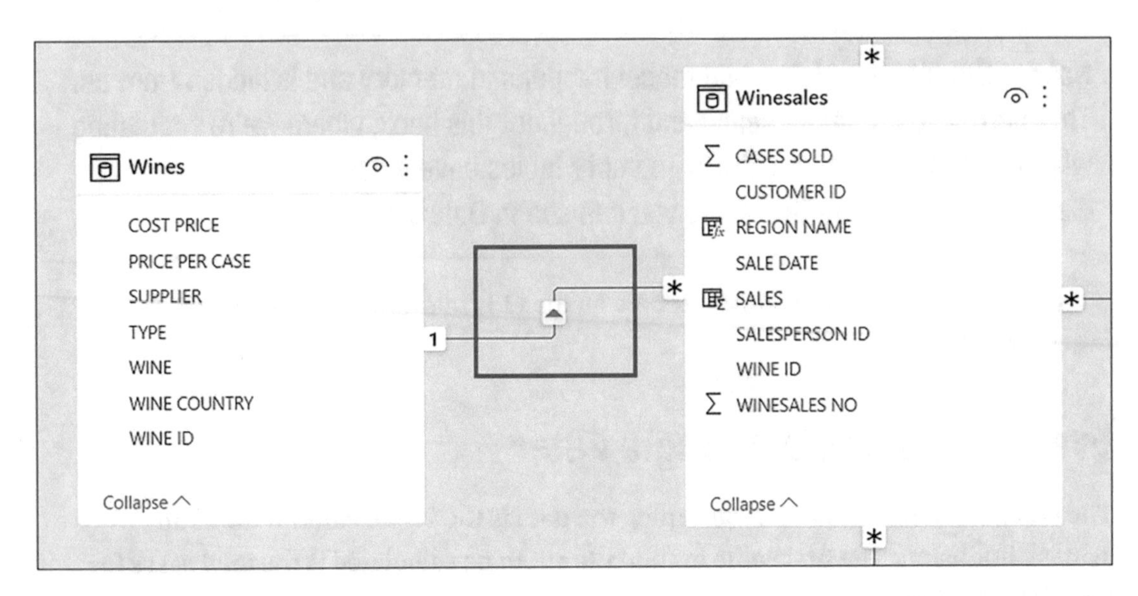

***Figure 4-3.** Filters propagate from the Wines dimension to the Winesales fact table*

Therefore, the Winesales fact table is now *cross-filtered* to only contain sales for "Bordeaux" wine that has the WINE ID that equals 1; see Figure 4-4. Notice there is no filter in the WINE ID column in the Winesales table because the filter on the Winesales table is a cross-filter that is generated only through filter propagation.

SALE DATE	WINESALES NO	SALESPERSON ID	CUSTOMER ID	WINE ID	CASES SOLD
03/07/2021	1903	6	30	1	323
24/04/2021	1769	6	55	1	208
21/03/2021	1703	6	5	1	421
29/01/2021	1618	6	34	1	236
07/01/2021	1571	6	1	1	234
26/12/2020	1557	6	3	1	322
19/11/2020	1482	6	35	1	249
08/11/2020	1455	6	35	1	485
06/10/2020	1411	6	21	1	365
05/07/2020	1245	6	21	1	491
01/06/2020	1182	6	33	1	358
01/05/2020	1124	6	46	1	280
18/04/2020	1100	6	24	1	409

***Figure 4-4.** The fact table is cross-filtered via the dimension*

This is the only filter affecting this visual, so the measure now sums the CASES SOLD column for "Bordeaux" wines and returns **54,070**.

The evaluation of the measure then moves on to "Champagne" and repeats the process of filtering the Wines dimension and cross-filtering the Winesales fact table using a *different filter context* each time. In the next evaluation, for instance, the WINE column from the Wines dimension now equals "Champagne" and so now returns **49,158.**

Note Experienced DAX users will know that this explanation of the filter context in action is a close approximation of what happens in memory and not exactly what happens. However, this explanation is easily understood at this stage of your knowledge and will serve you well for the time being. We will reveal what really happens under the hood later in this book.

And so on for all the wines in the WINE column of the Table visual. *Every evaluation of the "Total Cases" measure is evaluated in a different filter context.*

There is a way that we can prove that our Wines dimension, in memory, is filtered to one row on the evaluation of a measure that analyzes each wine. We can create this measure that counts the rows of the Wines dimension:

```
No. of Wines = COUNTROWS ( Wines )
```

If we put this measure into a Table visual containing the WINE column from the Wines dimension, the measure will return **1** for the evaluation of each wine; see Figure 4-5.

WINE	No. of Wines
Bordeaux	1
Champagne	1
Chardonnay	1
Chenin Blanc	1
Chianti	1
Grenache	1
Lambrusco	1
Malbec	1
Merlot	1
Piesporter	1
Pinot Grigio	1
Rioja	1
Sauvignon Blanc	1
Shiraz	1
Total	**14**

Figure 4-5. *The "No. of Wines" measure returns 1 because the Wines dimension has been filtered down to one row for each evaluation*

Notice too how "Lambrusco" wine returns a value because this measure filters only the Wines dimension and no other tables are involved.

Calculation in the Total Row

This now brings us to the calculation for the Total row of the visual, which returns **423,224**; see Figure 4-6.

Pinot Grigio	23,449
Rioja	33,951
Sauvignon Blanc	47,415
Shiraz	17,497
Total	**423,224**

Figure 4-6. *The Total row is evaluated in a different filter context*

This value is *not* the sum of the total values for each wine shown in the visual. When the measure is evaluated for the Total row, the filter is removed from the WINE column of the Wines dimension, so the expression is evaluated for *all* wines. In other words, it's our expression *"= SUM (Winesales[CASES SOLD])"* calculated in yet *another different filter context.*

Evaluations Using Multiple Filters

Let's create some more filters that affect the Table visual. For instance, we could include a slicer using the SALESPERSON column from the SalesPeople dimension[1] and also have the REGION column from the Regions dimension in a page-level filter[2]; see Figure 4-7.

WINE	Total Cases	Total Stores
Bordeaux	265	79
Champagne		79
Chardonnay	209	79
Chenin Blanc		79
Chianti	242	79
Grenache		79
Lambrusco		79
Malbec	256	79
Merlot	449	79
Piesporter	254	79
Pinot Grigio	112	79
Rioja	386	79
Sauvignon Blanc	261	79
Shiraz	131	79
Total	**2,565**	**79**

SALESPERSON
Abel
Blanchet
Charron
Denis
Leblanc
Reyer

Filters on this page ...
REGION
is Argentina
Filter type
Basic filtering
Search
Select all
Argentina 1
Australia 1
Canada 1
China 1
Czech Republic 1
England 1
Require single selection

Figure 4-7. *Filters are now placed on the Table visual from the slicer and the page-level filter*

[1] For information on working with slicers, visit `https://docs.microsoft.com/en-us/power-bi/visuals/power-bi-visualization-slicers`

[2] For information on working with the Filters pane, visit `https://docs.microsoft.com/en-us/power-bi/create-reports/power-bi-report-filter?tabs=powerbi-desktop`

We've filtered salesperson "Abel" and region "Argentina". You can see that the Total Cases value for "Bordeaux" is now **265** because the filter context has changed; WINE equals "Bordeaux", SALESPERSON equals "Abel", and REGION equals "Argentina". Again, we can imagine how these tables might look in memory; see Figure 4-8.

SALESPERSON ID	SALESPERSON	FIRSTNAME
1	Abel	Claude

REGION ID	REGION
100	Argentina

***Figure 4-8.** The in-memory tables filtering the Table visual*

You will notice, however, that the Total Stores measure is still returning the same value for every wine (i.e., 79). We will explain why presently.

Again, we can examine the data model (Figure 4-9) and can see how these filters propagate through the model and always arrive at the Winesales fact table, which is then cross-filtered accordingly.

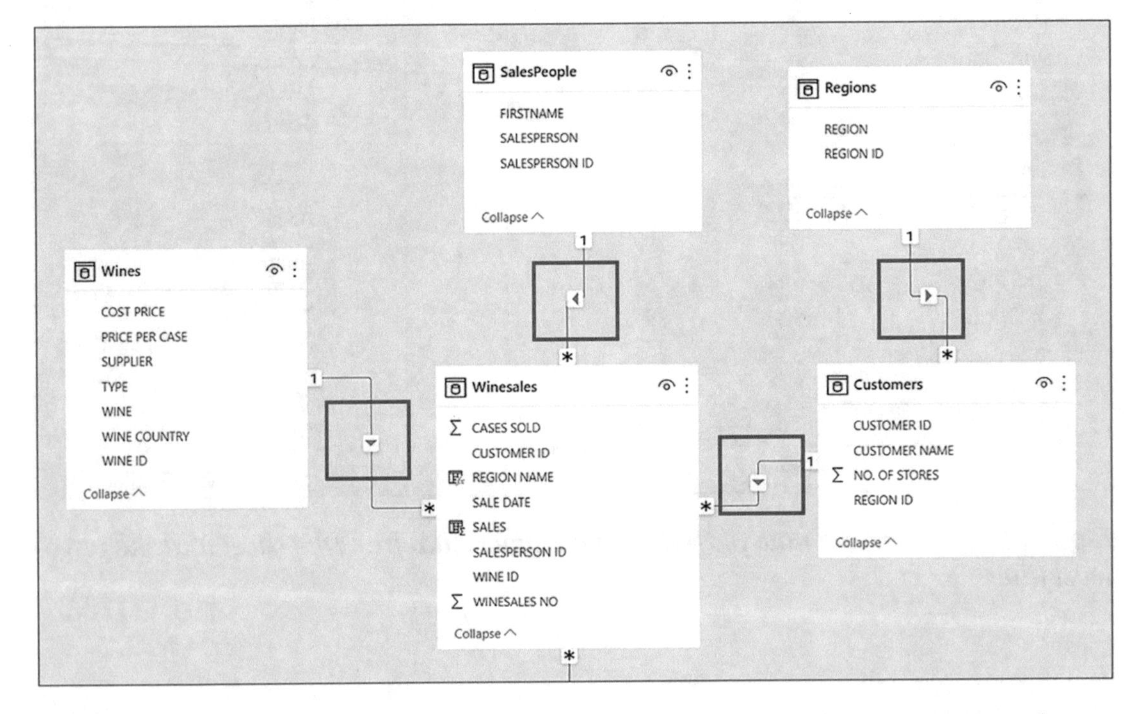

***Figure 4-9.** Filters propagate through the data model and always arrive on the fact table*

Notice how the Regions table creates a "snowflake" in the schema because it's *indirectly* related to the fact table via the Customers dimension. You can see how this arrangement of tables works; if the Regions table is filtered, for example, for "Argentina", this filter is propagated through to the Customers dimension, so customers in Argentina are now filtered in memory. This filter is then propagated onward to the fact table.

Depending on how the visual is constructed and what filters affect the visual will determine the outcome of the measure. This now brings us to the "Total Stores" measure shown in Figure 4-1. Notice it returns the same value of **1,181** for every wine and also in the Total row. This measure is summing the NO. OF STORES column in the Customers dimension. The Customers dimension has no filter on it when this measure is evaluated. The only filter is on the Wines dimension. Therefore, for the evaluation of every wine, the measure sums the values in the NO. OF STORES column in the Customers table for ***all*** the customers.

Looking again at the data model (Figure 4-10), we can see that if the Wines dimension is filtered, this filter is propagated to the fact table (shown by the tick), but the filter is ***not*** propagated onward to the Customers dimension (shown by the cross), as the arrow always points from the one side of the relationship into the many.

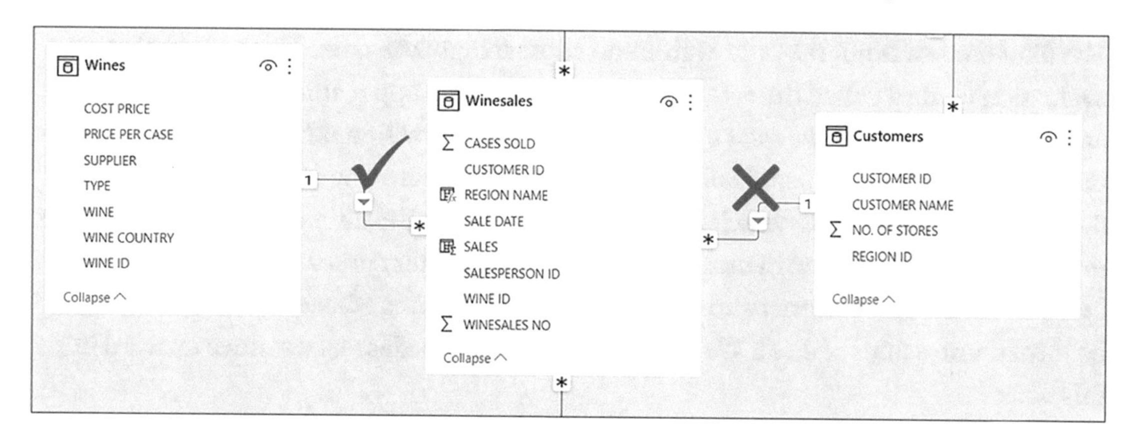

Figure 4-10. *Filters do not flow from the fact table to dimensions*

Note Well, how do you correctly calculate the number of stores in which each wine has been sold? One thing *not* to do, tempting though it is, is to edit the relationship to a "bidirectional" filter. Instead, you can use the DAX function CROSSFILTER to programmatically reverse the direction of the filter propagation. We look at the CROSSFILTER function later in this book.

The filter context underpins all DAX measures and is the reason why it's so important to distinguish between the two different types of table, dimension tables and fact tables, because they play two different roles in the evaluation of the measure:

- The role of dimension tables is to group the data and to propagate filters through the data model into the fact table.
- The role of fact tables is to summarize subsets of data that have been cross-filtered from dimensions.

DAX measures typically summarize data in the fact table that's been cross-filtered by dimension tables.

So next time you're wondering "why is my measure returning incorrect values," it's probably not the expression that's at fault; it's more likely because you haven't understood the current filter context in which the measure has been evaluated.

The Row Context

The filter context is not the only evaluation context that DAX uses. There is another evaluation context called the *row context.* Row context is applicable in any DAX expression that *iterates the rows of a table* where the expression is bound to the values in the current row. All calculated columns are evaluated in the row context and this is how they differ from measures, which are always evaluated in the filter context. However, just to make life difficult, some measures will use both the filter context *and* the row context in their evaluation. Also, there are some calculated columns whose row context can be turned into a filter context. We will be exploring these ideas as we move forward in this book.

To understand the row context, let's again refer to what we know about Excel formulas. In an Excel table, the formula is "copied down" where it is calculated for every row in the column. An "@" character is used in the formula to denote using the values in the current row. This is essentially what the row context is in DAX. When using the row context, the DAX expression iterates over every row in the table, and the values used in the expression are *the values sitting in the current row;* see Figure 4-11.

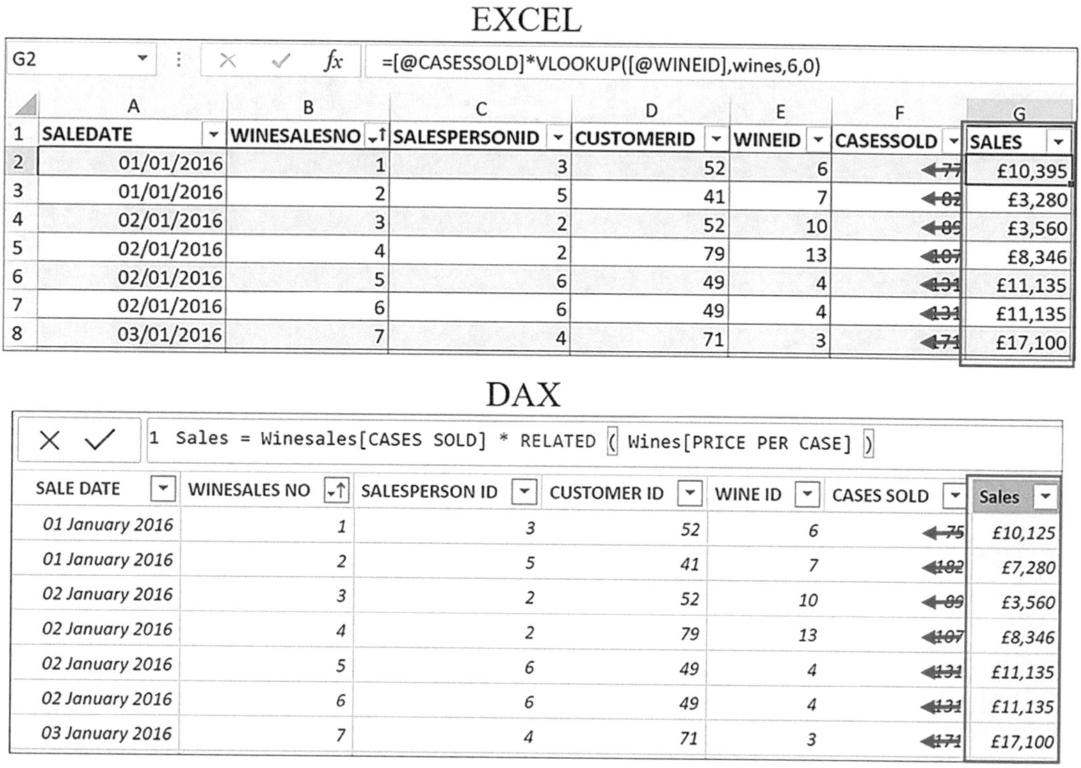

EXCEL

G2 fx =[@CASESSOLD]*VLOOKUP([@WINEID],wines,6,0)

	A	B	C	D	E	F	G
1	SALEDATE	WINESALESNO	SALESPERSONID	CUSTOMERID	WINEID	CASESSOLD	SALES
2	01/01/2016	1	3	52	6	77	£10,395
3	01/01/2016	2	5	41	7	82	£3,280
4	02/01/2016	3	2	52	10	89	£3,560
5	02/01/2016	4	2	79	13	107	£8,346
6	02/01/2016	5	6	49	4	131	£11,135
7	02/01/2016	6	6	49	4	131	£11,135
8	03/01/2016	7	4	71	3	171	£17,100

DAX

```
1 Sales = Winesales[CASES SOLD] * RELATED ( Wines[PRICE PER CASE] )
```

SALE DATE	WINESALES NO	SALESPERSON ID	CUSTOMER ID	WINE ID	CASES SOLD	Sales
01 January 2016	1	3	52	6	75	£10,125
01 January 2016	2	5	41	7	182	£7,280
02 January 2016	3	2	52	10	89	£3,560
02 January 2016	4	2	79	13	107	£8,346
02 January 2016	5	6	49	4	131	£11,135
02 January 2016	6	6	49	4	131	£11,135
03 January 2016	7	4	71	3	171	£17,100

Figure 4-11. *Both Excel table formulas and DAX calculated columns use values from the current row, known as the "row context" in DAX*

We can understand that calculated columns would normally use the row context, but measures can also use the row context in their evaluation. But surely the nature of all DAX measures is to group and summarize data, not to perform row-level calculations. Well, measures can perform row-level calculations too, and this is where the behavior of *iterators* comes in, a concept we will explore in the next chapter.

However, let's now summarize what you have learned in this chapter, and that is that all measures use the filter context in their evaluation. The filter context refers to filters that will be placed on the data model by the evaluation of the measure and depends on the construct of the visual in which the measure will be calculated and on any filters that impact on the visual. You now know also that there is a second evaluation context, the row context, where the DAX expression scans a table and performs row-level calculations as in the case of the calculated column. Understanding the two evaluation contexts that differentiate measures from calculated columns is the first major DAX concept that you have learned. Some people who have been using DAX, perhaps for some length of time, are often not able to explain this fundamental difference between measures and calculated columns.

CHAPTER 5

Iterators

There is a group of functions in DAX that are referred to as *iterators,* and from their name, we can infer that these functions iterate tables in the evaluation of a DAX expression. Any DAX function that ends in an "X" is an iterator, such as the "X" aggregators: SUMX, AVERAGEX, MAXX, MINX, COUNTAX. There are also "X" iterating functions that aren't aggregators such as CONCATENATEX and RANKX. Just to make life even more confusing, there are iterating functions that don't end in "X" such as FILTER and ADDCOLUMNS.

We will explore the FILTER, CONCATENATEX, and RANKX functions later. The ADDCOLUMNS function is beyond the remit of the book, but hopefully it will be something you self-explore as your knowledge of DAX increases. The focus of this chapter will be the aggregating iterators: SUMX, AVERAGEX, MAXX, MINX, and COUNTAX.

Aggregating iterators have two arguments: the table to be iterated and the expression that is to be evaluated for each row of the table, the result of which will then be aggregated. These functions create a row context inside the measure by iterating the table referenced by the function, and each row in the table is "visited" in memory by the measure. Remember that the measure will have generated a filter context first, so the table being iterated may have a filter or cross-filter on it.

© Alison Box 2022
A. Box, *Up and Running with DAX for Power BI,* https://doi.org/10.1007/978-1-4842-8188-8_5

DAX Measure

```
1 10 PC Increase Total =
2     SUMX ( Winesales,Winesales[CASES SOLD]* 1.1 )
```

SALES NO	SALESPERSON ID	CUSTOMER ID	WINE ID	CASES SOLD	10 Percent Increase
1903	6	30	1	323	355
1769	6	55	1	208	229
1703	6	5	1	421	463
1618				236	260
1571				234	257
1557				322	354
1482				249	274
1455				485	534
1411				365	402
1245				491	540
1182				358	394
1124				280	308

WINE	Total Cases	10 PC Increase Total
Bordeaux	54,070	59,477.00
Champagne	49,158	54,073.80
Chardonnay	42,030	46,233.00
Chenin Blanc	24,739	27,212.90
Chianti	27,323	30,055.30
Grenache	35,965	39,561.50
Malbec	34,290	37,719.00
Merlot	23,084	25,392.40
Piesporter	10,253	11,278.30
Pinot Grigio	23,449	25,793.90
Rioja	33,951	37,346.10
Sauvignon Blanc	47,415	52,156.50
Shiraz	17,497	19,246.70
Total	**423,224**	**465,546.40**

Figure 5-1. *The SUMX function iterates the cross-filtered fact table and performs a row-level calculation that is then summed by the measure*

For example, consider the measure "10 PC Increase Total" being evaluated in Figure 5-1. Here, we are using the aggregating iterator SUMX in the measure to multiply *in memory* the CASES SOLD value in each row of the fact table by 1.1. The results of these row-level calculations are then aggregated to return a scalar value returned by the measure, for example, **59,477.00** for "Bordeaux" wine. In a similar way, we could have used AVERAGEX or MAXX or any of the iterating aggregators.

Measures that include iterating functions use the row context in their iteration and then use the filter context to generate the scalar value.

Let's move forward now and explore these aggregating iterators in more detail, starting with SUMX.

The SUMX Function (and Other "X" Functions)

Now that you understand the purpose of DAX iterating aggregators, let's get to know one of the major iterating functions in DAX, and that's the SUMX function. We can then move on to explore other "X" aggregators.

SUMX returns the sum of an expression evaluated for each row in a table and has the following syntax:

= SUMX (table, expression)

where:

table is the table where you want to perform the calculation.

expression is the calculation you want to be performed for each row in that table.

Here's an example of the SUMX syntax:

= SUMX (Winesales, Winesales[CASES SOLD] * 0.1)

To illustrate the use of SUMX, let's start with this rather unrealistic but easy-to-understand scenario. We have been asked to find any CASES SOLD value that is greater than 100 and increase this value by 20%; otherwise, we only increase the value by 10%. Perhaps this is some strange way of predicting next year's volume of cases sold, so we'll call this calculation "Next Yr Cases". We then want to see what the "Next Yr Cases" value would be for each of our wines.

If we didn't know how to use SUMX, we would probably do this calculation in a clumsy way using both a calculated column and an implicit measure. We might create this calculated column using the IF function as shown in the following and then use an implicit measure by dragging the calculated column into the Values bucket of a Table visual; see Figure 5-2.

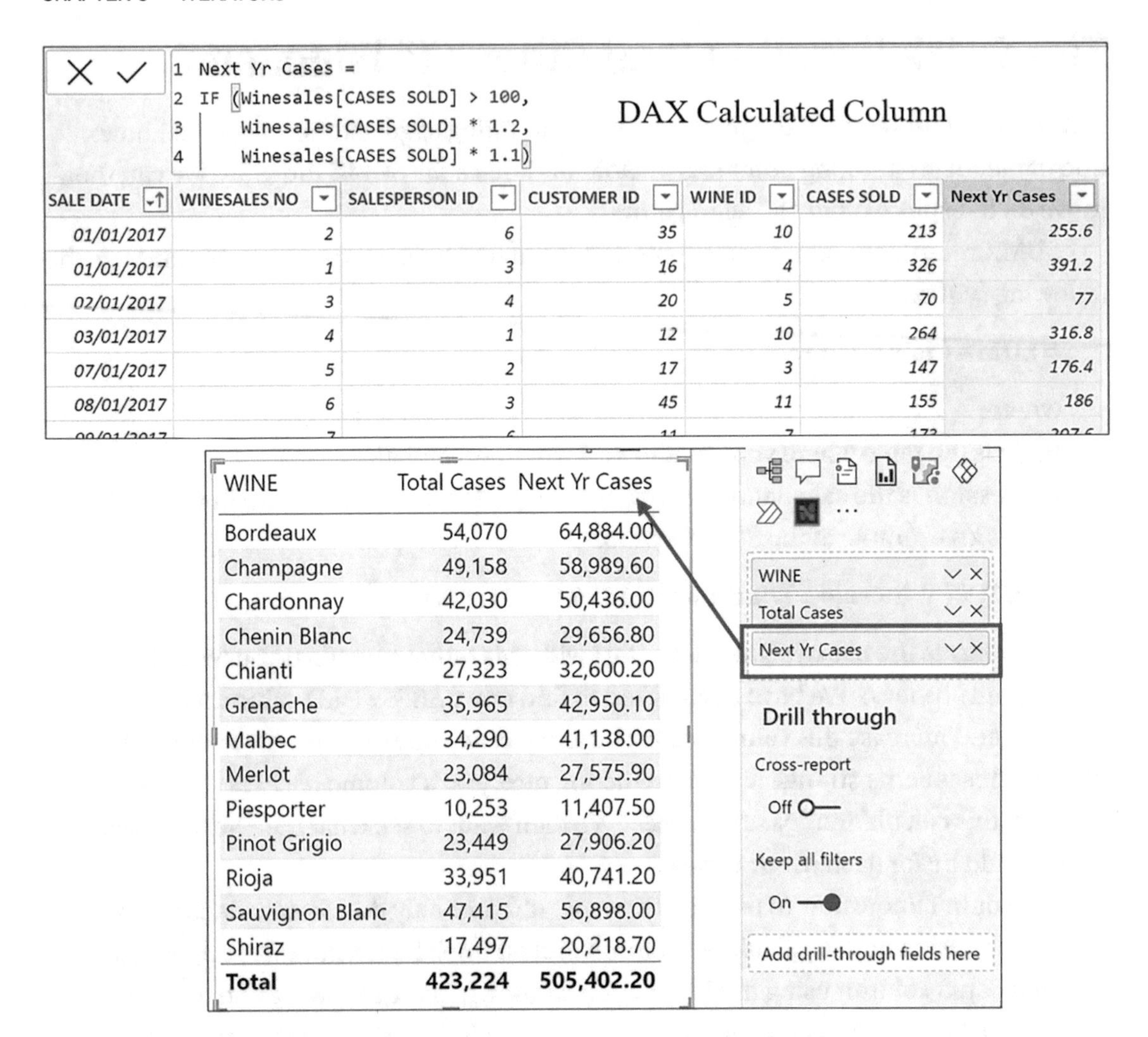

***Figure 5-2.** Creating a calculated column to be used as in implicit measure isn't efficient*

But let's think this through. We don't need to first create a calculated column to see the increased value for each row and then in *another step* sum this value for each wine, using an implicit measure. We can do it all in one go using SUMX. If we do this, the requirement for the calculated column is redundant; we can just use the measure. This is the real benefit because measures are always more efficient than calculated columns.

This is the explicit measure using SUMX that we can use instead of the calculated column/implicit measure combination:

```
Next Yr Cases Measure =
```

```
SUMX (
    Winesales,
    IF ( Winesales[CASES SOLD] > 100,
        Winesales[CASES SOLD] * 1.2,
        Winesales[CASES SOLD] * 1.1
    )
)
```

How does the SUMX measure work?

We know that SUMX sums the expression evaluated for *each row in the table.* The first argument in SUMX references the table where the calculation will be performed, in our case, Winesales. The second argument is the calculation you want to be done in memory *for each row* in this table. This is our expression using IF that SUMX calculates in memory by iterating every row. It then sums the results of this calculation, in this case, for each wine (because that's the current filter context for the evaluation of the measure).

Now that we have discovered the SUMX function, we can revisit a calculation we learned to author in Chapter 3, and that's the "Sales" calculation that's currently sitting in a *calculated column*; see Figure 3-6. Do you remember we created this *column* using the RELATED function?

```
Sales =
Winesales[CASES SOLD] * RELATED ( Wines[PRICE PER CASE] )
```

We then used this column in an implicit measure to find the total sales, but this wasn't the most efficient way of accomplishing this task. Well, now we can write a *measure* that will be our definitive "Total Sales" calculation using SUMX, as follows:

```
Total Sales =
SUMX (
    Winesales,
    Winesales[CASES SOLD] * RELATED ( Wines[PRICE PER CASE] )
)
```

This is a much cleaner way to calculate our Total Sales. The SUMX function iterates the Winesales table, and for every row in the current filter context, it multiplies the value in the CASES SOLD column with value in the PRICE PER CASE column of the Wines table (using RELATED to find the price of the wine in the current row context). It then sums the results of these row-level calculations for each wine.

Tip Select a measure and use the Measure Tools tab and the Formatting group to format your measures in the currency of your choice.

In a similar way, if you want to find the maximum sales or the average sales, the DAX measures would be these respectively:

```
Max Sales =
MAXX (
    Winesales,
    Winesales[CASES SOLD] * RELATED ( Wines[PRICE PER CASE] )
)

Avg Sales =
AVERAGEX (
    Winesales,
    Winesales[CASES SOLD] * RELATED ( Wines[PRICE PER CASE] )
)
```

You can see the results of these measures in Figure 5-3.

WINE	Total Sales	Avg Sales	Max Sales
Bordeaux	$4,055,250	$22,529	$37,500
Champagne	$7,373,700	$55,861	$75,000
Chardonnay	$4,203,000	$22,476	$25,000
Chenin Blanc	$1,236,950	$6,185	$7,500
Chianti	$1,092,920	$7,385	$11,960
Grenache	$1,078,950	$5,928	$10,500
Malbec	$2,914,650	$17,145	$27,710
Merlot	$900,276	$5,734	$7,800
Piesporter	$1,384,155	$12,036	$21,870
Pinot Grigio	$703,470	$4,187	$6,000
Rioja	$1,527,795	$7,755	$9,000
Sauvignon Blanc	$1,896,600	$11,289	$14,000
Shiraz	$1,364,766	$6,723	$11,700
Total	**$29,732,482**	**$13,472**	**$75,000**

Figure 5-3. *Measures using SUMX, AVERAGEX, and MAXX*

Let's explore another example of using AVERAGEX by calculating the average price that our customers have paid for their wines. We need to first find the price of every transaction (in the current filter context) in the Winesales table and then calculate the average of these prices, so the measure would look like this:

```
Average Price =
AVERAGEX (
    Winesales,
    RELATED ( Wines[PRICE PER CASE] )
)
```

The "Average Price" measure uses the RELATED function to calculate the price of each transaction in memory, and AVERAGEX then averages these prices. You can see the results of this measure in Figure 5-4.

CUSTOMER NAME	Average Price
Back River & Co	$66.20
Ballard & Sons	$95.00
Barstow Ltd	$46.71
Beaverton & Co	$82.14
Black Ltd	$61.83
Bluffton Bros	$59.73
Branch Ltd	$63.40
Brooklyn & Co	$45.64
Brooklyn Ltd	$73.00
Brown & Co	$45.75
Burlington Ltd	$62.80

***Figure 5-4.** Calculating the average price that customers paid for their wines*

This may seem a simple measure, but even some experienced DAX users struggle to get it right, so let's explain its evaluation. We can see in Figure 5-4 that the filter context is on the CUSTOMER NAME column of the Customers dimension and "Black River & Co" is the first instance. The Winesales fact table is cross-filtered to contain only this customer's sales. The RELATED function, nested inside AVERAGEX, in memory calculates the PRICE PER CASE value from the Wines dimension for each row in the Winesales table for "Black River & Co." The AVERAGEX function then finds the average of these prices (it sums the prices and divides by the number of rows in the Winesales table for this customer).

Total Row Grief

This brings us to another common problem for people who are new to DAX: understanding the calculation on the Total row of a Table or Matrix visual. People often complain that it's not correct. This is probably because they've used the SUM function when they should have used SUMX.

Consider the measures in Figure 5-5 that compare the "Total Sales" measure to the "Total Sales Wrong" measure. You can see that for each wine, the "Total Sales" and the "Total Sales Wrong" measures both return correct results. But when the measures evaluate the Total row, the "Total Sales Wrong" measure shows an incorrect result.

WINE	Sum Price	Total Sales	Total Sales Wrong
Bordeaux	$75	$4,055,250	$4,055,250
Champagne	$150	$7,373,700	$7,373,700
Chardonnay	$100	$4,203,000	$4,203,000
Chenin Blanc	$50	$1,236,950	$1,236,950
Chianti	$40	$1,092,920	$1,092,920
Grenache	$30	$1,078,950	$1,078,950
Lambrusco	$20		
Malbec	$85	$2,914,650	$2,914,650
Merlot	$39	$900,276	$900,276
Piesporter	$135	$1,384,155	$1,384,155
Pinot Grigio	$30	$703,470	$703,470
Rioja	$45	$1,527,795	$1,527,795
Sauvignon Blanc	$40	$1,896,600	$1,896,600
Shiraz	$78	$1,364,766	$1,364,766
Total	**$917**	**$29,732,482**	**$388,096,408**

Figure 5-5. *The Total row calculation is incorrect for the "Total Sales Wrong" measure*

So what is the problem with "Total Sales Wrong" when it evaluates the Total row? The problem, as is often the case, will be found within the filter context. Remember what we learned earlier; the Total row calculation is *not* the sum of the total values you see in the visual. In the Total row, the measure is evaluated in a different filter context where the filter has been removed from the WINE column. So let's look at how things can easily go awry. This is the measure for "Total Sales Wrong":

```
Total Sales Wrong =
SUM ( Winesales[CASES SOLD] ) * SUM ( Wines[PRICE PER CASE] )
```

Let's also extract the two constituent expressions into their own separate measures:

```
Total Cases =
SUM ( Winesales[CASES SOLD] )

Sum Price =
SUM ( Wines[PRICE PER CASE] )
```

Our "Total Sales Wrong" measure is multiplying the results of these two expressions; refer to Figure 5-5.

You can see that the problem lies in using SUM, particularly in trying to sum the PRICE PER CASE values. What is the sum of these values? It's the sum of the price in the current filter context. So for the evaluation of each wine, it's simply the price of the wine, for example, **$75** for "Bordeaux"; see Figure 5-6.

WINE ID	WINE	SUPPLIER	TYPE	WINE COUNTRY	PRICE PER CASE	COST PRICE
1	Bordeaux	Laithwaites	Red	France	$75.00	$25.00

Figure 5-6. *The sum of the price on the evaluation of each wine is the price of the wine*

Multiplying this value by the sum of the cases sold for each wine gives the correct total sales value when evaluating each wine. But for the evaluation of the Total row, the filter has been released from the WINE column, so the "*SUM (Wines[PRICE PER CASE])*" expression sums the prices for all the wines and returns **$917**, see Figure 5-7. It is this value that is multiplied by the sum of the cases sold.

WINE ID	WINE	SUPPLIER	TYPE	WINE COUNTRY	PRICE PER CASE	COST PRICE
1	Bordeaux	Laithwaites	Red	France	$75.00	$25.00
2	Champagne	Laithwaites	White	France	$150.00	$100.00
3	Chardonnay	Alliance	White	France	$100.00	$75.00
4	Malbec	Laithwaites	Red	Germany	$85.00	$40.00
5	Grenache	Redsky	Red	France	$30.00	$10.00
6	Piesporter	Redsky	White	Germany	$135.00	$50.00
7	Chianti	Redsky	Red	Germany	$40.00	$10.00
8	Pinot Grigio	Majestic	White	Italy	$30.00	$5.00
9	Merlot	Majestic	Red	France	$39.00	$15.00
10	Sauvignon Blanc	Majestic	White	Italy	$40.00	$20.00
11	Rioja	Majestic	Red	Italy	$45.00	$15.00
12	Chenin Blanc	Alliance	White	France	$50.00	$10.00
13	Shiraz	Alliance	Red	France	$78.00	$30.00
14	Lambrusco	Alliance	White	Italy	$20.00	$15.00

Figure 5-7. *The sum of the prices on the evaluation of each Total row*

So in the "Total Sales Wrong" measure, the Total row calculation is the sum of the prices for all the wines multiplied by the sum of cases for all the wines: **917** x **423,224** = **388,096,408**.

The incorrect expression using SUM *sums and then multiplies*. The correct expression using SUMX *multiplies and then sums.*

The SUM function should only be used in the simplest of measures to sum the values in a single column and never when you want to sum the results of multiplications or other calculations. In fact, even when you use the SUM function in a DAX expression, this is converted internally by the DAX engine into SUMX.

So for instance, this expression

```
=SUM ( Winesales[CASES SOLD] )
```

is converted internally to this:

```
=SUMX ( Winesales, Winesales[CASES SOLD] )
```

In learning about iterators and how to use SUMX and the other "X" iterating functions, we're progressing well into more difficult areas of DAX. We've also shed light on other challenging areas of DAX, such as the filter context and the nature of measures. You've learned some other important concepts too, understanding the role of fact tables and dimensions within the data model, but we're still only starting out.

The real power behind DAX is still waiting in the wings for us to discover, and that's the use of the function called CALCULATE.

CHAPTER 6

The CALCULATE Function

CALCULATE is the most important function in DAX. Quite a sweeping statement you might think but as soon as you get to grips with CALCULATE, you'll quickly realize that there won't be many expressions you author in DAX where this function won't be required, even though you might think we've done pretty well up to now. In this chapter, you will learn how to construct expressions using CALCULATE which you will find relatively straightforward. It's understanding when and why you must use CALCULATE, and its purpose inside the measure, that will be more challenging to grasp, and so this will be the true focus of this chapter.

Why You Need CALCULATE

Let's look at solving a scenario that will explain how CALCULATE can help us. In our data model, we have our DateTable dimension that is related to the Winesales fact table by the DateTable[DATEKEY}column and the Winesales[SALE DATE] column as shown in Figure 6-1.

© Alison Box 2022
A. Box, *Up and Running with DAX for Power BI*, https://doi.org/10.1007/978-1-4842-8188-8_6

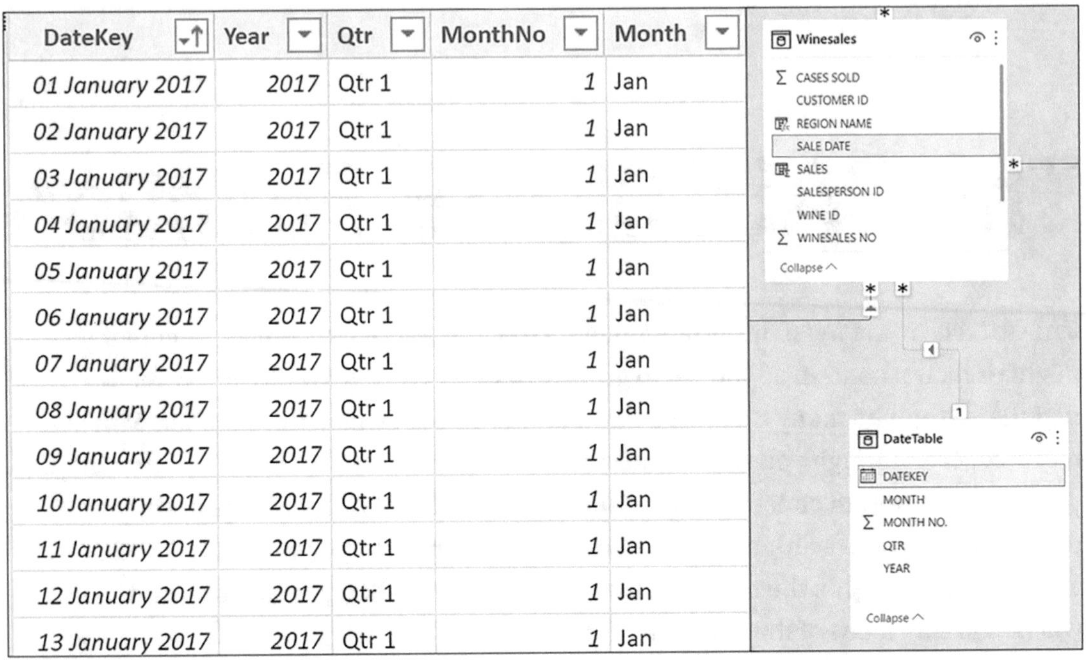

DateKey	Year	Qtr	MonthNo	Month
01 January 2017	2017	Qtr 1	1	Jan
02 January 2017	2017	Qtr 1	1	Jan
03 January 2017	2017	Qtr 1	1	Jan
04 January 2017	2017	Qtr 1	1	Jan
05 January 2017	2017	Qtr 1	1	Jan
06 January 2017	2017	Qtr 1	1	Jan
07 January 2017	2017	Qtr 1	1	Jan
08 January 2017	2017	Qtr 1	1	Jan
09 January 2017	2017	Qtr 1	1	Jan
10 January 2017	2017	Qtr 1	1	Jan
11 January 2017	2017	Qtr 1	1	Jan
12 January 2017	2017	Qtr 1	1	Jan
13 January 2017	2017	Qtr 1	1	Jan

Winesales
CASES SOLD
CUSTOMER ID
REGION NAME
SALE DATE
SALES
SALESPERSON ID
WINE ID
WINESALES NO
Collapse

DateTable
DATEKEY
MONTH
MONTH NO.
QTR
YEAR
Collapse

Figure 6-1. *The DateTable dimension is related to the fact table*

If we filter on the YEAR column in the DateTable, the filter will propagate to the Winesales table to filter the sales for that year. We've been asked to carry out a specific analysis of our data. For each of our wines, we would like to calculate what percentage the total cases sold for 2021 is of the total cases sold for all years as shown in Figure 6-2.

WINE	Total Cases	2021 Cases	2021 PC
Bordeaux	54,070	14,940	27.63%
Champagne	49,158	11,461	23.31%
Chardonnay	42,030	11,302	26.89%
Chenin Blanc	24,739	6,952	28.10%
Chianti	27,323	8,535	31.24%
Grenache	35,965	9,702	26.98%
Malbec	34,290	10,543	30.75%
Merlot	23,084	8,158	35.34%
Piesporter	10,253	4,080	39.79%
Pinot Grigio	23,449	6,040	25.76%
Rioja	33,951	10,161	29.93%
Sauvignon Blanc	47,415	14,689	30.98%
Shiraz	17,497	4,137	23.64%
Total	**423,224**	**120,700**	**28.52%**

Figure 6-2. *A Table visual showing what percentage the cases sold for 2021 are of the total for all years*

In other words, *in the same visual,* we need to have both the "Total Cases" measure for all years and the "Total Cases" measure filtered for the year 2021. We can then divide "Total Cases" for all years into "2021 Cases" and express this as a percentage.

If you look at the example in Figure 6-3, you can see that we have copied and pasted the "Total Cases" measure and named it "2021 Cases".

WINE	Total Cases	2021 Cases
Bordeaux	54,070	54,070
Champagne	49,158	49,158
Chardonnay	42,030	42,030
Chenin Blanc	24,739	24,739
Chianti	27,323	27,323
Grenache	35,965	35,965
Malbec	34,290	34,290
Merlot	23,084	23,084
Piesporter	10,253	10,253
Pinot Grigio	23,449	23,449
Rioja	33,951	33,951
Sauvignon Blanc	47,415	47,415
Shiraz	17,497	17,497
Total	**423,224**	**423,224**

Figure 6-3. *We can copy a measure and attempt to apply filters to the copied measure*

We want to see if we can filter the "2021 Cases" measure to show values for 2021, *while at the same time* the "Total Cases" measure shows values for all years. However, we have a problem. If we use a slicer to filter the YEAR column from the DateTable, it filters *both* measures; see Figure 6-4.

WINE	Total Cases	2021 Cases
Bordeaux	14,940	14,940
Champagne	11,461	11,461
Chardonnay	11,302	11,302
Chenin Blanc	6,952	6,952
Chianti	8,535	8,535
Grenache	9,702	9,702
Malbec	10,543	10,543
Merlot	8,158	8,158
Piesporter	4,080	4,080
Pinot Grigio	6,040	6,040
Rioja	10,161	10,161
Sauvignon Blanc	14,689	14,689
Shiraz	4,137	4,137
Total	**120,700**	**120,700**

YEAR
2017
2018
2019
2020
2021

Figure 6-4. *Filters are applied to all measures in the visual*

It seems that we must find a way to apply different filter contexts for different measures *in the same visual.* However, as yet, any filters being used by a visual, be they from the visual itself, from slicers, or from filters in the filters pane, apply the *same* filter to all the measures in the visual. We can't yet pick and choose which filters affect which measures. At the moment it's all or nothing. This is where the CALCULATE function can help us.

Note At this juncture, the slicer filtering the YEAR column can be removed from the canvas as it does not impact the data as required and is now redundant.

CALCULATE evaluates an expression in a *modified* filter context and has the following syntax:

= **CALCULATE (expression , filter1 , filter2 etc.)**

where:

expression is what you want calculated. This can be a DAX **expression** or a **measure** that defines an expression.

filter1, filter2, etc. is how you want to filter the **expression** or **measure**. You can have multiple filters, and these are combined in an "AND" logical statement.

Here are two examples of the CALCULATE syntax:

= CALCULATE (SUM (Winesales[CASES SOLD]), Wines[WINE] = "Bordeaux")

= CALCULATE ([Total Cases], Wines[WINE] = "Bordeaux")

The first example uses an ***expression*** in the expression argument, and the second uses a ***measure*** in the expression argument (highlighted in gray).

This is the first time that we have nested a measure inside a "parent" measure. Note that when you type your expression in the DAX editor, if you type a square bracket " [", IntelliSense will list only measures; see Figure 6-5.

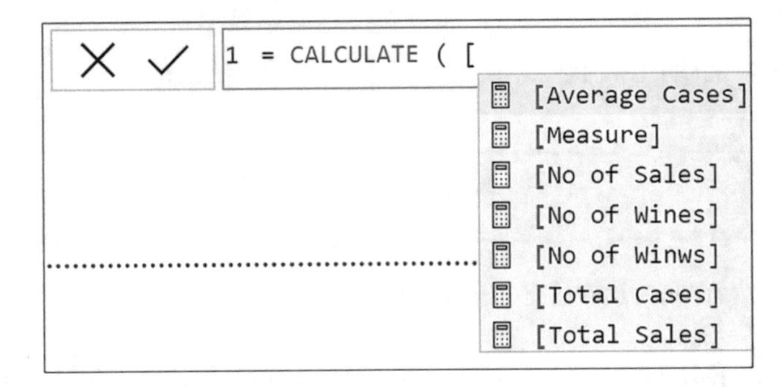

Figure 6-5. *Typing a square bracket " [" into the DAX editor, lists all your measures*

CALCULATE takes *an* ***expression*** or a ***measure*** and evaluates it in a different filter context from the active filters coming through from the visual, slicers, or the filters pane. The end result of this new filter context generated by CALCULATE depends on the current state of the active filters. This is what is meant by the filter context being "modified" in the description of CALCULATE. However, rather than trying to explain what CALCULATE does, perhaps it's easier to work through some examples.

Using Single Filters

Back to our scenario. In the Table visual in Figure 6-4, how do we generate a measure to calculate the Total Cases for 2021 while retaining the measure that calculates Total Cases for all years? This is the measure, using CALCULATE that will do the job:

```
2021 Cases =
CALCULATE ( [Total Cases], DateTable[Year] = 2021)
```

Now we can create the final measure that will calculate the percentage that each wine's total cases for 2021 are of the total for all years and format it as percent:

```
2021 Percentage =
 [2021 Cases] / [Total Cases]
```

Or better still:

```
2021 Percentage =
DIVIDE ( [2021 Cases], [Total Cases] )
```

Note The DIVIDE function returns a blank value by default if there is a divide by zero error, so using DIVIDE is the preferred method of performing divisions.

When we put either of these measures into a Table visual, we can see that the total cases for "Bordeaux" wine in 2021 comprised **27.63%** of the total cases for "Bordeaux" wine for all years (2017 to 2021).

Let's look more closely at the evaluation of the "2021 Cases" measure in Figure 6-2. We can see that coming through from the Table visual, we have a filter on the WINE column in the Wines dimension. However, using the "2021 Cases" measure, CALCULATE in memory *also* filters the DateTable dimension so that YEAR equals 2021. This filter is then applied to the Winesales fact table alongside the filter coming through from the Wines dimension; see Figure 6-6.

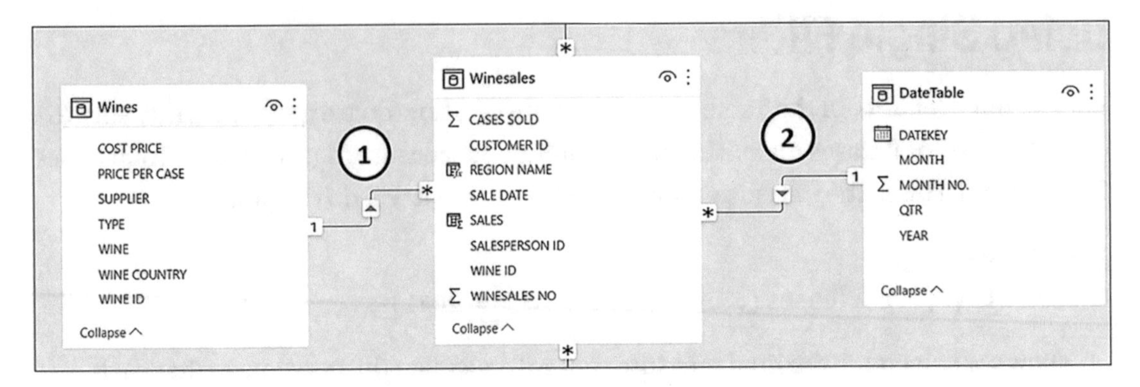

Figure 6-6. *How filters propagate for the "2021 Cases" measure:*

1. The current filter context filters each WINE in the Wines dimension. This filter is propagated to the Winesales table and cross-filters each wine.

2. The filter provided by CALCULATE programmatically filters the DateTable for the year "2021". This filter is also propagated to the Winesales table and so applies a second cross-filter on Winesales.

The thing to note here is that the 2021 filter on the DateTable dimension only affects the evaluation of *this* measure and no other measures in the visual. You can think of CALCULATE as being a way to *programmatically* generate a filter context in memory that interacts with active filters coming through from the visual, slicers, and the filters pane.

Let's now explore some more examples of using CALCULATE. For example, let's calculate the total sales where the CASES SOLD value is greater than 350.

```
Total Sales for Cases Sold Greater than 350 =
CALCULATE (
    [Total Sales],
    Winesales[CASES SOLD]  > 350
)
```

Using Multiple Filters

CALCULATE accepts multiple filter arguments that are combined in an AND logical statement. If you require an OR statement, you can use the OR operator or the OR function within a single filter argument inside CALCULATE, and we will be exploring both of these scenarios. You can also use more complex filters that require aggregate expressions inside the filter arguments of CALCULATE, and we will be moving forward to understand these expressions too.

AND and OR Filters

The following are three more examples of measures that use CALCULATE to modify the filter context. Notice how all the filter arguments to CALCULATE are combined in an "AND".

```
Total Cases in May 2021 =
CALCULATE (
    [Total Cases],
    DateTable[YEAR] = 2021,
    DateTable[MONTH] = "may"
)

Total Cases for Abel in Argentina =
CALCULATE (
    [Total Cases],
    SalesPeople[SALESPERSON] = "abel",
    Regions[REGION] = "argentina"
)

Average Cases for Black Ltd in 2021 =
CALCULATE (
    AVERAGE ( Winesales[CASES SOLD] ),
    DateTable[Year] = 2021,
    Customers[CUSTOMER NAME] = "black ltd"
)
```

You can see the results of these expressions using the WINE column from the Wines dimension in Figure 6-7.

WINE	Total Cases	Total Cases in May 2021	Total Cases for Abel in Argentina	Average Cases for Black Ltd in 2021
Rioja	33,951	657	386	164
Piesporter	10,253	447	254	87
Bordeaux	54,070	1,031	265	
Champagne	49,158	313		
Chardonnay	42,030	1,390	209	
Chenin Blanc	24,739	128		
Chianti	27,323	1,032	242	
Grenache	35,965	645		
Malbec	34,290	1,296	256	
Merlot	23,084	1,027	449	
Pinot Grigio	23,449	413	112	
Sauvignon Blanc	47,415	603	261	
Shiraz	17,497	252	131	
Total	**423,224**	**9,234**	**2,565**	**126**

Figure 6-7. *Measures using multiple filters generated by CALCULATE*

In the preceding examples, filter arguments in CALCULATE are combined in an "AND" statement, for example, cases sold for 2021 ***AND*** May. However, what if you require a filter that uses "OR", for example, 2021 ***OR*** 2020. Using CALCULATE, filtering using "OR" on the same column is straightforward. Filtering using "OR" on *different* columns is a little more challenging, and this is where our calculations will get a little trickier. Let's take the simpler calculations first.

To use "OR" on the same column, you can use the double pipe (||) operator within the same filter argument, as in these examples:

```
Total Cases 2020 or 2021 =
CALCULATE ( [Total Cases],
    DateTable[YEAR] = 2021
        || DateTable[YEAR] = 2020
)
```

```
Average Cases Argentina or Australia =
CALCULATE (
    AVERAGE ( Winesales[CASES SOLD] ),
    Regions[REGION] = "argentina"
        || Regions[REGION] = "australia"
)
```

You can also use the OR function, but unlike Excel, you can only put two parameters into the DAX OR function as in this example:

```
Average Cases Argentina or Australia =
CALCULATE (
    AVERAGE ( Winesales[CASES SOLD] ),
    OR ( Regions[REGION] = "argentina",
        Regions[REGION] = "australia")
)
```

Complex Filters

Let's take another example of an "OR" filter. For example, we may want to find Total Sales for red wines OR French wines using the TYPE and WINECOUNTRY columns in the Wines table, respectively, and use this to analyze our salespeople's performance of these wines. This would be the expression:

```
Sales for Red or French #1=
CALCULATE (
    [Total Sales],
    Wines[TYPE] = "red"
        || Wines[WINE COUNTRY] = "France"
)
```

This measure appears to work just fine as you can see in Figure 6-8.

SALESPERSON	Total Sales	Sales for Red or French #1
Abel	$5,265,266	$4,647,576
Blanchet	$4,860,044	$4,193,329
Charron	$5,147,366	$4,583,416
Denis	$5,431,390	$4,518,335
Leblanc	$4,792,407	$4,220,947
Reyer	$4,236,009	$3,584,654
Total	**$29,732,482**	**$25,748,257**

Figure 6-8. *The "Sales for Red or French #1" measure evaluated for each salesperson*

However, experienced DAX users would be surprised that this expression was valid and would expect an error message as shown in Figure 6-9 that states

> ***"The expression contains multiple columns, but only a single column can be used in a True/False expression that is used as a table filter expression."***

```
1 Sales for Red or French = CALCULATE ( [Total Sales], Wines[TYPE]= "red" || Wines[WINE COUNTRY] = "france" )
```

The expression contains multiple columns, but only a single column can be used in a True/False expression that is used as a table filter expression.

Figure 6-9. *This error message was removed in the March 2021 update of Power BI*

This message tells us that referencing two columns from the same table in a single filter is not allowed. In fact, the expression using "OR" on different columns has only become legitimate since the March 2021 update of Power BI.

However, although it appears to now be valid, there is still an inherent problem with it. This expression doesn't respond correctly to specific filter selections. To show this, we have written an alternative measure, "Sales for Red or French #2", and can now compare the two versions of this expression in a Table visual where we are filtering "Red" wines via the slicer; see Figure 6-10.

SALESPERSON	Total Sales	Sales for Red or French #1	Sales for Red or French #2
Abel	$2,050,276	$4,647,576	$2,050,276
Charron	$2,029,616	$4,583,416	$2,029,616
Denis	$2,711,085	$4,518,335	$2,711,085
Leblanc	$2,232,097	$4,220,947	$2,232,097
Blanchet	$1,734,279	$4,193,329	$1,734,279
Reyer	$2,177,254	$3,584,654	$2,177,254
Total	**$12,934,607**	**$25,748,257**	**$12,934,607**

TYPE
- Red
- White

Figure 6-10. *Using "OR" on different columns from the same table doesn't respond correctly to specific filters*

You will see that the measure "Sales for Red or French #1" doesn't respond to filters from the slicer that uses the TYPE column from the Wines dimension. It continues to calculate sales for both red or French wines disregarding the slicer. The second measure, "Sales for Red or French #2", however, does show just sales for red wines. We will look at the details of this measure in the chapter on the FILTER function, but for the moment, we have to ask this question: Why has an expression that filters two different columns from the same table been invalid until recently, and now that we are allowed to do it, why doesn't it calculate correctly with a filter on the TYPE column?

Let's look more closely at the problem. When you have a filter on just one column, the rows of the table are filtered in memory where the filter criterion on the column equates to true. But when you place filters on *multiple* columns, you can only further reduce the rows. For example, once you've filtered out the red wines, you can only then filter the red wines that are French.

How can we solve this predicament? One way is to ensure that there are no filters on either the Wines[TYPE] column or the Wines[WINE COUNTRY] column so that in every evaluation, values in *both* columns are considered. This is the route that DAX takes in the expression "Sales for Red or French #1".

Note It's beyond the scope of this chapter to elaborate on the details of the "Sales for Red or French #1" expression or why it returns errors in the presence of certain filters. However, we do uncover the problem in Chapter 18. All we need to note at this stage is that the expression doesn't always return the correct result.

Is there an alternative approach? Perhaps we could try this; rather than applying filters directly to columns, we could *filter out the rows* that we want to evaluate instead. For example, we could iterate the rows in the Wines dimension, and if we find a red wine, filter the row out, or if we find a French wine, filter that row out too. We could then, in memory, build a new *virtual table* comprising just the rows for wines that are red or French. This in-memory virtual Wines table that has been filtered to just the rows we need could then propagate that filter to the Winesales fact table, just like the "real" Wines dimension filtering the Winesales table. Would that work?

Well yes, it would because in DAX, there is a group of functions called "table" functions that generate *in-memory virtual* tables that, when used inside CALCULATE, will propagate filters just like "real" tables. Now that we know this, all that remains for us to discover is the name of the table function that will generate our virtual table containing just the rows for red or French wines.

Before we find this function, however, there's a little more learning to be done. We need to look more closely at the *different types* of DAX functions and particularly to understand what we mean by "table functions." Then we can solve our "red or French" conundrum.

CHAPTER 7

DAX Table Functions

A skill that will serve you well when working with DAX is a good imagination. You've already learned to construct a picture in your mind of the current filters that are propagating through the data model. The scanning of tables by iterators can only be envisaged, and designing the correct CALCULATE expression is done through inferring what filters must be changed. There is yet another aspect of DAX that is hidden from us, and that therefore must be imagined. That is the generation of virtual tables. Much of your DAX code will involve building in-memory tables that are used in the evaluation of the measure. In this chapter, we are going to explore this concept, how we create table expressions through the use of table functions, and their purpose in manipulating the data model. In doing so, we will be focusing on the most ubiquitous of the table functions, and that is the FILTER function.

Types of DAX Functions

In DAX, we can divide functions into three categories depending on the type of value the functions return; see Table 7-1.

© Alison Box 2022
A. Box, *Up and Running with DAX for Power BI*, https://doi.org/10.1007/978-1-4842-8188-8_7

Table 7-1. *Types of DAX functions*

Function Type	Example	Description
Scalar Functions	Return scalar values. e.g., SUM, COUNTROWS, SUMX, CALCULATE	These functions return a scalar or single value and are used in all measures.
Table Functions	Return virtual tables. e.g., FILTER, VALUES, PREVIOUSMONTH, ALL	Table functions are used to generate "virtual" tables that propagate filters through the data model in the same way as "real" tables. The virtual tables are typically subsets of rows or subsets of columns of the original table, but they can expand the number of rows in the case of the ALL function. Because measures must always return scalar values and not tables, table functions are always nested inside scalar functions.
CALCULATE Modifiers	Modify the filter arguments of CALCULATE. e.g., CROSSFILTER, USERELATIONSHIP, KEEPFILTERS, ALL	We'll meet this type of function later. These functions change the behavior of any filters generated by CALCULATE and so are always nested inside CALCULATE. These functions don't return any value.

Note The ALL function is both a Table function and a CALCULATE modifier. We'll look more closely at this later.

How do you know what type of function you are using? The best way is to consult the DAX Function Library here: `https://docs.microsoft.com/en-us/dax/dax-function-reference`, and it will tell you what a DAX function returns; for example, the FILTER function returns "*a table* containing only the filtered rows," see Figure 7-1.

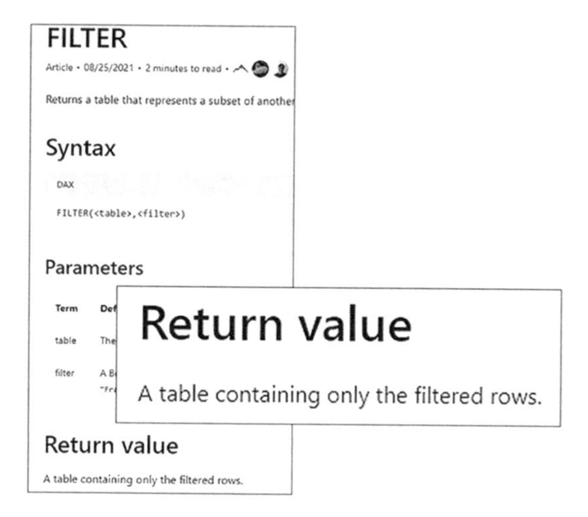

***Figure 7-1.** Use the DAX Function Library to check the function type*

We've already explored some scalar functions, and we'll meet some more in later chapters such as the SELECTEDVALUE function. We will also delve into CALCULATE modifiers like CROSSFILTER, USERELATIONSHIP and ALL later on. In this chapter, we will focus only on table functions.

Table Functions

Table functions create *table expressions* and can be used for two purposes:

1. To generate additional tables in your data model using the **New Table** button. These are referred to as *calculated tables*. If this is your requirement, the recommendation is that new tables are generated using Power Query, not DAX.
2. To generate in-memory virtual tables as part of the evaluation of measures.

In this chapter, we will only be considering the latter of these, the generation of virtual tables using table expressions inside DAX measures.

Table expressions can be used in measures wherever a function accepts a "table" as one of its arguments or as the filter argument inside CALCULATE. Up to now, we've always referenced an actual table inside functions like COUNTROWS or SUMX, but we can use a table expression instead. Inside CALCULATE, we've created Boolean expressions as column filters, but we could also use table expressions. When creating measures, table expressions are *always* nested inside functions that return scalar values and are never used on their own.

Examples of Table Expressions

Consider the expressions in Figure 7-2 where, in place of referencing a table, we're using a table expression instead.

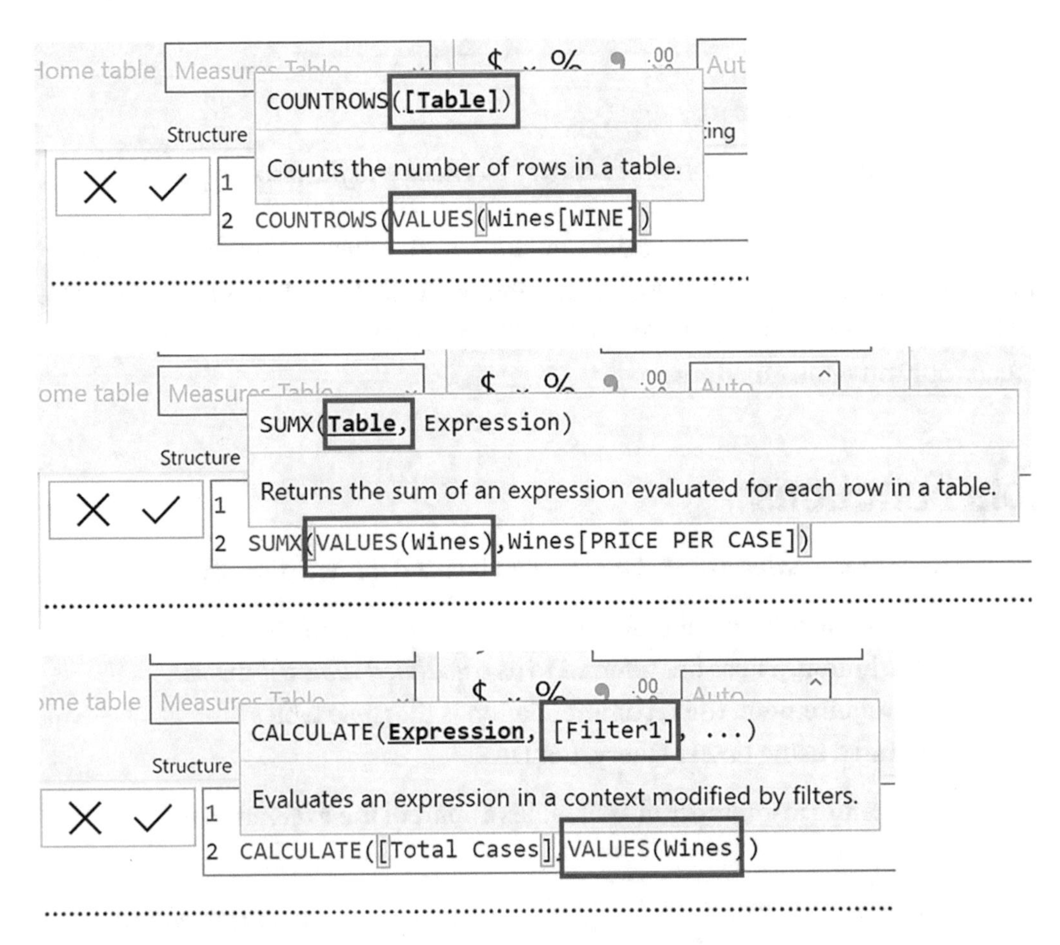

Figure 7-2. *Examples of table expressions*

These examples use a table function called VALUES. You don't need to know at this stage what the VALUES function is doing (we'll meet VALUES in a later chapter). You just need to understand that it's a *table expression* being used as the "table" argument or as the "filter" argument inside CALCULATE.

Why Do We Need Table Expressions?

There are two very different reasons why we use table expressions inside DAX measures.

Nested inside any other function other than CALCULATE, table expressions supply the "table" argument and often create subsets of the original table, either subsets of rows or subsets of columns. For example, FILTER nested inside SUMX will normally generate a table with fewer rows for SUMX to iterate. As we will discover in later chapters, some table functions are also used to generate "hybrid" tables that comprise combinations of columns from different tables.

On the other hand, as filter arguments inside CALCULATE, table expressions generate virtual tables that are used as *filters*. Understanding the use of table expressions as filter arguments inside CALCULATE is a challenging concept to new DAX users, and we'll be exploring this concept in detail as we move through this chapter.

However, we will begin our journey through table functions by understanding the use of the most common table function in DAX, and that is FILTER.

The FILTER Function

The FILTER function returns a *table* that is a subset of another table and has the following syntax:

= FILTER (table , filter)

where:

table is the table that you want to filter. The table can also be supplied by another table function.

filter is the filter you want to apply to the **table** as a Boolean expression, for example, *"Wines[TYPE]= "red"*.

Here is an example of the FILTER function syntax:

= COUNTROWS (FILTER (Wines, Wines[TYPE]= "red"))

FILTER as a table function can be used to generate table expressions as explained in "Why Do We Need Table Expressions?" section. We've learned that these functions have different behaviors depending on whether they are used to change the shape of the data, such as reducing the rows considered by an expression, or whether they are used inside CALCULATE. The FILTER function is no exception, so let's now consider these two behaviors.

FILTER Used to Reduce Rows

For instance, we could calculate the number of high-volume sales where high volume is any transaction where the CASES SOLD value is greater than 300. To do this, we can use FILTER nested inside COUNTROWS to count the rows of the filtered Winesales table as in the following expression:

```
No. of High Volume Sales =
COUNTROWS ( FILTER ( Winesales, Winesales[CASES SOLD] > 300 ) )
```

You can see the result of this measure in Figure 7-3. FILTER can also be nested inside SUMX, whereby the number of rows in the table iterated by SUMX will be reduced by FILTER. For example, the "Total Sales" measure that we authored in Chapter 5 could be extended to filter the sales where the volume of cases is greater than 300 (shown in Figure 7-4):

```
Cases GT 300 =
SUMX (
    FILTER ( Winesales, Winesales[CASES SOLD] > 300 ),
    Winesales[CASES SOLD] * RELATED (Wines [PRICE PER CASE] ))
```

WINE	No. of High Volume Sales
Bordeaux	89
Champagne	100
Grenache	32
Malbec	1
Sauvignon Blanc	64
Total	**286**

Figure 7-3. *Using FILTER nested inside COUNTROWS to calculate the number of high-volume sales*

WINE	Cases GT 300
Bordeaux	$2,658,150
Champagne	$6,075,150
Grenache	$310,530
Malbec	$27,710
Sauvignon Blanc	$835,280
Total	**$9,906,820**

Figure 7-4. *Using FILTER nested inside SUMX to calculate the sales value where cases sold is greater than 300*

However, if you want to use this calculation, this is not the best expression for doing the job. We will be discovering that FILTER is an iterator, and in this respect, it will scan the Winesales fact table that may contain many millions of rows. We will be exploring later in this chapter more efficient ways of performing this task.

FILTER as the Filter Argument of CALCULATE

If FILTER is used in a filter argument of CALCULATE, FILTER generates an in-memory table that is used to filter the data model, just as dimensions filter the data model.

Before March 2021, *it was a requirement* to use the FILTER function inside CALCULATE in the following two situations:

1. When the filter includes *more than one column* from the same table
2. When the filter includes *an expression*

However, it is now possible to omit the FILTER function when filtering two or more columns in the same table, but depending on slicer selections, the measure can still fail. It is also now possible to omit FILTER if the expression is a simple Boolean test using an aggregate function, such as AVERAGE, but using any other expression in the filter argument still requires the use of FILTER.

For people new to DAX, it is very important to understand that the new syntax, where FILTER is no longer required in the situations outlined before, is a recent development

(DAX was first introduced in 2009). Any DAX resources you browse or any code you copy and paste will most probably be using the old syntax using FILTER.

With this in mind, let's return to our "Sales for Red or French #1" measure we authored when exploring the CALCULATE function in the previous chapter. This was the measure:

```
Sales for Red or French #1 =
CALCULATE (
    [Total Sales],
    Wines[TYPE] = "red"
        || Wines[WINE COUNTRY] = "France" )
```

This expression returns incorrect results if there is a filter on the TYPE column or the WINE COUNTRY column, assuming that if you are slicing, you now want to calculate sales only for red wines and French wines, not red *or* French wines, which is the current calculation. If so, the correct values are shown in the "Total Sales" measure on the left in Figure 7-5 as this measure is responding to the filters in the slicers.

SALESPERSON	Total Sales	Sales for Red or French #1	Sales for Red or French #2
Abel	$2,050,276	$4,647,576	£2,050,276
Blanchet	$1,734,279	$4,193,329	£1,734,279
Charron	$2,029,616	$4,583,416	£2,029,616
Denis	$2,711,085	$4,518,335	£2,711,085
Leblanc	$2,232,097	$4,220,947	£2,232,097
Reyer	$2,177,254	$3,584,654	£2,177,254
Total	**$12,934,607**	**$25,748,257**	**£12,934,607**

TYPE
■ Red
□ White

Figure 7-5. *Omitting FILTER can return incorrect results*

Note You will learn later in this book the precise details as to why the "#1" measure returns incorrect results when there is a filter on either TYPE or WINECOUNTRY.

We established that the root of the problem lies in the fact that we're using two *different* columns in our filter and indeed in earlier days, we were prevented from authoring such code. To resolve this problem, we need to use the table function FILTER inside CALCULATE. So let's now get to grips with how we can use FILTER in this context and use it to author the correct version of the measure, "Sales for Red or French #2":

```
Sales for Red or French #2=
CALCULATE (
    [Total Sales],
         FILTER ( Wines, Wines[TYPE] = "red"
      || Wines[WINE COUNTRY] = "France" )
)
```

In Figure 7-6, you can see the measure evaluated when put into a Table visual. We've also included the "Total Sales" measure to provide context and clarity on the evaluation.

SALESPERSON	Total Sales	Sales for Red or French #1	Sales for Red or French #2
Abel	$1,244,481	$4,647,576	$1,244,481
Blanchet	$967,929	$4,193,329	$967,929
Charron	$1,116,846	$4,583,416	$1,116,846
Denis	$1,573,080	$4,518,335	$1,573,080
Leblanc	$1,224,252	$4,220,947	$1,224,252
Reyer	$1,272,654	$3,584,654	$1,272,654
Total	**$7,399,242**	**$25,748,257**	**$7,399,242**

TYPE
- ■ Red
- □ White

WINE COUNTRY
- ■ France
- □ Germany
- □ Italy

Figure 7-6. *The calculation of the "Sales for Red or French #2" measure*

See Figure 7-7 for a step-by-step guide through the evaluation of this measure. In the "Sales for Red or French #2" measure, FILTER is nested inside CALCULATE to provide the filter argument. The FILTER function is an *iterator*. We met iterators when we looked at the SUMX function in Chapter 5. These are functions that scan a table on a row-by-row basis and in the case of FILTER perform a test on each row. If the test applied by FILTER is true for a row, that row is extracted to a virtual table of its own. This virtual table, used as the filter argument to CALCULATE, is then used to propagate filters through the model just like a "real" table.

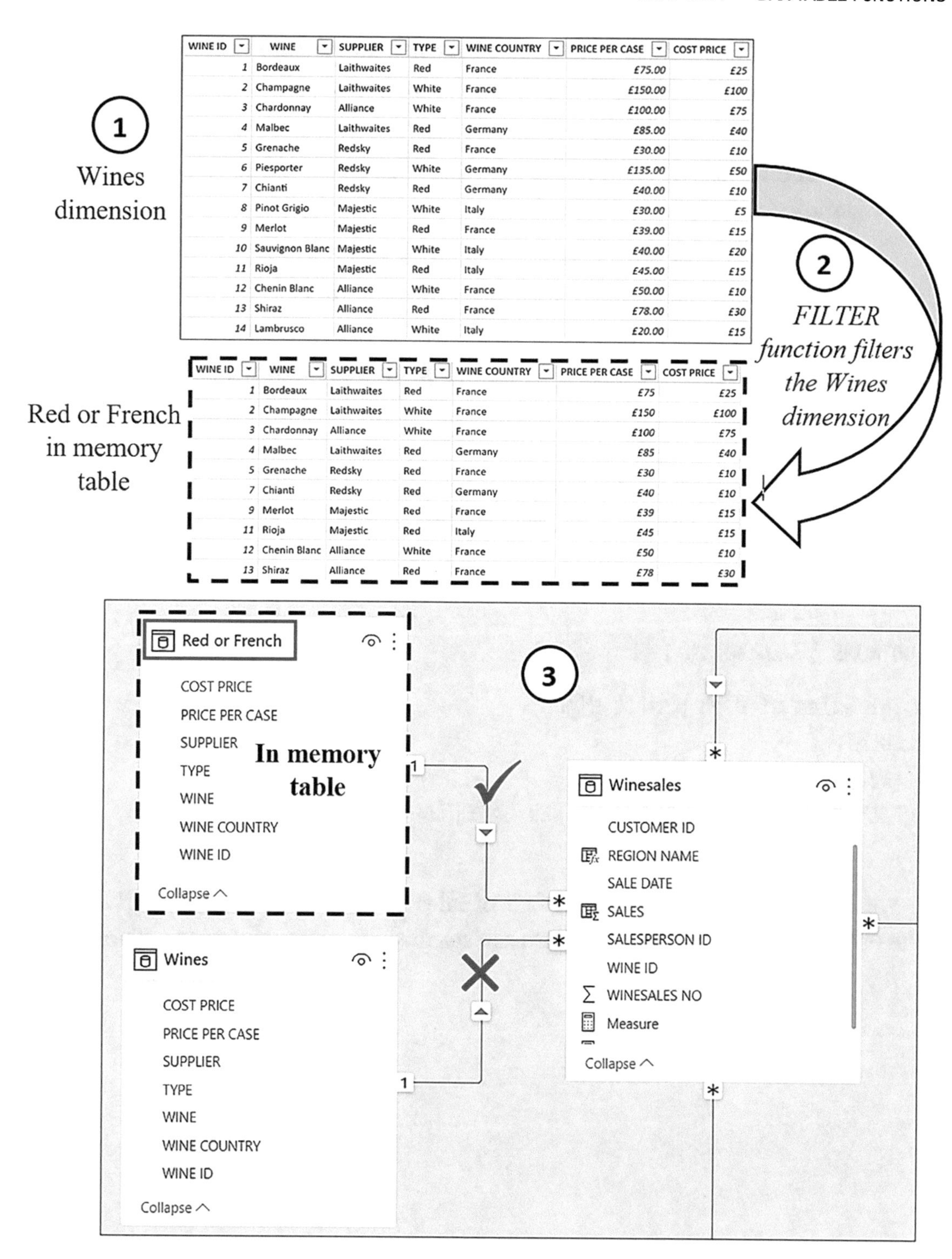

WINE ID	WINE	SUPPLIER	TYPE	WINE COUNTRY	PRICE PER CASE	COST PRICE
1	Bordeaux	Laithwaites	Red	France	£75.00	£25
2	Champagne	Laithwaites	White	France	£150.00	£100
3	Chardonnay	Alliance	White	France	£100.00	£75
4	Malbec	Laithwaites	Red	Germany	£85.00	£40
5	Grenache	Redsky	Red	France	£30.00	£10
6	Piesporter	Redsky	White	Germany	£135.00	£50
7	Chianti	Redsky	Red	Germany	£40.00	£10
8	Pinot Grigio	Majestic	White	Italy	£30.00	£5
9	Merlot	Majestic	Red	France	£39.00	£15
10	Sauvignon Blanc	Majestic	White	Italy	£40.00	£20
11	Rioja	Majestic	Red	Italy	£45.00	£15
12	Chenin Blanc	Alliance	White	France	£50.00	£10
13	Shiraz	Alliance	Red	France	£78.00	£30
14	Lambrusco	Alliance	White	Italy	£20.00	£15

WINE ID	WINE	SUPPLIER	TYPE	WINE COUNTRY	PRICE PER CASE	COST PRICE
1	Bordeaux	Laithwaites	Red	France	£75	£25
2	Champagne	Laithwaites	White	France	£150	£100
3	Chardonnay	Alliance	White	France	£100	£75
4	Malbec	Laithwaites	Red	Germany	£85	£40
5	Grenache	Redsky	Red	France	£30	£10
7	Chianti	Redsky	Red	Germany	£40	£10
9	Merlot	Majestic	Red	France	£39	£15
11	Rioja	Majestic	Red	Italy	£45	£15
12	Chenin Blanc	Alliance	White	France	£50	£10
13	Shiraz	Alliance	Red	France	£78	£30

Figure 7-7. *Stepping through the "Sales for Red or French #2" measure*

1. The FILTER function iterates the Wines table in memory and filters any rows where TYPE = “red” or WINECOUNTRY = “France”.
2. FILTER generates an in-memory virtual table containing only those rows where the test is true.
3. The virtual table generated by FILTER is used as the filter argument to CALCULATE to filter the Winesales table.

We’ve been examining the use of the FILTER function to perform an “OR” test on two different columns of the same table. Let’s look at another example with the same issue.

Consider the scenario where you want to find the number of sales (i.e., the number of rows in the Winesales table) for high profit wines. High profit wines are where wines have a price that is three times the cost price. This test involves two columns in the Wines dimension, PRICE PER CASE and COST PRICE, and therefore, it’s recommended that you use FILTER. These are the measures you can use:

```
No. of Sales =
COUNTROWS ( Winesales )

No. of Sales of High profit Wines =
CALCULATE (
   [No. of Sales],
   FILTER ( Wines, Wines[PRICE PER CASE] >= Wines[COST PRICE] * 3 )
)
```

We’ve included the expression for “No. of Sales” that we will nest inside the “No. of Sales of High profit Wines” measure. You can see this measure calculated in Figure 7-8.

WINE	No of Sales	No of Sales of High-profit Wines
Bordeaux	180	180
Champagne	132	
Chardonnay	187	
Chenin Blanc	200	200
Chianti	148	148
Grenache	182	182
Malbec	170	
Merlot	157	
Piesporter	115	
Pinot Grigio	168	168
Rioja	197	197
Sauvignon Blanc	168	
Shiraz	203	
Total	**2,207**	**1,075**

SALESPERSON	No of Sales	No of Sales of High-profit Wines
Abel	376	173
Blanchet	343	160
Charron	347	168
Denis	435	228
Leblanc	355	166
Reyer	351	180
Total	**2,207**	**1,075**

Figure 7-8. *Finding high profit wines*

You will notice again that FILTER can be omitted here because expressions using different columns from the same table are now valid. However, take note that if you had a filter on either the PRICE PER CASE column or the COST PRICE column, you would not see correct values being returned. Therefore, it is recommended that you use FILTER nested inside CALCULATE whenever more than one column is being referenced.

However, we also need FILTER whenever we need to use an *expression* in the filter argument in CALCULATE. We've set out two examples of this requirement where we are calculating the following:

1. The number of sales where the total sales values are greater than 20,000. We are using the "Total Sales" measure in the filter test.

2. The number of sales that are greater than the average sales value. To calculate the average sales, we are using the AVERAGEX expression that you learned in Chapter 5.

First, we have authored the "wrong" version of the measures that omits the FILTER function. These expressions will return error messages. We have then authored the correct expressions using FILTER. Therefore, it's important that you understand that

FILTER is required when you use any *expression* in the filter test of CALCULATE. We have highlighted in gray the FILTER expressions to help clarify the code used:

```
Sales Greater than 20K Wrong =
CALCULATE ( [No. of Sales], [Total Sales] > 20000 )

Sales Greater than 20K =
CALCULATE ( [No. of Sales],
       FILTER ( Winesales, [Total Sales] > 20000 ) )

Sales Greater than Avg Wrong =
CALCULATE (
    [No. of Sales],
    [Total Sales]
        > AVERAGEX (
            Winesales,
            Winesales[CASES SOLD] *
                        RELATED ( Wines[PRICE PER CASE] )
        )
)

Sales Greater than Avg =
CALCULATE (
    [No. of Sales],
    FILTER (
        Winesales,
        [Total Sales]
            > AVERAGEX (
               Winesales,
               Winesales[CASES SOLD] *
                        RELATED ( Wines[PRICE PER CASE] )
            )
    )
)
```

However, if the requirement is to calculate the number of sales that are greater than the average cases sold, this expression does not require FILTER because it's using the simple aggregate function AVERAGE. Since September 2021, we are now allowed to author code that uses the simple aggregate functions, such as AVERAGE or MAX in the predicate as follows:

```
Cases GT Avg =
CALCULATE (
    [No. of Sales],
    Winesales[CASES SOLD] > AVERAGE ( Winesales[CASES SOLD] )
)
```

However, experienced DAX users would probably prefer to see this measure expressed using FILTER:

```
Cases GT Avg =
CALCULATE (
    [No. of Sales],
    FILTER ( Winesales, Winesales[CASES SOLD]
            > AVERAGE ( Winesales[CASES SOLD] ) )
)
```

In this section on the FILTER function, you have learned that FILTER generates a virtual table that can be used in the filter argument of CALCULATE. This virtual table is used to filter the data model just like "real" tables do.

This leads us to another aspect of the FILTER function (and indeed table functions generally) that we need to explore in more detail, and that's the difference between using a *table expression* as a filter inside CALCULATE and using a simple *column* filter instead.

Column Filters vs. Table Filters

What you have learned is that in the "filter" argument to CALCULATE, you can supply two types of filter: a filter using a *column* and/or a filter using a *table*. In short, within CALCULATE, there are two ways to modify the filter context: using columns or using tables. What you need to understand now is that there will be a considerable difference in the evaluation of a measure depending on which type of filter you choose.

So far in this book, the only table function we've met is the FILTER function, so we'll use FILTER to illustrate the difference between column filters and table filters but to appreciate that it's relevant to all table expressions used as filters inside CALCULATE.

Note We'll be exploring a number of other table functions as we move through this book such as ALL, VALUES, and the functions known as "time intelligence."

Why do we need to distinguish between table filters and column filters? There are essentially two reasons why this difference is important:

1. Because the DAX engine has to generate the virtual tables, table filters take longer to process.
2. Your measure may return a different result depending on the filter type.

We will now explore these two scenarios. In the first example, we look at how table filters increase the processing weight of the measure. In the second example, we will see that table filters can produce different results from column filters.

Table Filters Are Less Efficient

In this example, let's take two similar expressions using CALCULATE. The first uses a column filter and the second, a table expression as the filter argument. In both expressions, we are filtering the rows in the Winesales fact table that contain cases sold greater than 300.

```
Cases GT 300 #1 =
CALCULATE ([Total Sales], Winesales[CASES SOLD] > 300 )

Cases GT 300 #2 =
CALCULATE (
    [Total Sales],
    FILTER ( Winesales, Winesales[CASES SOLD] > 300 )
)
```

You can see in Figure 7-9 that both these measures return the same result, so how does the table filter differ from the column filter? To answer this question, we must look more carefully at the evaluation of each of these measures, taking the evaluation of "Grenache" wine that returns **$310,530** as our example.

WINE	Cases GT 300 #1	Cases GT 300 #2
Bordeaux	$2,658,150	$2,658,150
Champagne	$6,075,150	$6,075,150
Grenache	$310,530	$310,530
Malbec	$27,710	$27,710
Sauvignon Blanc	$835,280	$835,280
Total	**$9,906,820**	**$9,906,820**

Figure 7-9. *The measures return the same result*

When measure "Cases GT 300 #1" is evaluated, a filter is placed on the CASES SOLD *column* to filter values greater than 300. The Total Sales values are then calculated for the filtered rows of the Winesales table; see Figure 7-10.

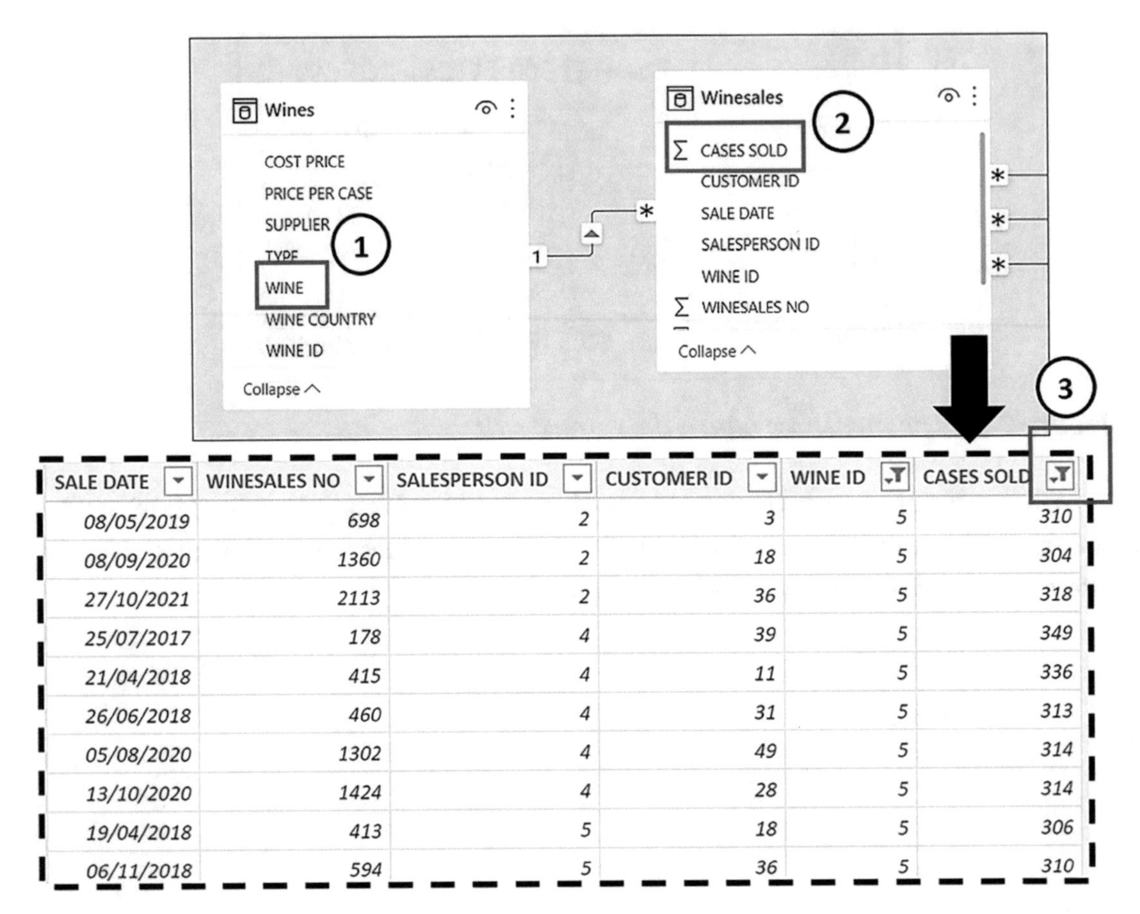

SALE DATE	WINESALES NO	SALESPERSON ID	CUSTOMER ID	WINE ID	CASES SOLD
08/05/2019	698	2	3	5	310
08/09/2020	1360	2	18	5	304
27/10/2021	2113	2	36	5	318
25/07/2017	178	4	39	5	349
21/04/2018	415	4	11	5	336
26/06/2018	460	4	31	5	313
05/08/2020	1302	4	49	5	314
13/10/2020	1424	4	28	5	314
19/04/2018	413	5	18	5	306
06/11/2018	594	5	36	5	310

***Figure 7-10.** Stepping through the "Cases GT 300 #1" measure*

1. The wine "Grenache" is filtered in the WINE column of the Wines table and is cross-filtered to the Winesales table that only now contains rows for "Grenache".

2. The CASES SOLD column in the Winesales table is further filtered in memory to contain only the rows for this wine that are greater than 300. The "Total Sales" measure is then calculated for just these rows.

3. Note the filter on the CASES SOLD column.

When the measure "Cases GT 300 #2" is evaluated, the FILTER function iterates the Winesales table to extract rows where the CASES SOLD is greater than 300 into a virtual table (remembering that Winesales is filtered to just contain "Grenache" wines). The total sales for the virtual table generated by FILTER are then calculated; see Figure 7-11.

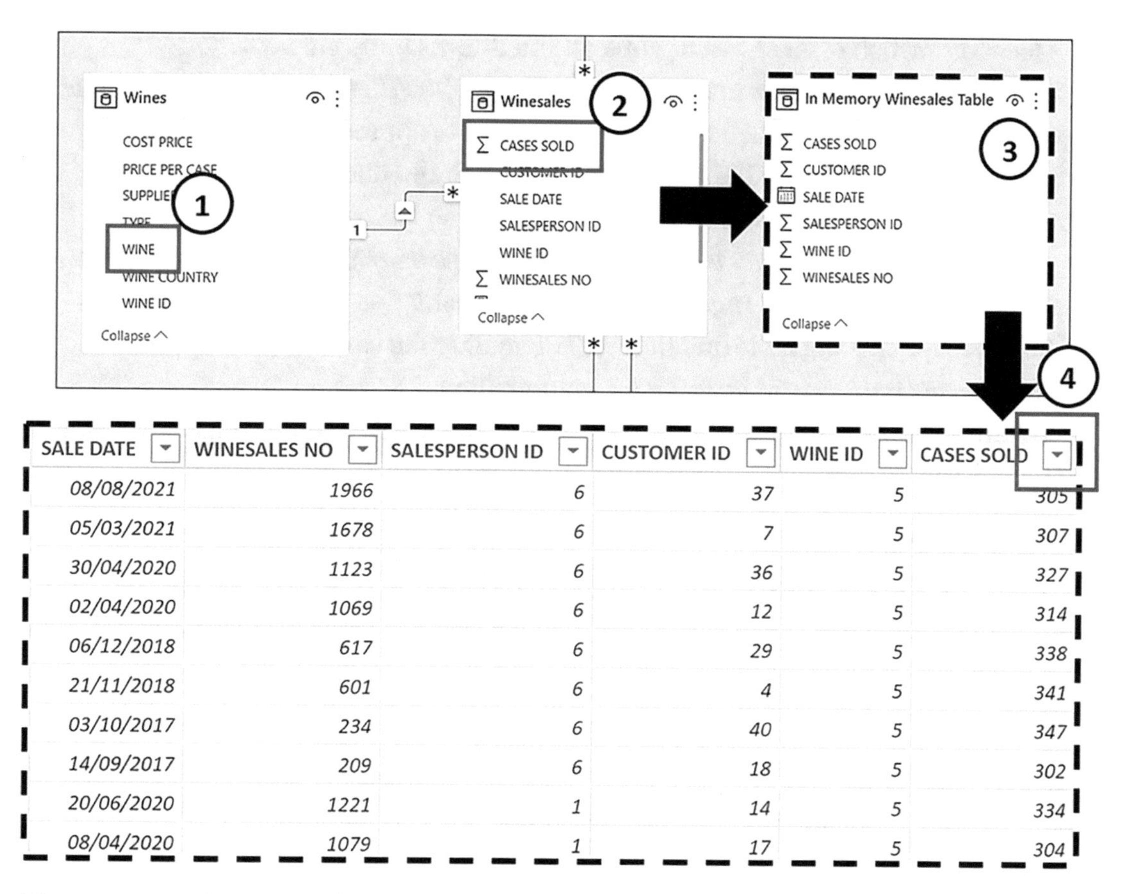

SALE DATE	WINESALES NO	SALESPERSON ID	CUSTOMER ID	WINE ID	CASES SOLD
08/08/2021	1966	6	37	5	305
05/03/2021	1678	6	7	5	307
30/04/2020	1123	6	36	5	327
02/04/2020	1069	6	12	5	314
06/12/2018	617	6	29	5	338
21/11/2018	601	6	4	5	341
03/10/2017	234	6	40	5	347
14/09/2017	209	6	18	5	302
20/06/2020	1221	1	14	5	334
08/04/2020	1079	1	17	5	304

Figure 7-11. *Stepping through the "Cases GT 300 #2" measure*

1. The wine "Grenache" is filtered in the WINE column of the Wines table and is cross-filtered to the Winesales table that now only contains rows for "Grenache".
2. The FILTER function iterates the Winesales table to filter CASES SOLD greater than 300 and generates a virtual table.
3. The "Total Sales" measure is calculated for the rows in the virtual Winesales table in memory.
4. Note there is no filter on the CASES SOLD column because it's the virtual table that has generated the filtered rows.

Question: Which of these evaluations do you think is more efficient?

If your fact table contains many millions of rows, the FILTER function must iterate these rows to build the virtual table. We're sure you can appreciate that you pay a heavy processing price if you use table filters rather than column filters. Marco Russo and Alberto Ferrari explain this in more technical terms:

"A side effect of a table filter is that it requires a large materialization to the storage engine to enable the formula engine to compute the result."[1]

This is why using the table function FILTER to filter the cases sold is not good practice because you should be using the column filter.

To further make the point, in this video, Marco Russo takes you through why using the FILTER function unnecessarily is not a good idea:

My Power BI report is slow: what should I do? by Marco Russo

Before we leave the subject of the problematic table filters, there is a third version of the "Cases GT 300" measure that "newbies" might consider authoring. The expression "Cases GT 300 #3" uses SUMX and returns the same values as the previous two versions of the measure discussed before:

```
Cases GT 300 #3 =
SUMX ( FILTER ( Winesales, Winesales[CASES SOLD] > 300 ),
                [Total Sales] )
```

What is the problem with this expression? You of course now know. The answer is it's inefficient. First, FILTER iterates the fact table to generate a table containing the rows to be considered. Then SUMX iterates the table generated by FILTER. That's a lot of iterations!

The recommended expression is always to use a simple filter on the CASES SOLD column in the filter argument of CALCULATE.

Table Filters Return Different Results

To understand this aspect of the table filters, let's consider these two measures, the first using a column filter and the second using a table filter:

[1] Marco Russo and Alberto Ferrari (2020), *The Definitive Guide to DAX*, 2nd ed, p. 699 [Microsoft Press]

```
Bordeaux Wines #1 =
CALCULATE (
    SUM ( Winesales[CASES SOLD] ),
    Wines[WINE] = "Bordeaux" )

Bordeaux Wines #2 =
CALCULATE (
    SUM ( Winesales[CASES SOLD] ),
    FILTER ( Wines, Wines[WINE] = "Bordeaux" )
)
```

You might think that these two measures should return the same result. However, if we put these measures into a Table visual that contains the WINE column from the Wines dimension (Figure 7-12), we get different results. The measure #1 gives the value for "Bordeaux" for every wine, but in #2, we get blanks for any wine other than Bordeaux.

WINE	Bordeaux Wines #1	Bordeaux Wines #2
Bordeaux	54,070	54,070
Champagne	54,070	
Chardonnay	54,070	
Chenin Blanc	54,070	
Chianti	54,070	
Grenache	54,070	
Lambrusco	54,070	
Malbec	54,070	
Merlot	54,070	
Piesporter	54,070	
Pinot Grigio	54,070	
Rioja	54,070	
Sauvignon Blanc	54,070	
Shiraz	54,070	
Total	**54,070**	**54,070**

Figure 7-12. *"Bordeaux Wines #1" and "Bordeaux Wines #2" in a Table visual with the WINE column from the Wines dimension*

So why the difference? Let's look more closely at the "Bordeaux Wines #1" measure. In the first evaluation of this measure, the active filter context is on the WINE column of the Wines dimension and is filtering "Bordeaux" in the first instance. This filter is now cross-filtered to the Winesales table to sum the CASES SOLD for "Bordeaux". On the next evaluation, "Champagne" is in the filter context. But *CALCULATE modifies the filter context* and *replaces* the filter on the WINE column from "Champagne" to "Bordeaux". It's this filter that is now cross-filtered to the Winesales table to sum the CASES SOLD for "Bordeaux". And so on for every evaluation of each wine and also the Total row evaluation; see Figure 7-13.

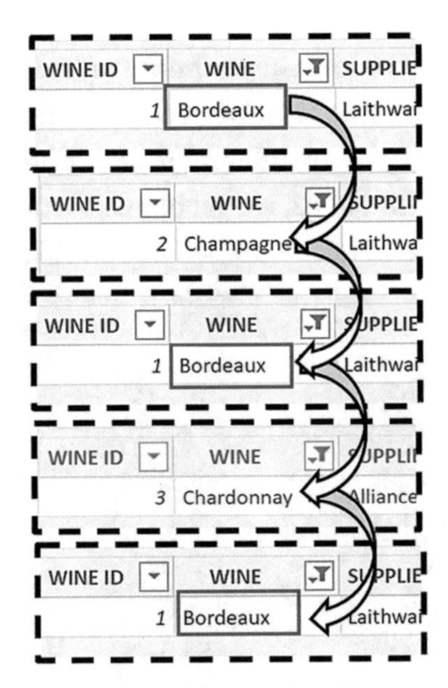

Figure 7-13. *CALCULATE replaces the filter on the WINE column so it always filters "Bordeaux"*

Note As mentioned earlier, at this stage in your knowledge of DAX, this explanation of how the filters work is not yet complete, but it will stand you in good stead for the time being. We will get to a more accurate explanation later in Chapter 18.

This is why the total cases for "Bordeaux" are always returned because CALCULATE *replaces* the filter on the WINE column to "Bordeaux" for the evaluation of each wine.

Let's now look at the second measure using FILTER where we get a value returned for "Bordeaux" but not for the other wines. This measure uses a table filter:

```
Bordeaux Wines #2 =
CALCULATE (
    SUM ( Winesales[CASES SOLD] ),
    FILTER ( Wines, Wines[WINE] = "Bordeaux" )
)
```

The current filter context is on the WINE column of the Wines dimension, filtering "Bordeaux" in memory in the first instance. The FILTER function inside CALCULATE scans this table looking for the value "Bordeaux" and generates a virtual table containing just the "Bordeaux" row. It's this table filter that is now cross-filtered to the Winesales table to sum the CASES SOLD for "Bordeaux". On the evaluation for "Champagne", the WINE column in the Wines dimension is filtered accordingly. However, the *FILTER function does not modify the filter context,* so the FILTER function inside CALCULATE scans this one-row table containing "Champagne" looking for the value "Bordeaux". It won't find it, and so there is nothing to filter. There is now an empty filter generated by FILTER, and an empty filter returns no value; see Figure 7-14. This is why there are no values returned by the measure other than for "Bordeaux".

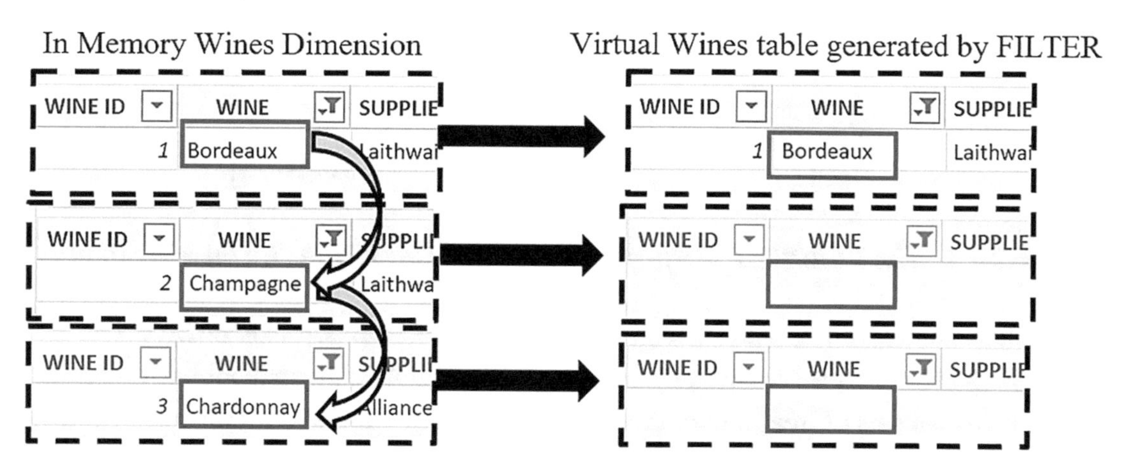

Figure 7-14. *FILTER can't replace the filter on the WINE column to equal "Bordeaux", so there are no rows filtered other than for "Bordeaux"*

The important thing to remember about the FILTER function is that it's a weak function. Unless you use the ALL function that we explore in the next chapter, FILTER will only filter the rows *that are in the current filter context* and will therefore typically return a *subset* of the original filter. Using column filters inside CALCULATE, on the other hand, will replace filters where required.

Using the KEEPFILTERS Function

This behavior of CALCULATE whereby a *column* filter is always *replaced* is, by all accounts, rather odd and unintuitive, giving you the same value for every evaluation. The filter generated by FILTER, even though it's a table filter, looks more "normal." As we're learning, it's always best to use column filters if possible, so to make the column expression behave more intuitively, we can use a function called KEEPFILTERS as in this example:

```
Bordeaux Wines #1 =
CALCULATE (
    SUM ( Winesales[CASES SOLD] ),
    KEEPFILTERS ( Wines[WINE] = "Bordeaux" )
)
```

This function modifies the behavior of CALCULATE and prevents it from replacing filters. In Figure 7-15, you can see that we now only get a value return for "Bordeaux" for the "Bordeaux Wines #1" measure and no value is returned for the other wines.

WINE	Bordeaux Wines #1	Bordeaux Wines #2
Bordeaux	54,070	54,070
Total	**54,070**	**54,070**

Figure 7-15. *The KEEPFILTERS function prevents CALCULATE replacing filters*

In this chapter, you've learned to generate virtual tables as part of your DAX expressions. These tables are used by measures to manipulate the data model, either by returning subsets of "real" tables or to act as in-memory dimensions that propagate filters through the data model. You've also been warned of the different behaviors of table filters and column filters, particularly with respect to using the FILTER function. As we move forward and tackle more challenging calculations, this difference will become more important. For the moment, however, let's just remember this:

Always use column filters where you can. Only use table filters where necessary.

CHAPTER 8

The ALL Function and All Its Variations

In previous chapters, we have explored the filter context and how the construct of the visual, slicers, and filters all come together to filter the data model on the evaluation of a measure. You have learned that with the CALCULATE function, you can modify these filters programmatically. What you don't yet know is how to *remove* filters so you can calculate your own totals and subtotals. But better still, knowing how to remove filters means you can programmatically reapply totally different filters than those that are currently defining the filter context. Let me introduce you to the ALL function that allows you to take control of this aspect of the evaluation of your measures.

On the face of it, the ALL function appears to be an easy function to understand. The ALL function returns all the rows of a table, or all the distinct values in a column, ignoring any filters that might have been applied. However, what you will be discovering in this chapter is that the simplicity of the ALL function belies the fact that it's one of the most challenging DAX functions with which to come to terms. In this chapter, we will be delving into this "wolf in sheep's clothing" function; the objective is to teach you every aspect of ALL and all the variations on the ALL function. This will enable you to move forward and author more complex measures.

There are at least two reasons why the ALL function is challenging to understand. Firstly, there are a number of variations of the ALL function:

- ALLSELECTED
- ALLEXCEPT
- ALLCROSSFILTERED
- ALLNOBLANKROW

You need to know which of these to use and when.

© Alison Box 2022
A. Box, *Up and Running with DAX for Power BI*, https://doi.org/10.1007/978-1-4842-8188-8_8

Note The ALLCROSSFILTERED and ALLNOBLANKROW functions are outside the remit of this book.

Secondly, ALL (and its variations) has a dual face; it can be used either as a table function or as a modifier to CALCULATE, as described in the following:

- **ALL as a table function** - When used as a table function, ALL behaves as described before; that is, it returns all the rows of a table or all the distinct values in a column or columns.
- **ALL as a modifier** - When ALL is used as a top-level filter argument in CALCULATE, it acts as a modifier to CALCULATE and *removes* the filters from tables or columns. In other words, it doesn't generate a virtual table.

In fact, ALL is two completely different functions. This is something that many inexperienced users of DAX don't appreciate. This is because mostly, the ALL function behaves the way you would expect, whether you use it as a top-level filter argument in CALCULATE or nested inside other functions such as FILTER or COUNTROWS. It removes filters whether by generating virtual tables containing all the rows or by removing filters from tables and columns. However, we will explore later how understanding this difference is crucial in understanding the ALL function.

Although I've been referring solely to the ALL function here, we will also be exploring the ALLSELECTED and ALLEXCEPT functions.

The ALL Function

The ALL function has the following syntaxes:

= ALL (table)

where:

table is the table from where you want to clear the filters.

Here is an example of the ALL function syntax, referencing a table:

= ALL (Winesales)

Unlike other functions that use tables as arguments, you can't nest another table function inside the ALL function; you can only use base tables.

Or you can reference a column.

= ALL (column 1, column 2, etc.)

where:

column(s) is the column or columns from where you want to clear the filters.

Here is an example of the ALL function syntax referencing a column:

= ALL (Wines[TYPE])

The ALL function will have a different impact on the filtering of the data model depending on the syntax you use, whether ALL is removing filters from tables or removing filters from columns. It will have a different impact yet again if you use ALL to remove filters from fact tables or dimensions. Therefore, to make it easier to understand the behavior of ALL, we'll take these three different objects from where ALL can remove filters and explore them separately, as follows:

1. Fact tables
2. Dimensions
3. A column or columns

Applied to the Fact Table

Let's again consider a scenario. In the visual in Figure 8-1, we're using this measure to calculate the number of sales:

```
No. of Sales =
COUNTROWS ( Winesales )
```

We have then calculated the "Grand Total No. of Sales" to act as a denominator to calculate the percentage shown in "No. of Sales as Percent of Grand Total"; see Figure 8-1.

WINE	No. of Sales	Grand Total No. of Sales	No. of Sales as Percent of Grand Total
Bordeaux	180	2,207	8.16%
Champagne	132	2,207	5.98%
Chardonnay	187	2,207	8.47%
Chenin Blanc	200	2,207	9.06%
Chianti	148	2,207	6.71%
Grenache	182	2,207	8.25%
Lambrusco		2,207	
Malbec	170	2,207	7.70%
Merlot	157	2,207	7.11%
Piesporter	115	2,207	5.21%
Pinot Grigio	168	2,207	7.61%
Rioja	197	2,207	8.93%
Sauvignon Blanc	168	2,207	7.61%
Shiraz	203	2,207	9.20%
Total	**2,207**	**2,207**	**100.00%**

Figure 8-1. *Using ALL to calculate the percentage of the Grand Total*

To arrive at these calculations, first, we need to author a measure that ignores the filters coming through from the Wines dimension so we can calculate the number of sales for *all the wines*, **2,207**. To do this, we can use ALL as a table function to generate a virtual table containing all the rows of the Winesales fact table and then use COUNTROWS to count the rows in this table. This is the expression:

```
Grand Total No. of Sales =
COUNTROWS ( ALL ( Winesales ) )
```

Finally, we can then divide the Grand Total into each wine's total to find the percentage, as in the following measure:

```
No. of Sales as Percent of Grand Total =
DIVIDE ( [No. of Sales] , [Grand Total No. of Sales] )
```

Let's explore the impact of adding more filters to the report. In Figure 8-2, we have placed a filter on the SalesPeople dimension using a slicer, but you can see that the measure using the ALL function always returns the Grand Total regardless of the filter.

WINE	No. of Sales	Grand Total No. of Sales	No. of Sales as Percent of Grand Total
Bordeaux	30	2,207	1.36%
Champagne	29	2,207	1.31%
Chardonnay	36	2,207	1.63%
Chenin Blanc	25	2,207	1.13%
Chianti	24	2,207	1.09%
Grenache	30	2,207	1.36%
Lambrusco		2,207	
Malbec	25	2,207	1.13%
Merlot	31	2,207	1.40%
Piesporter	25	2,207	1.13%
Pinot Grigio	30	2,207	1.36%
Rioja	34	2,207	1.54%
Sauvignon Blanc	19	2,207	0.86%
Shiraz	38	2,207	1.72%
Total	**376**	**2,207**	**17.04%**

SALESPERSON
- ■ Abel
- ☐ Blanchet
- ☐ Charron
- ☐ Denis
- ☐ Leblanc
- ☐ Reyer

***Figure 8-2.** The ALL function ignores filters from dimensions*

To understand the behavior of ALL in this example, we must again consider the filter context. On the evaluation for "Bordeaux" wine, there are two active filters: one on the Wines dimension filtering "Bordeaux" and one on the SalesPeople dimension filtering

"Abel". However, the ALL function always counts the rows of the virtual table containing *all the rows* of the Winesales fact table ignoring any filters propagating from dimensions; see Figure 8-3.

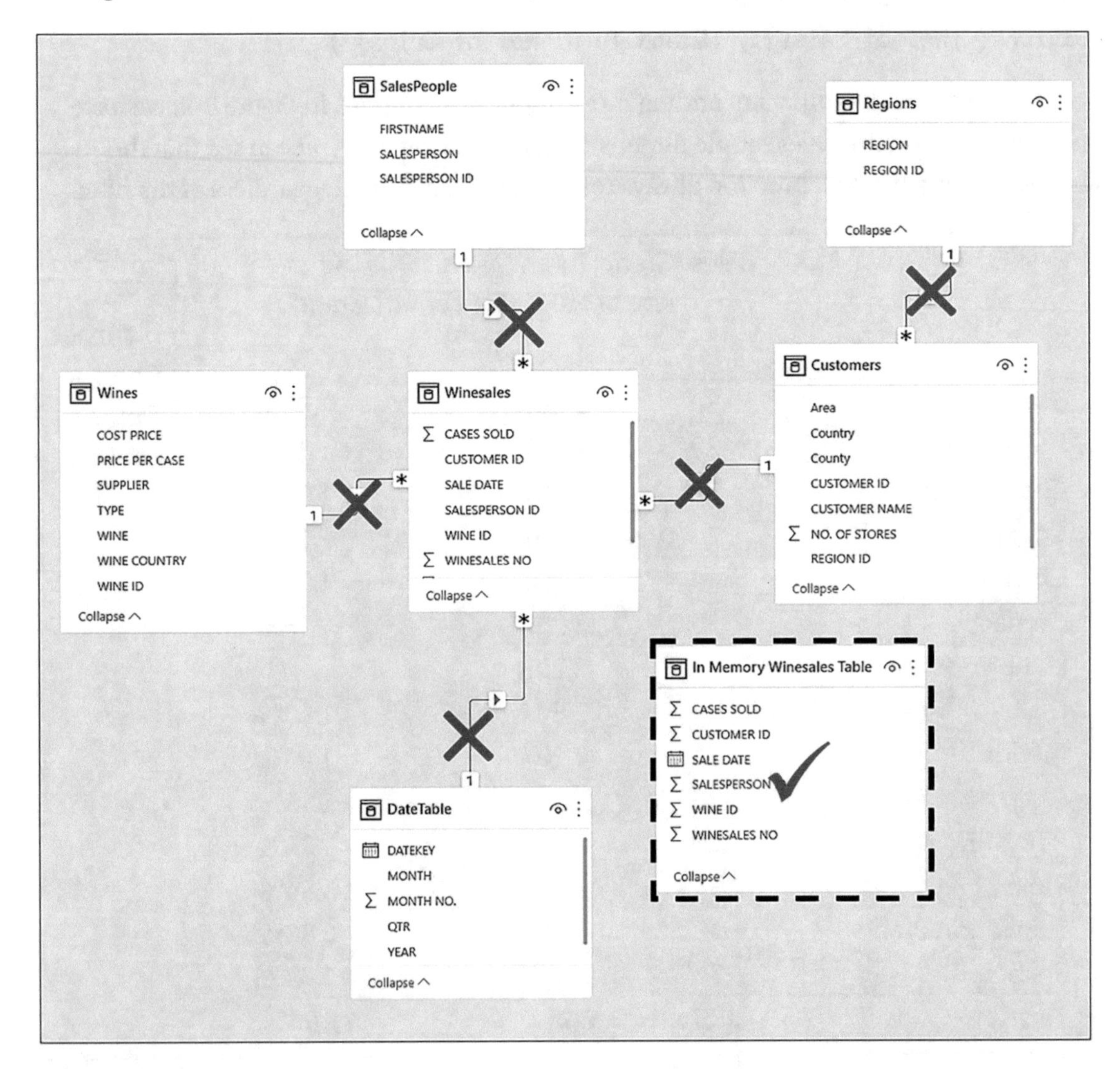

Figure 8-3. *The ALL function passed to the fact table generates a virtual fact table that is used for all evaluations, and any filters from dimensions are ignored*

Let's look at another example of using the ALL function on the fact table, but this time nesting ALL inside CALCULATE. For example, you may want to find the grand total of cases sold, again so you could use this value as a denominator to find percentages; see Figure 8-4. This would be the measure:

```
Total Cases All Winesales =
CALCULATE ( [Total Cases], ALL( Winesales ) )
```

WINE	Total Cases	Total Cases All Winesales
Bordeaux	54,070	423,224
Champagne	49,158	423,224
Chardonnay	42,030	423,224
Chenin Blanc	24,739	423,224
Chianti	27,323	423,224
Grenache	35,965	423,224
Lambrusco		423,224
Malbec	34,290	423,224
Merlot	23,084	423,224
Piesporter	10,253	423,224
Pinot Grigio	23,449	423,224
Rioja	33,951	423,224
Sauvignon Blanc	47,415	423,224
Shiraz	17,497	423,224
Total	**423,224**	**423,224**

Figure 8-4. *The ALL function nested inside CALCULATE to find the grand total of cases sold*

You can see how this measure again ignores any filters on the data model.

However, let's now focus on an expression that you may require that calculates the average cases sold *for all wines* so you can compare this average to the average cases sold for *each* wine. This would be the measure that would find this average:

```
Avg Cases All Winesales =
CALCULATE( AVERAGE ( Winesales[CASES SOLD] ), ALL ( Winesales ) )
```

You could then author the following measure using FILTER to calculate the number of sales where the cases sold value is greater than the average for all the wines:

```
No. of Sales Where Cases is GT Avg All Wines =
CALCULATE (
```

```
    [No. of Sales],
    FILTER ( Winesales, Winesales[CASES SOLD]
                            >= [Avg Cases All Winesales] )
)
```

In the code for "No. of Sales Where Cases is GT Avg All Wines" the FILTER function iterates the Winesales table to filter any rows where the value in the CASES SOLD column is greater than the value calculated by "Avg Cases All Winesales". However, to fully appreciate the details of this expression, you need to understand the concept of context transition that we will be exploring in a later chapter.

You can see the results of these expressions in Figure 8-5.

WINE	Total Cases	Total Cases All Winesales	Avg Cases	Avg Cases All Winesales	No. of Sales Where Cases is GT Avg All Wines
Bordeaux	54,070	423,224	300.39	191.76	145
Champagne	49,158	423,224	372.41	191.76	131
Chardonnay	42,030	423,224	224.76	191.76	185
Chenin Blanc	24,739	423,224	123.70	191.76	
Chianti	27,323	423,224	184.61	191.76	70
Grenache	35,965	423,224	197.61	191.76	91
Lambrusco		423,224		191.76	
Malbec	34,290	423,224	201.71	191.76	88
Merlot	23,084	423,224	147.03	191.76	12
Piesporter	10,253	423,224	89.16	191.76	
Pinot Grigio	23,449	423,224	139.58	191.76	14
Rioja	33,951	423,224	172.34	191.76	28
Sauvignon Blanc	47,415	423,224	282.23	191.76	168
Shiraz	17,497	423,224	86.19	191.76	
Total	**423,224**	**423,224**	**191.76**	**191.76**	**932**

Figure 8-5. *Calculating the grand total cases sold and the average cases for all wines*

What you have to understand here is that when ALL is nested inside CALCULATE, it doesn't behave as a table function. Instead, ALL is *removing* all the cross-filters on the fact table and therefore evaluating all the rows of the fact table. We will be exploring this behavior in detail as we move through this chapter.

Using ALL in this way, we've been able to find the percentages of the Grand Total. However, you may have a different requirement, and that is to calculate percentages across *filtered* items. This brings us to the second place where we can use ALL, and that is when it's passed onto dimensions.

Using ALL on Dimension Tables

For example, in Figure 8-2, we've filtered salesperson "Abel" in the slicer and can see the total number of sales for Abel for *all the wines* is **376.** We want to know what the individual wine totals are for "Abel" as the percentage of *this value*. In other words, we need to remove the filter on the Wines dimension while retaining the filter on the SALESPERSON column in the SalesPeople dimension that is filtering "Abel".

If we remove a filter from a specific dimension, filters propagating from other dimensions into the fact table will be unaffected. Therefore, if we remove the filter from the Wines dimension, the filter on the SalesPeople dimension will be preserved, therefore calculating the number of sales *for all the wines* for the filtered salesperson.

However, this measure using ALL on the Wines dimension isn't correct:

```
No. of Sales All Wines Wrong =
COUNTROWS ( ALL ( Wines ) )
```

This measure would generate a table containing *all the rows in the Wines dimension* and then count the number of rows in this table, returning **14** because there are 14 rows in the Wines dimension. Remember that the table whose rows we want to count is that of the Winesales fact table, filtered to show the sales of all the wines for the salesperson selected in the slicer. Therefore, we need to calculate the number of sales in the Winesales table which we've already done a number of times:

```
No. of Sales = COUNTROWS ( Winesales)
```

Because we want to modify the filter context to remove the filter from the Wines dimension, we can use the "No. of Sales" measure inside CALCULATE, and then using the ALL function as the filter argument in CALCULATE, we can modify the filter context as follows:

```
No. of Sales All Wines =
CALCULATE ( [No. of Sales], ALL ( Wines ) )
```

Finally, we can divide to arrive at the percentage:

```
No. of Sales as Percent of Filtered Value =
DIVIDE ( [No. of Sales] , [No. of Sales All Wines] )
```

Let's focus on the measure "No. of Sales All Wines" shown in Figure 8-6. We can see it calculates the same value that is sitting in the Total row of the "No. of Sales" measure, **376**.

WINE	No. of Sales	No. of Sales All Wines	No. of Sales as Percent of Filtered Value
Bordeaux	30	376	7.98%
Champagne	29	376	7.71%
Chardonnay	36	376	9.57%
Chenin Blanc	25	376	6.65%
Chianti	24	376	6.38%
Grenache	30	376	7.98%
Lambrusco		376	
Malbec	25	376	6.65%
Merlot	31	376	8.24%
Piesporter	25	376	6.65%
Pinot Grigio	30	376	7.98%
Rioja	34	376	9.04%
Sauvignon Blanc	19	376	5.05%
Shiraz	38	376	10.11%
Total	**376**	**376**	**100.00%**

SALESPERSON
- [x] Abel
- [ ] Blanchet
- [ ] Charron
- [ ] Denis
- [ ] Leblanc
- [ ] Reyer

***Figure 8-6.** Removing the filter from a dimension using ALL*

In this Table visual, initially, filters are on both the Wines dimension and the SalesPeople dimension, but when the "No. of Sales All Wines" measure is evaluated for each wine, all the filters are *removed* from the Wines dimension (because we are using CALCULATE), therefore always returning the value for *all the wines*. Filters from any other dimensions, for example, the SalesPeople dimension, are retained; see Figure 8-7.

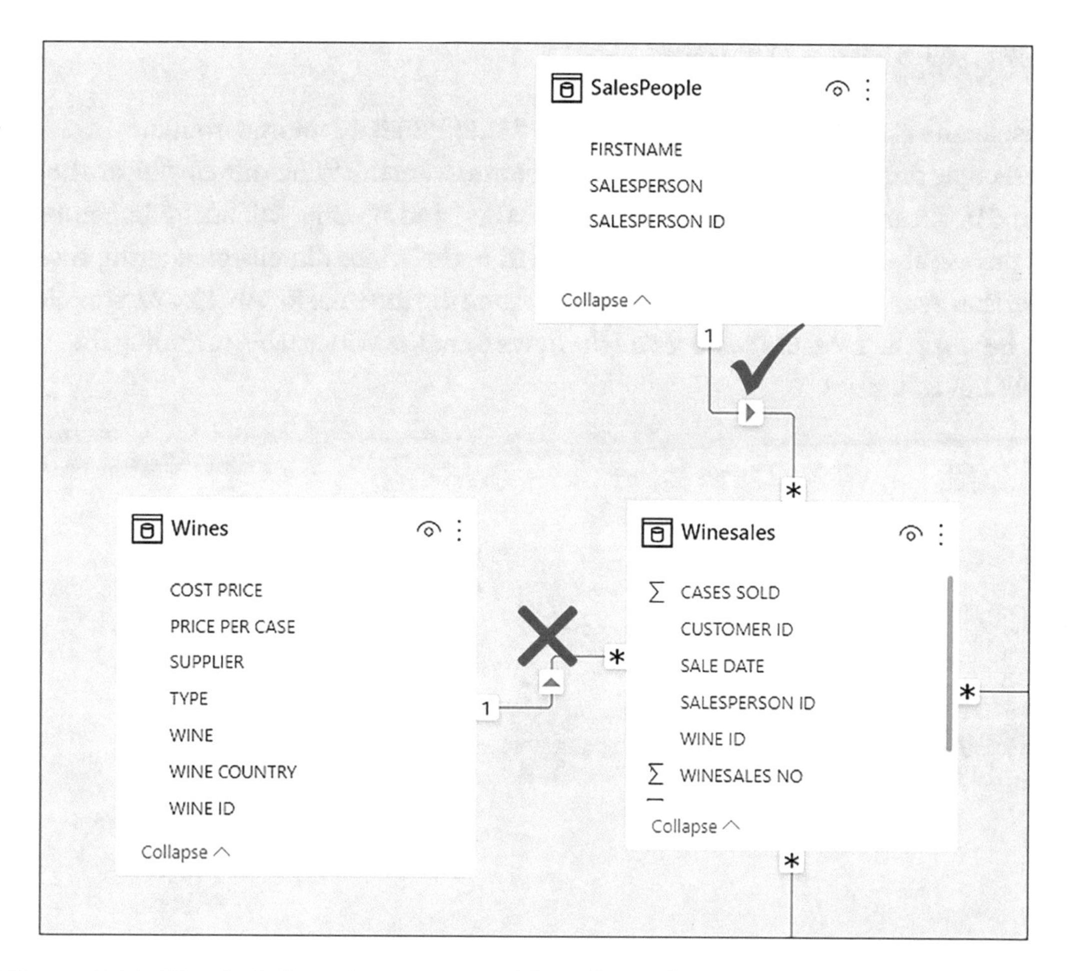

Figure 8-7. *The ALL function removes filters from the Wines dimension, but other filters are preserved*

Perhaps we're beginning to appreciate that there's much to understanding the ALL function! We're getting there, but we're not quite there yet. For instance, consider the measure we've just been working with:

```
No. of Sales All Wines =
CALCULATE ( [No. of Sales], ALL ( Wines ) )
```

It may not be the calculation that you want. The problem is that it removes *all* the filters in the Wines dimension. There will come a time when we need to be more specific regarding from which columns in a dimension we need to remove the filters.

Using ALL on a Column

Consider the example in Figure 8-8. Both the SALESPERSON column from the SalesPeople dimension *and* the SUPPLIER column from the Wines dimension are being filtered by slicers. We're filtering salesperson "Abel" and supplier "Alliance". Remember that there is also a filter on the WINE column from the Wines dimension filtering each wine. However, the "No. of Sales All Wines" is showing the total for Abel for *all* suppliers, **376**, because the measure removes *all the filters* from the Wines table including the SUPPLIER column.

WINE	No. of Sales	No. of Sales All Wines	No. of Sales as Percent of Filtered Value
Chardonnay	36	376	9.57%
Chenin Blanc	25	376	6.65%
Lambrusco		376	
Shiraz	38	376	10.11%
Total	**99**	**376**	**26.33%**

SALESPERSON
- [x] Abel
- [ ] Blanchet
- [ ] Charron
- [ ] Denis
- [ ] Leblanc
- [ ] Reyer

SUPPLIER
- [x] Alliance
- [ ] Laithwaites
- [ ] Majestic
- [ ] Redsky

Figure 8-8. *ALL that references a table will remove filters from all columns in a table, which may be incorrect*

Therefore, the percentage in "No. of Sales as Percent of Filtered Value" would be correct if you want to show the percentage "Abel's" sales of "Alliance" are of "Abel's" total sales for all suppliers (**376**). However, this would be incorrect if you want to show the percentage "Abel's" sales are of the total sales only for "Alliance" (**99**). If the latter is the goal, we must calculate "Abel's" total for all the wines that are supplied by "Alliance" (or whatever supplier has been filtered), which is **99**.

Let's look more closely at the problem. The current filter context uses filters on *two* columns in the Wines dimension: WINE and SUPPLIER. If we could see the in-memory Wines table for the evaluation of "Chardonnay", it might look something like Figure 8-9.

WINE ID	WINE	SUPPLIER	TYPE	WINE COUNTRY	PRICE PER CASE	COST PRICE
3	Chardonnay	Alliance	White	France	$100.00	$75.00

Figure 8-9. *Filters are on both the WINE column and the SUPPLIER column*

The measure "No. of Sales All Wines" removes both these filters and so calculates the number of sales for Abel for *all* wines and *all* suppliers. Using the ALL function with a table name as its argument, whether it's a fact table or a dimension, will remove *all* the filters from that table. We can, however, use ALL to remove filters from just *specific columns.*

To remedy the problem in Figure 8-8, we need to remove the filter from the WINE column but retain the filter on the SUPPLIER column. This is the measure we can create to do this:

```
No. of Sales All Wines #2 =
CALCULATE ( [No. of Sales] , ALL ( Wines[WINE] ) )
```

You can see that in this measure, we've used a reference to the WINE column inside ALL, and so ALL removes the filter from this column only. Figure 8-10 shows what is happening in memory, and you can see that the filter is retained on the SUPPLIER column.

WINE ID	WINE	SUPPLIER	TYPE	WINE COUNTRY	PRICE PER CASE	COST PRICE
14	Lambrusco	Alliance	White	Italy	$20.00	$15.00
13	Shiraz	Alliance	Red	France	$78.00	$30.00
12	Chenin Blanc	Alliance	White	France	$50.00	$10.00
3	Chardonnay	Alliance	White	France	$100.00	$75.00

Figure 8-10. *Using ALL on a column removes the filter from that column only*

We can now calculate the correct percentage and see this evaluated in Figure 8-11:

```
No. of Sales as Percent of Filtered Value #2 =
DIVIDE ( [No. of Sales] , [No. of Sales All Wines #2] )
```

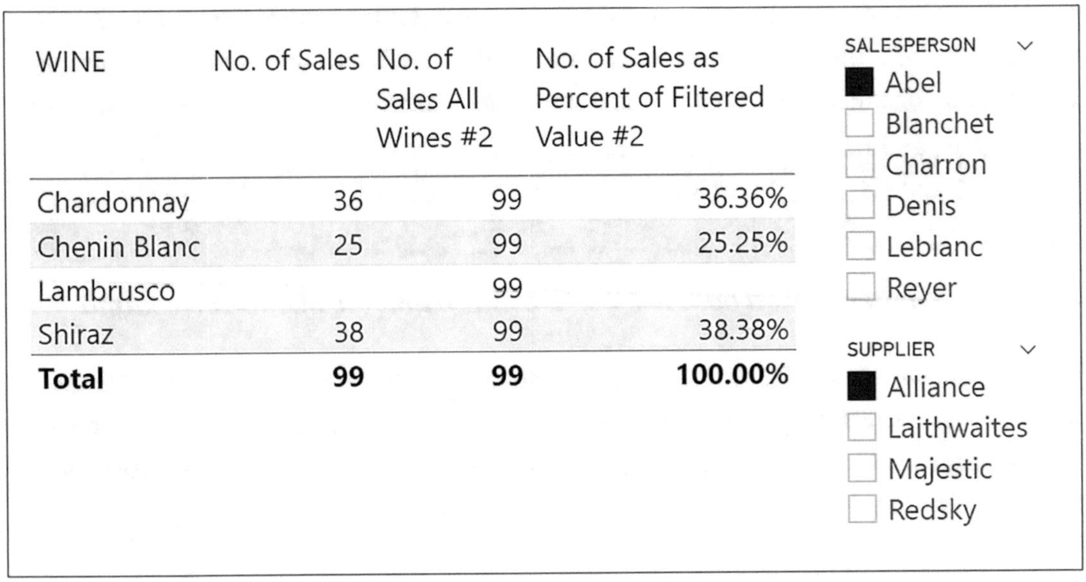

WINE	No. of Sales	No. of Sales All Wines #2	No. of Sales as Percent of Filtered Value #2
Chardonnay	36	99	36.36%
Chenin Blanc	25	99	25.25%
Lambrusco		99	
Shiraz	38	99	38.38%
Total	**99**	**99**	**100.00%**

SALESPERSON
- [x] Abel
- [ ] Blanchet
- [ ] Charron
- [ ] Denis
- [ ] Leblanc
- [ ] Reyer

SUPPLIER
- [x] Alliance
- [ ] Laithwaites
- [ ] Majestic
- [ ] Redsky

Figure 8-11. *The correct percentage for sales for "Abel" for "Alliance" supplier*

Let's consider another example where we must use the ALL function to remove the filter from a specific column. This is where the requirement is to calculate percentages across grouped items. For example, in the Matrix visual in Figure 8-12, there are two columns from the Wines dimension in the Rows bucket of the Matrix: WINE COUNTRY and TYPE. We've calculated the percentage the "Total Cases" values for each TYPE are of the "Total Cases" values for each WINE COUNTRY and can see that "White" wines constitute **47.02%** of "French" wines.[1]

[1] For information on constructing Matrix visuals, visit `https://www.burningsuit.co.uk/7-secrets-of-the-matrix-visual/`

WINE COUNTRY	TYPE	Total Cases	All Wines Type	Percentage of Wine Country
⊟ France	Red	130,616	246,543	52.98%
	White	115,927	246,543	47.02%
	Total	**246,543**	**246,543**	**100.00%**
⊟ Germany	Red	61,613	71,866	85.73%
	White	10,253	71,866	14.27%
	Total	**71,866**	**71,866**	**100.00%**
⊟ Italy	Red	33,951	104,815	32.39%
	White	70,864	104,815	67.61%
	Total	**104,815**	**104,815**	**100.00%**
Total		**423,224**	**423,224**	**100.00%**

Figure 8-12. *Calculating percentages across grouped data*

It would then be insightful to create a stacked column chart where we can show the total cases for each WINE COUNTRY and TYPE. We could then use the Tooltip, populated with our percentage measure to show the percentage breakdown across TYPE, as in Figure 8-13.

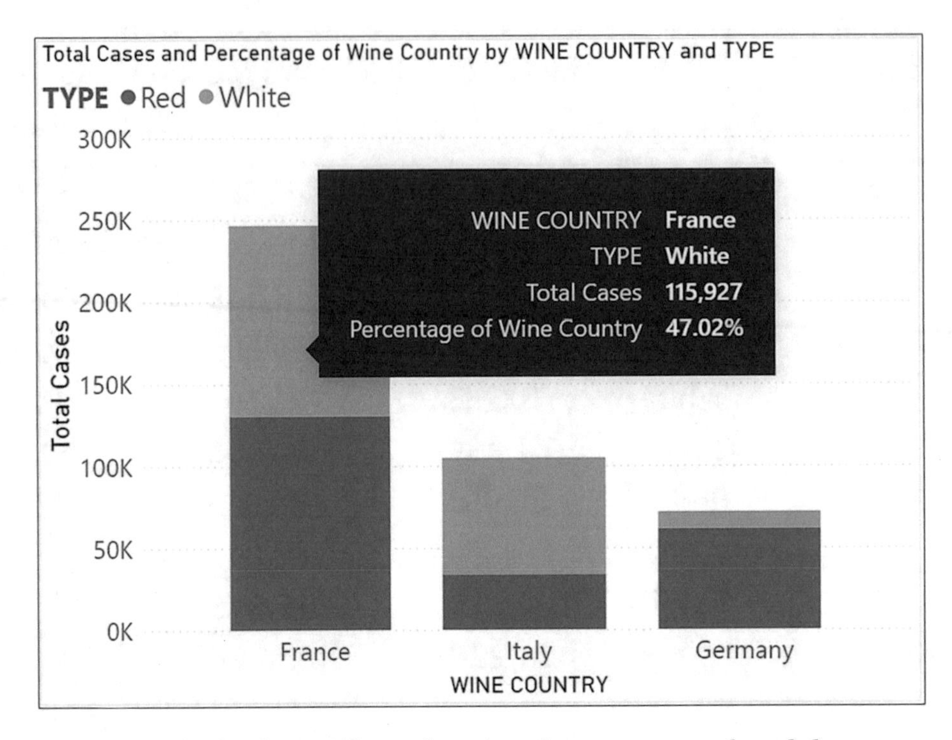

***Figure 8-13.** Stacked column chart showing the percentage breakdown across WINE COUNTRY in the Tooltip*

These are the measures that are calculated in Figure 8-12.

```
All Wines Type =
CALCULATE ( [Total Cases], ALL ( Wines[TYPE] ) )
```

```
Percentage of Wine Country =
DIVIDE ( [Total Cases] , [All Wines Type] )
```

Let's look more closely at how the "All Wines Type" measure is evaluated in the Matrix visual. The first evaluation starts with a filter on WINE COUNTRY of "France" and a filter on TYPE of "Red", and this is propagated to the fact table. However, to calculate the Total Cases for "France", the filter on TYPE must be removed so that the measure calculates Total Cases for *both* "Red" and "White" types for "France". If the filter from the TYPE column is removed using ALL, then "France" is the only filter propagated to the fact table; see Figure 8-14.

WINE ID	WINE	SUPPLIER	TYPE	WINE COUNTRY	PRICE PER CASE	COST PRICE
13	Shiraz	Alliance	Red	France	$78.00	$30.00
9	Merlot	Majestic	Red	France	$39.00	$15.00
5	Grenache	Redsky	Red	France	$30.00	$10.00
1	Bordeaux	Laithwaites	Red	France	$75.00	$25.00

WINE ID	WINE	SUPPLIER	TYPE	WINE COUNTRY	PRICE PER CASE	COST PRICE
13	Shiraz	Alliance	Red	France	$78.00	$30.00
12	Chenin Blanc	Alliance	White	France	$50.00	$10.00
9	Merlot	Majestic	Red	France	$39.00	$15.00
5	Grenache	Redsky	Red	France	$30.00	$10.00
3	Chardonnay	Alliance	White	France	$100.00	$75.00
2	Champagne	Laithwaites	White	France	$150.00	$100.00
1	Bordeaux	Laithwaites	Red	France	$75.00	$25.00

***Figure 8-14.** Using ALL on the TYPE column removes the filter from only that column*

Being able to identify which table and/or column you want to remove filters from is key to using ALL successfully. However, consider the example in Figure 8-15 where we have *four* columns in the Rows bucket. To calculate the percentage for each WINE COUNTRY, we need to remove the filters from *three* columns in the Wines dimension, that is, TYPE, SUPPLIER, and WINE.

WINE COUNTRY	TYPE	SUPPLIER	WINE	Total Cases	All Wines Type, Supplier & Wine	Percentage of Wine Country #2
France	Red	Laithwaites	Bordeaux	54,070	246,543	21.93%
		Redsky	Grenache	35,965	246,543	14.59%
		Majestic	Merlot	23,084	246,543	9.36%
		Alliance	Shiraz	17,497	246,543	7.10%
	White	Alliance	Chardonnay	42,030	246,543	17.05%
			Chenin Blanc	24,739	246,543	10.03%
		Laithwaites	Champagne	49,158	246,543	19.94%
	Total			**246,543**	**246,543**	**100.00%**
Germany	Red	Laithwaites	Malbec	34,290	71,866	47.71%
		Redsky	Chianti	27,323	71,866	38.02%
	White	Redsky	Piesporter	10,253	71,866	14.27%
	Total			**71,866**	**71,866**	**100.00%**
Italy	White	Majestic	Sauvignon Blanc	47,415	104,815	45.24%
			Pinot Grigio	23,449	104,815	22.37%
	Red	Majestic	Rioja	33,951	104,815	32.39%
	Total			**104,815**	**104,815**	**100.00%**

Figure 8-15. *Removing filters from multiple columns*

Inside the ALL function, you can reference multiple column names, so you could write this measure:

```
All Wines Type, Supplier & Wine =
CALCULATE ( [Total Cases],
    ALL ( Wines[TYPE], Wines[SUPPLIER], Wines[WINE] )
)
```

Note Because we are removing the filter from the WINE column, "Lambrusco" wine that has no data will appear in the visual. To fix this, use a visual-level filter to filter nonblank items.

However, you can appreciate how tedious this could get if you had many columns from which you must remove filters. This is where you could use the ALLEXCEPT function instead of ALL.

The ALLEXCEPT Function

ALLEXCEPT removes all filters in a table except filters that are applied to the columns you specify. This can be used for situations in which you want to remove the filters on many but not all of the columns in a table.

The ALLEXCEPT function has the following syntax:

= ALLEXCEPT (table, column1, colum2, etc.)

where:

table is the table where you want to clear the filters from *except* the filters on the columns specified in the next arguments.

column1, column2 are the columns where you want filters preserved.

Here is an example of the ALLEXCEPT syntax:

= ALLEXCEPT (Wines, Wines[WINE COUNTRY])

Note that in ALLEXCEPT, unlike ALL, you need to first supply the table name.

Therefore, in the Matrix visual in Figure 8-15, you could author an alternative version of the "All Wines Type, Supplier & Wine" measure as follows:

```
All Except Wine Country =
CALCULATE ( [Total Cases],
    ALLEXCEPT ( Wines, Wines[WINE COUNTRY] )
)
```

So now we can calculate the percentage:

```
Percentage of Wine Country #2=
DIVIDE ( [Total Cases] , [All Except Wine Country] )
```

You might think that surely we've exhausted all possible "ALL" variations! We've looked at removing filters from entire tables, either fact tables or dimensions. We've also seen how we can remove filters from specific columns and how to remove filters from several columns while retaining filters on others. However, there is still another scenario that we need to explore. Consider Figure 8-16.

WINE	No. of Sales	Grand Total No. of Sales
Bordeaux	180	2,207
Champagne	132	2,207
Chardonnay	187	2,207
Chenin Blanc	200	2,207
Total	**699**	**2,207**

WINE
- ■ Bordeaux
- ■ Champagne
- ■ Chardonnay
- ■ Chenin Blanc
- □ Chianti
- □ Grenache
- □ Lambrusco
- □ Malbec
- □ Merlot
- □ Piesporter
- □ Pinot Grigio
- □ Rioja
- □ Sauvignon Blanc
- □ Shiraz

Figure 8-16. *The "Grand Total No. of Sales" measure is not the total for the selected wines in the slicer*

Here, we have a Table visual into which the WINE column from the Wines dimension has been placed. You can see that four wines have been filtered using the slicer. The "No. of Sales" measure calculates the number of sales for the selected wines. The "Grand Total No. of Sales" measure has also been included and has the following expression:

```
Grand Total No. of Sales =
COUNTROWS ( ALL ( Winesales ) )
```

This measure returns the total number of sales for all wines irrespective of the slicer selection. This would also be true if the wines filter was generated from a filter placed in the Filters pane. If we want to calculate percentages of the total only for the selected wines (**699** in this case), this "Grand Total No. of Sales" measure is not going to work.

The problem is that the values selected in the slicer come from the same column that is put into the Table visual, which is the WINE column. We're using the slicer to *reduce* the wines shown in the visual. Therefore, we need to find a function that specifically finds grand totals *for the items that have been filtered in the visual.* The function we need is called ALLSELECTED.

The ALLSELECTED Function

The syntax for the ALLSELECTED function is the same as for the ALL function:

= ALLSELECTED (table)

or

= ALLSELECTED (Column 1, Column 2, etc.)

However, if you were to look at the function description in the DAX Function Library, you may be a little bemused:

> *"ALLSELECTED removes context filters from columns and rows in the current query, while retaining all other context filters or explicit filters. The ALLSELECTED function gets the context that represents all rows and columns in the query, while keeping explicit filters and contexts other than row and column filters. This function can be used to obtain visual totals in queries."*

To be fair, it is very difficult to explain what this function does. It's much easier to look at an example of using it. Therefore, let's return to our problem of calculating the grand total for only the wines selected in the slicer. This is the DAX expression we need:

```
Grand Total No. of Sales for Selected Wines =
CALCULATE ( [No. of Sales], ALLSELECTED ( Wines[WINE] ) )
```

You can see this measure and the percentage calculated from it in Figure 8-17.

WINE	No. of Sales	Grand Total No. of Sales for Selected Wines
Bordeaux	180	699
Champagne	132	699
Chardonnay	187	699
Chenin Blanc	200	699
Total	**699**	**699**

WINE
- ■ Bordeaux
- ■ Champagne
- ■ Chardonnay
- ■ Chenin Blanc
- ☐ Chianti
- ☐ Grenache
- ☐ Lambrusco
- ☐ Malbec
- ☐ Merlot
- ☐ Piesporter
- ☐ Pinot Grigio
- ☐ Rioja
- ☐ Sauvignon Blanc
- ☐ Shiraz

Figure 8-17. *The ALLSELECTED function calculates the correct grand total*

How does this expression work? Well again, let's consider the current filter context for the first evaluation of this measure, that is, "Bordeaux" in the WINE column of the Wines dimension. However, ALLSELECTED replaces the filter on the WINES column with the filter from the slicer. Therefore, the Wines dimension is filtered to reflect the slicer selection; see Figure 8-18.

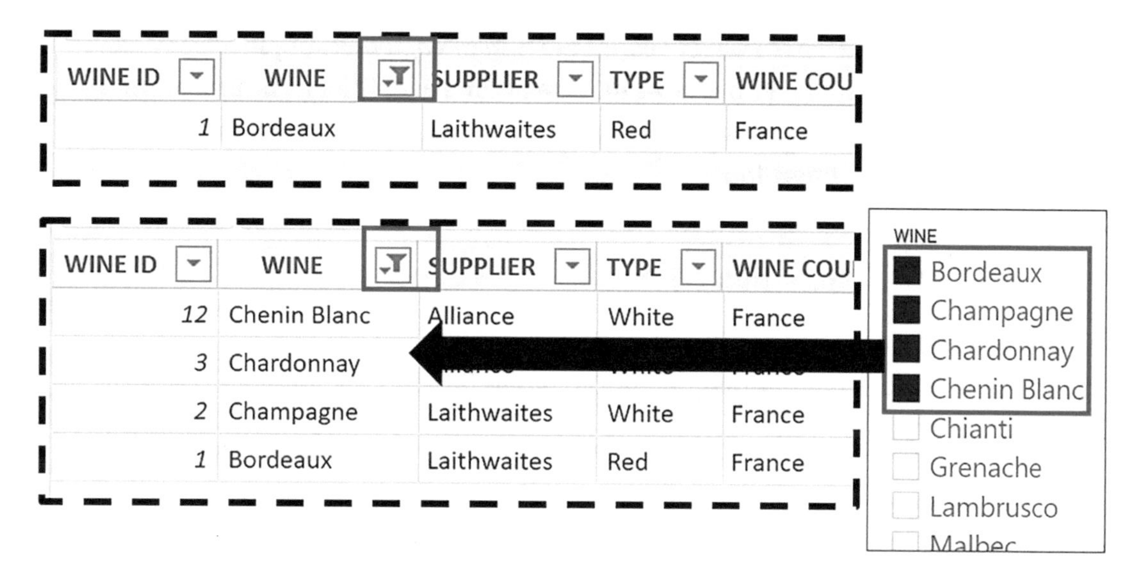

Figure 8-18. *The ALLSELECTED function replaces the filter to reflect the slicer selections*

Mostly you can use ALLSELECTED in place of ALL because often you're using slicers or the Filters pane to reduce the number of items shown in visuals. If there are no selections from slicers or from the Filters pane, ALLSELECTED will remove all filters, just like ALL.

Up to now, we've been using the ALL function (and its variations) while not considering whether it's being used as a *table function* or is being used as a *modifier* to CALCULATE. The "ALL" functions seem to be doing their job, and we're thankful for that. We know that ALL removes filters whether by removing filters from tables and columns or by generating virtual tables containing all the rows. However, we are now going to focus our attention on the difference between ALL as a table function and ALL as a modifier to CALCULATE. Remember how in Chapter 1 we said that when working with DAX, the devil is in the detail? Understanding this difference in these two behaviors of ALL is a fine example of paying attention to this detail.

ALL as a Modifier to CALCULATE

To understand this aspect of ALL, let's consider a scenario that we've looked at before, which is removing the filter from the WINE column in the Wines dimension while still retaining the filter on the SUPPLIER column (see Figure 8-11). This was to calculate the *number of sales* for the selected supplier.

However, this time we're going to count the *number of wines* supplied by "Alliance" by counting the rows in the Wines table that are filtered accordingly, using a slicer. To do this, we've created two similar measures that both use the ALL function on the WINE column in the Wines dimension:

```
No. of Wines #1 =
COUNTROWS ( ALL ( Wines[WINE] ) )

No. of Wines #2 =
CALCULATE ( COUNTROWS ( Wines ), ALL ( Wines[WINE] ) )
```

However, only one of these measures returns the correct result; see Figure 8-19.

WINE	No. of Wines #1	No. of Wines #2
Chardonnay	14	4
Chenin Blanc	14	4
Lambrusco	14	4
Shiraz	14	4
Total	**14**	**4**

SUPPLIER
- ■ Alliance
- □ Laithwaites
- □ Majestic
- □ Redsky

Figure 8-19. *Using ALL on a column can return different results*

The "No. of Wines #1" measure uses ALL as a *table* function and generates a one-column table of *all* the distinct values in the WINE column. The measure then counts the number of rows in this virtual table and returns 14 rows as shown in Figure 8-20.

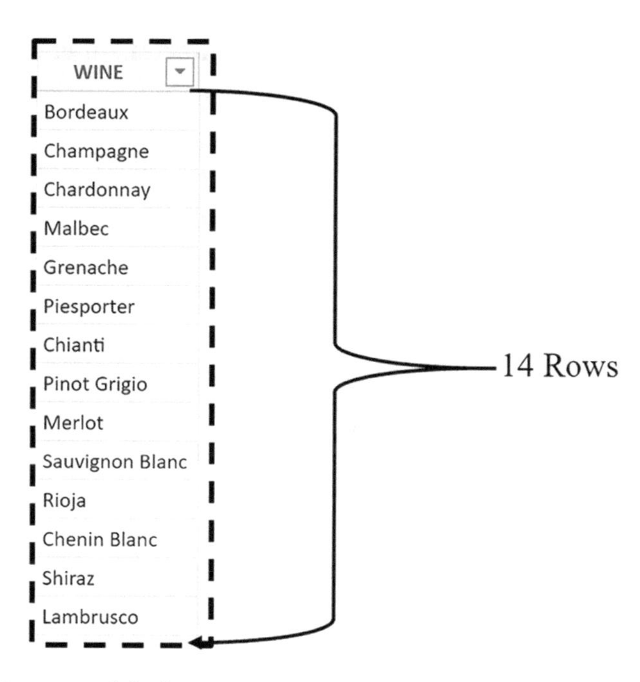

Figure 8-20. *ALL as a table function generates a virtual table of distinct values*

The "No. of Wines #2" measure uses ALL inside CALCULATE as a *modifier* and therefore *removes* the filter from the WINES column but preserves the filter on the SUPPLIER column. This measure then counts the number of rows in the Wines dimension and returns four rows; see Figure 8-21.

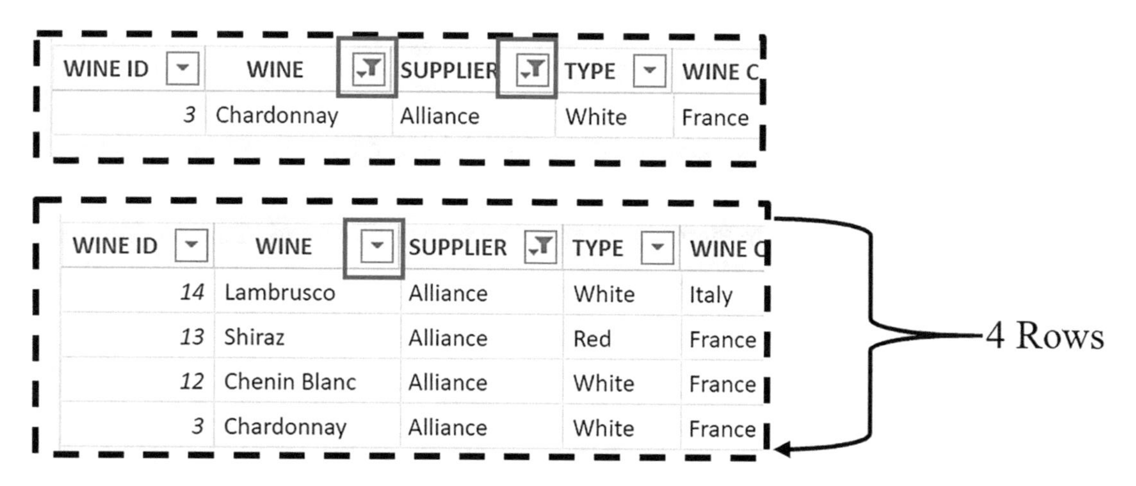

Figure 8-21. *The ALL function as a CALCULATE modifier removes the filter on the WINE column*

This example has been easy to explain. However, the ALL function acting as a modifier to CALCULATE can be more challenging to understand, and this is certainly the case in the next example we're going to explore.

We've built three measures that calculate the number of sales where the cases sold is greater than 300. They're all using the expression *"ALL (Winesales)"* (highlighted in gray), and the expressions look much the same. You might therefore expect them to return the same result:

```
No. of Sales Where Cases GT 300 #1 =
CALCULATE ( [No. of Sales],
    ALL ( Winesales ),
   Winesales[CASES SOLD] > 300
)

No. of Sales Where Cases GT 300 #2 =
CALCULATE (
    [No. of Sales],
    FILTER (
    ALL ( Winesales ), Winesales[CASES SOLD] > 300 )
)

No. of Sales Where Cases GT 300 #3 =
CALCULATE (
    [No. of Sales],
    ALL ( Winesales ),
    FILTER ( Winesales, Winesales[CASES SOLD] >300 )
)
```

However, as you can see in Figure 8-22, whereas measures #1 and #2 return the same value, measure #3 returns a different value.

WINE	No. of Sales Where Cases GT 300 #1	No. of Sales Where Cases GT 300 #2	No. of Sales Where Cases GT 300 #3
Bordeaux	286	286	89
Champagne	286	286	100
Chardonnay	286	286	
Chenin Blanc	286	286	
Chianti	286	286	
Grenache	286	286	32
Lambrusco	286	286	
Malbec	286	286	1
Merlot	286	286	
Piesporter	286	286	
Pinot Grigio	286	286	
Rioja	286	286	
Sauvignon Blanc	286	286	64
Shiraz	286	286	
Total	**286**	**286**	**286**

Figure 8-22. *Similar expressions can return different results*

These three measures all use ALL on the Winesales table so they should ignore any filters on the Winesales table. This is true for measures #1 and #2 (there are **286** rows in the Winesales table where CASES SOLD is greater than 300), but what about measure #3? In this measure, the ALL function appears to be ignored, and the cross-filter propagated from the Wines dimension is retained. Therefore, this measure returns the number of sales *for each wine* where CASES SOLD is greater than 300.

Question: Which of these measures is the odd one out?

You might think measure #3 is the odd one out because it returns a different value. However, you could argue that measure #2 is the odd one out because it's the only measure where ALL is being used as a *table function*. In the other two measures, ALL is acting as a CALCULATE *modifier*.

To understand this, let's look at the evaluation of each of these measures in more detail.

In this measure

```
No. of Sales Where Cases GT 300 #1 =
CALCULATE ( [No. of Sales],
    ALL ( Winesales ), Winesales[CASES SOLD] > 350
)
```

there are two filter arguments in CALCULATE. The first one using ALL is *modifying* the filter to remove filters from the Winesales table. This is evaluated first and produces an empty filter. The second filter is a *column* filter on the CASES SOLD column, filtering cases sold greater than 300; see Figure 8-23.

Evaluation of the ALL argument that modifies CALCULATE is done first

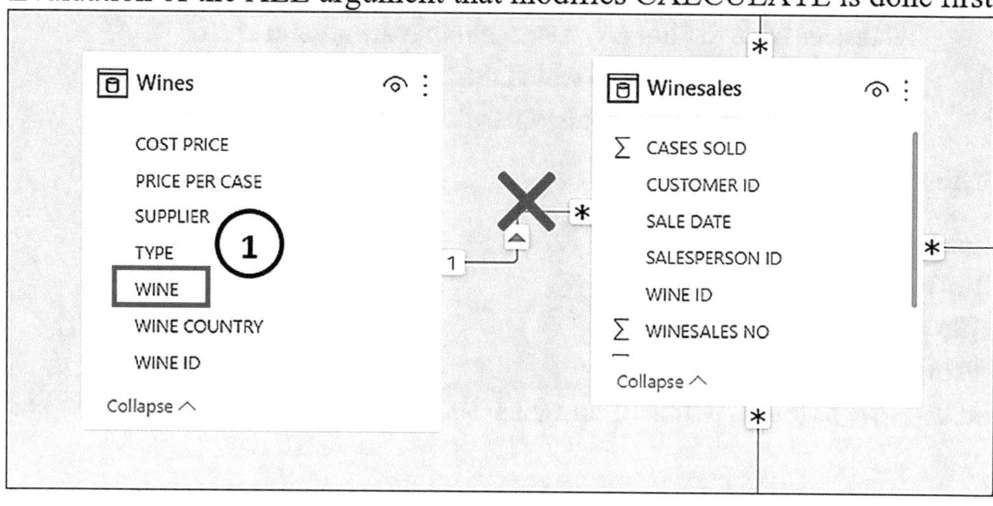

Evaluation of the filter argument is done next

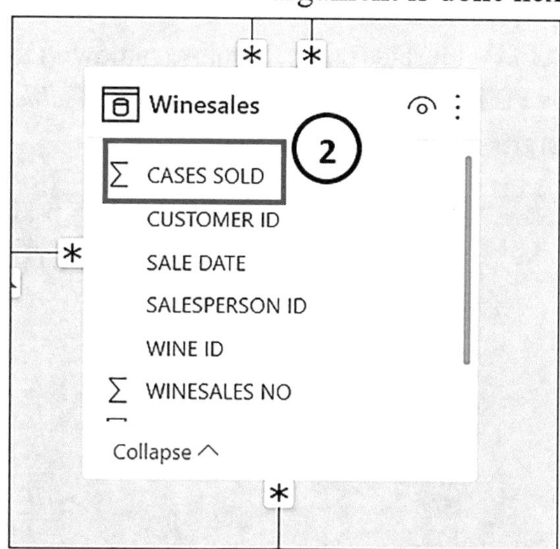

Figure 8-23. *The evaluation of "No. of Sales Where Cases GT 300 #1"*

1. ALL behaves as a modifier to CALCULATE and removes any filters or cross-filters on Winesales, including the filter coming through from the Wines dimension. This results in an empty filter, and therefore, the Winesales table now has no filters on it.

2. A new filter is then placed on the CASES SOLD column of the Winesales table to filter any cases sold that are greater than 300. This is the new filter in which the "No. of Sales" measure is evaluated and the rows of the Winesales table are counted.

In this measure

```
No. of Sales Where Cases GT 300 #2 =
CALCULATE (
    [No. of Sales],
    FILTER (
    ALL ( Winesales ), Winesales[CASES SOLD] >350 )
)
```

there is just one filter argument in CALCULATE supplied by the FILTER function (highlighted in gray). Inside the FILTER function, the ALL function generates a virtual table of all the rows in the Winesales table, therefore removing the cross-filter from the Wines dimension. The FILTER function iterates this *virtual table* to return the rows where CASES SOLD is greater than 300; see Figure 8-24.

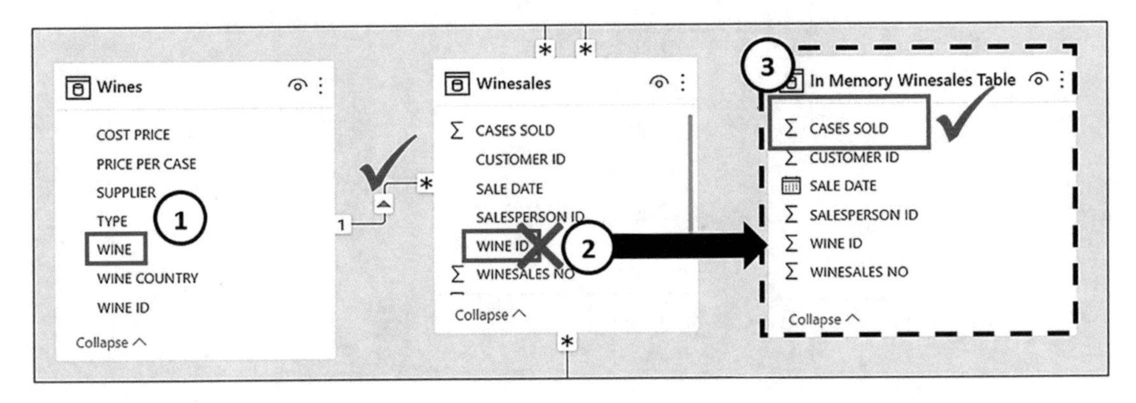

Figure 8-24. *The evaluation of "No. of Sales Where Cases GT 300 #2"*

1. The current filter context filters each WINE in the Wines dimension, and this is cross-filtered to the Winesales table.

2. The ALL function inside FILTER generates a virtual table of all the rows of Winesales, ignoring the wine filter.

3. The FILTER function iterates over the virtual Winesales table to filter out the rows where CASES SOLD is greater than 300. This is the new filter in which the "No. of Sales" measure is evaluated and the rows of the virtual Winesales table are counted.

The outcome of this measure is the same as in #1 before. However, you can appreciate that the generation of a virtual table is less efficient than simply placing a filter on a column. Here is yet another example of paying a heavy processing price when using a table filter inside CALCULATE (we've looked at this earlier when learning about the FILTER function).

In this measure

```
No. of Sales Where Cases GT 300 #3 =
CALCULATE (
    [No. of Sales],
    ALL ( Winesales ),
    FILTER ( Winesales, Winesales[CASES SOLD] >300 )
)
```

there are two filter arguments inside CALCULATE. The first, *"ALL (Winesales)"*, is a CALCULATE modifier. The second, *"FILTER (Winesales, Winesales[CASES SOLD] > 300"*, is a table filter. We need to understand that CALCULATE modifiers are evaluated first before any other filter arguments. Let's take the first argument that is *modifying* the filter context to remove the filters from the Winesales table. This is evaluated first and creates an empty filter because all filters on the Winesales table have been removed.

The second filter uses the FILTER function to create a *virtual Winesales table.* But which rows have been filtered in the virtual Winesales table generated by FILTER? We have asked FILTER to filter to the rows where CASES SOLD is greater than 300. However, remember what we know about the FILTER function. This function filters only the rows *in the current filter context.* So the table generated by FILTER still contains the rows *for each wine* (e.g., only rows for "Bordeaux" in the first evaluation), and these rows are further filtered to just rows where CASES SOLD is greater than 350; see Figure 8-25.

Evaluation of the ALL argument that modifies CALCULATE is done first

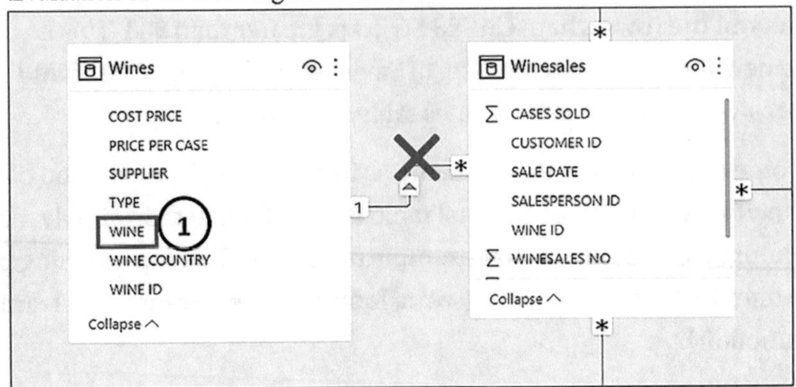

Evaluation of the filter argument using FILTER is done next

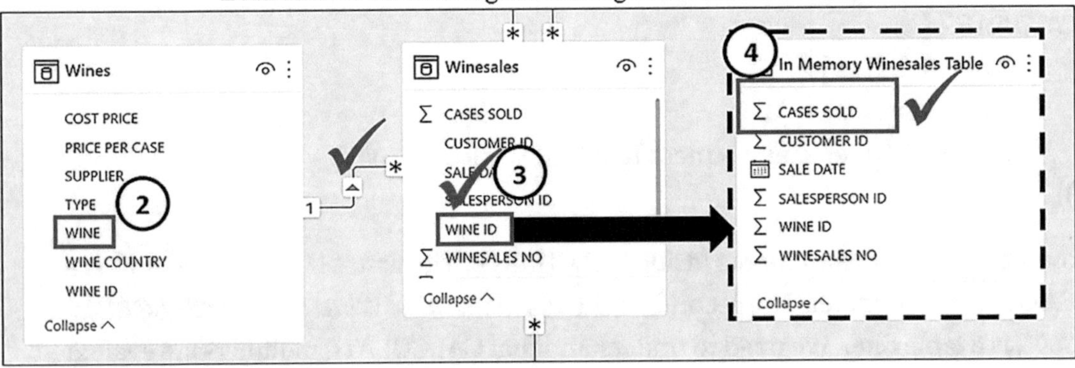

Figure 8-25. *The evaluation of "No. of Sales Where Cases GT 300 #3"*

1. The first argument uses ALL as a modifier to remove the filters from the Winesales table. This is evaluated first, and there is now an empty filter on the Winesales table.

2. The filter argument using FILTER is now evaluated separately. The Wines dimension is cross-filtered to the Winesales fact table filtering each wine.

3. The FILTER function iterates the Winesales table in the current filter context and generates a virtual table containing the rows for, for example, "Bordeaux" wine.

4. It then further filters these rows so only rows containing CASES SOLD that is greater than 300 for that wine remain in the table.

Because the first argument using ALL has produced an empty filter, this is the new filter in which the "No. of Sales" measure is evaluated and the rows of the virtual Winesales table are counted.

So let's summarize what we now know about ALL. The ALL function as a *table* function generates a virtual table containing all the rows from a table or all the distinct rows of a column or columns. This virtual table containing all the rows can be *refiltered* by FILTER, and this will then propagate filters through the model as in measure #2 before.

The ALL function as a modifier to CALCULATE is evaluated first before any filter arguments inside CALCULATE. ALL as a modifier *removes* any filters from a table or a column and generates an empty filter. Any other filter arguments of CALCULATE are then evaluated and generate the new filter context as in measures #1 and #3 before.

Because ALL has a different behavior when used as a top-level argument to CALCULATE, users believed it should have a different name when used in this context. As a result, in 2019, a new function was introduced into the DAX Function Library, REMOVEFILTERS. This function is synonymous with ALL, but it can be used only as a CALCULATE modifier and not as a table expression like ALL.

In this chapter we have explored the ALL, ALLEXCEPT and ALLSELECTED functions that are challenging functions with which to get to grips. Regardless of how long you've been using DAX, the examples described here will always be problematic to understand, but it's only by thinking through the evaluation of these measures, paying close attention to the details, can we come to truly understand how DAX works.

Having covered ALL and its variations, we can now move on to look at a group of functions called time intelligence functions where paradoxically, the ALL function has mostly been made redundant.

CHAPTER 9

Calculations on Dates: Using DAX Time Intelligence

Have you ever wanted to compare sales for the current month against sales for last month? Or perhaps something a little more ambitious, such as cumulative totals or even a rolling monthly average? If the answer is yes, and why wouldn't it be, calculations using date data such as these require the use of a group of DAX functions called "time intelligence" functions. Exploring these functions will be the focus of this chapter, and you will learn how to design expressions to enable you to evaluate data across different granularities of time such as financial years, quarters, months, and even down to the day grain. In doing so, you will be able to compare and contrast calculations over those periods to build insights into the data that's important to you, such as trends and patterns over time.

Note The term "time intelligence" is a little misleading. These are not *time* intelligence functions but *date* intelligence functions, so these functions will not help you with calculations on hours, minutes, or seconds, although we can do these calculations with the help of a Time dimension.

The starting point to using time intelligence functions is the creation of a date dimension. This is because most time intelligence functions are designed to work with a date table as an integral part of the data model. You may feel your data model doesn't require a date dimension, but you'll struggle to create the date-based calculations you need, and you certainly won't be able to reap the benefits of time intelligence measures.

© Alison Box 2022
A. Box, *Up and Running with DAX for Power BI*, https://doi.org/10.1007/978-1-4842-8188-8_9

However, people new to DAX often don't appreciate this aspect of date calculations and therefore don't have a date dimension in their model. If this is the case, Power BI will help you with your date analysis by generating built-in date hierarchies, and this is what we will explore first.

Power BI Date Hierarchies

In the absence of a date dimension in your model, if you have columns of a date data type in any tables, for every one of these columns, Power BI will generate an in-memory date table for you that also contains a date hierarchy. We have removed the DateTable dimension from our data model, and so the SALE DATE column is now expressed as a date hierarchy as shown in Figure 9-1.

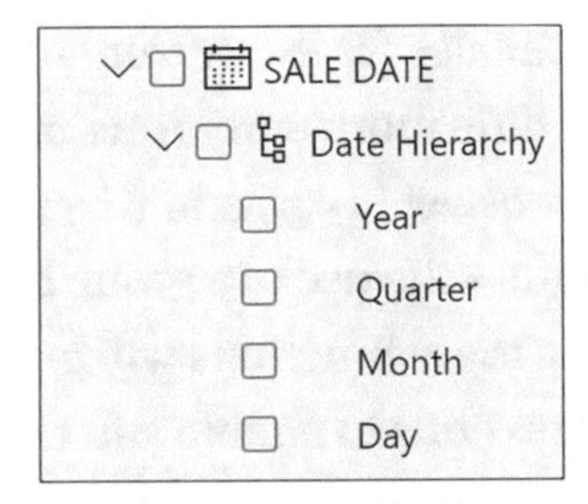

Figure 9-1. *A date type column with a date hierarchy generated by Power BI*

This feature is called "Auto date/time," but you can turn off this behavior either globally or only for the current file in the **Options** pane shown in Figure 9-2.

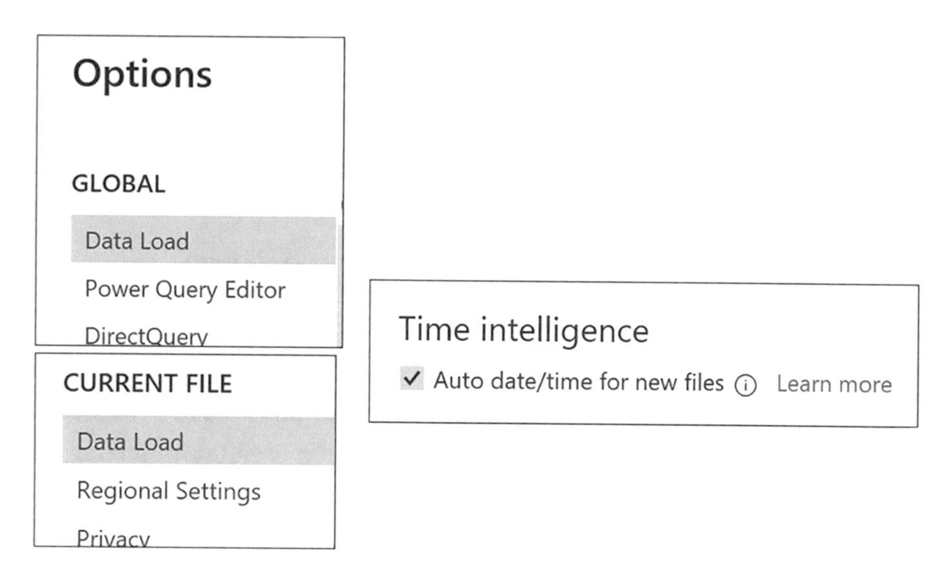

Figure 9-2. *You can turn off the generation of a date hierarchy using the Options pane*

If you have the "Auto date/time" feature turned on and you don't have a date dimension in your model, any fields of a date data type will be structured into hierarchies. These built-in date hierarchies are useful for drilling into different date granularities when put into Power BI visuals and also make it possible to slice by year, quarter, month, and day. For example, in Figure 9-3, we are using the SALE DATE hierarchy to drill into Month granularity in a Power BI line chart and slice by year.

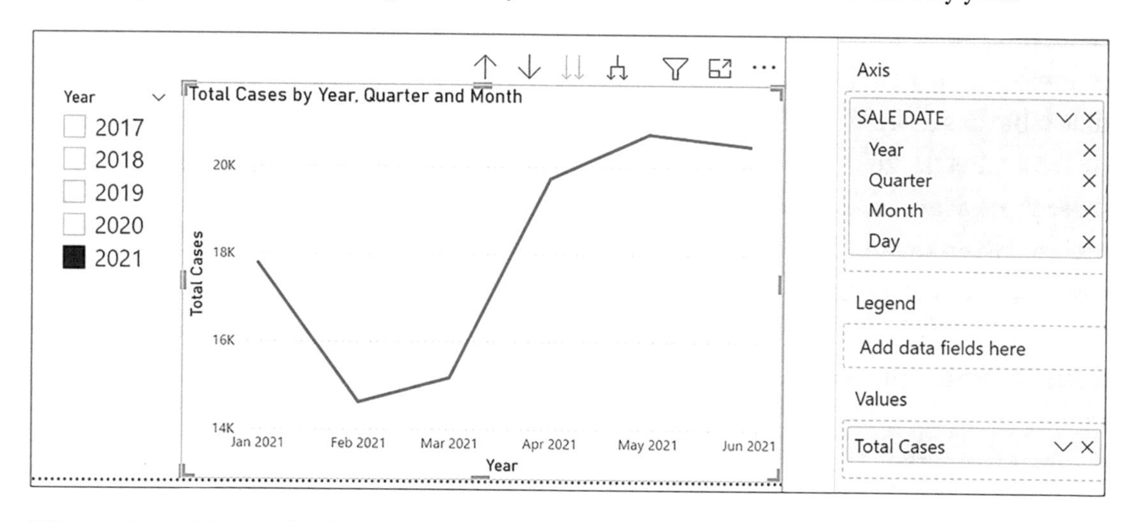

Figure 9-3. *Using the built-in date hierarchy to visualize date data*

However, there are a number of drawbacks to using these hierarchies:

- What if your financial year doesn't start in January?
- What if you want to analyze sales by week granularity? How would you add week numbers?
- What if you want to compare sales in 2020 with sales in 2021 in a clustered column chart?

All the preceding problems present a real challenge if you're using built-in date hierarchies, but if you have a date table dimension in your model, life becomes a lot easier as far as date calculations go. Therefore, the first step is generating your date dimension table and integrating it into your data model.

Creating a Date Table

To generate your date table, you can use DAX or Power Query as explained comprehensively in these two links:

```
www.sqlbi.com/articles/creating-a-simple-date-table-in-dax/
https://exceleratorbi.com.au/build-reusable-calendar-table-power-query/
```

Failing these two suggestions, you could use Excel to create a date table.

The only mandatory column in a date table is a column containing a list of sequential dates that includes *all the dates* that cover the time span of your data. For example, our wine sales begin in January 2017 and end in December 2021; therefore, our date table has a DATEKEY column with values starting on January 1, 2017, and ending on December 31, 2021 (the end of our financial year). You must include all the dates in these years even if there is no data for specific dates. The other columns in the date table are used to group and categorize these dates and are completely arbitrary. However, it would be normal to have columns for your financial year and quarters and columns for months, including month name and month number. You could also include different financial years and week numbers. To analyze by months, you need to include both month name *and* month number. This is so you can sort the month names correctly, and some measures will require referencing both month name and number.

We've now replaced our DateTable back into our data model. You can see the DateTable is related to the fact table using the SALE DATE and the DATEKEY columns as shown in Figure 9-4.

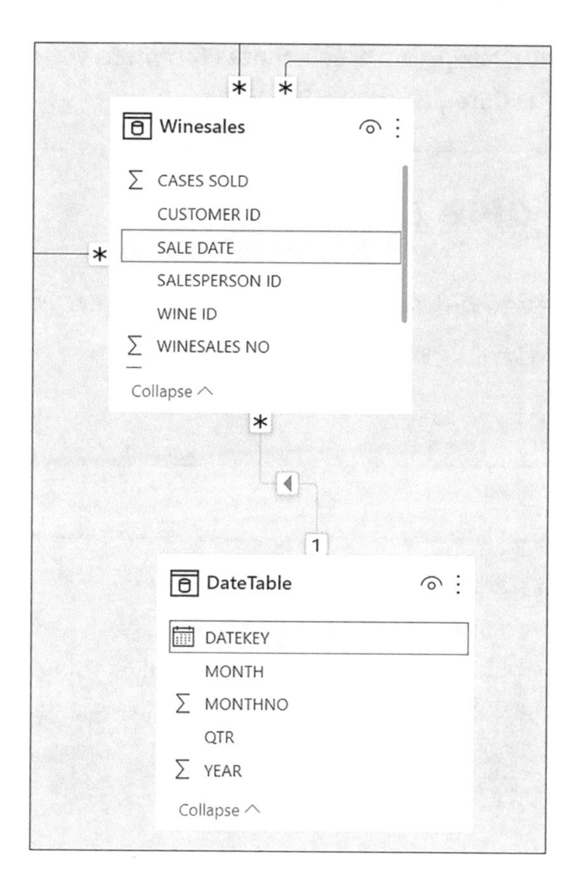

Figure 9-4. *The DateTable is related to the Winesales fact table using the DATEKEY column*

Note It's usual to use the column in your date table that contains the list of unique dates as the linking field or primary key, but it would be possible to use some other unique field in the date table as the linking field. However, you must always have a column containing a list of sequential dates in your date table even if you don't use this field to link to the fact table.

The next requirement regarding the date table is to ensure the model "knows" this is your date dimension. This is particularly true if you haven't used the field containing the list of unique dates as the primary key of the date table. You do this by marking the date dimension as a date table by selecting **Mark as date table** from the **Table Tools** tab.

Now, in the "Date column" drop-down, select the column in your date table that contains the list of unique dates, as shown in Figure 9-5.

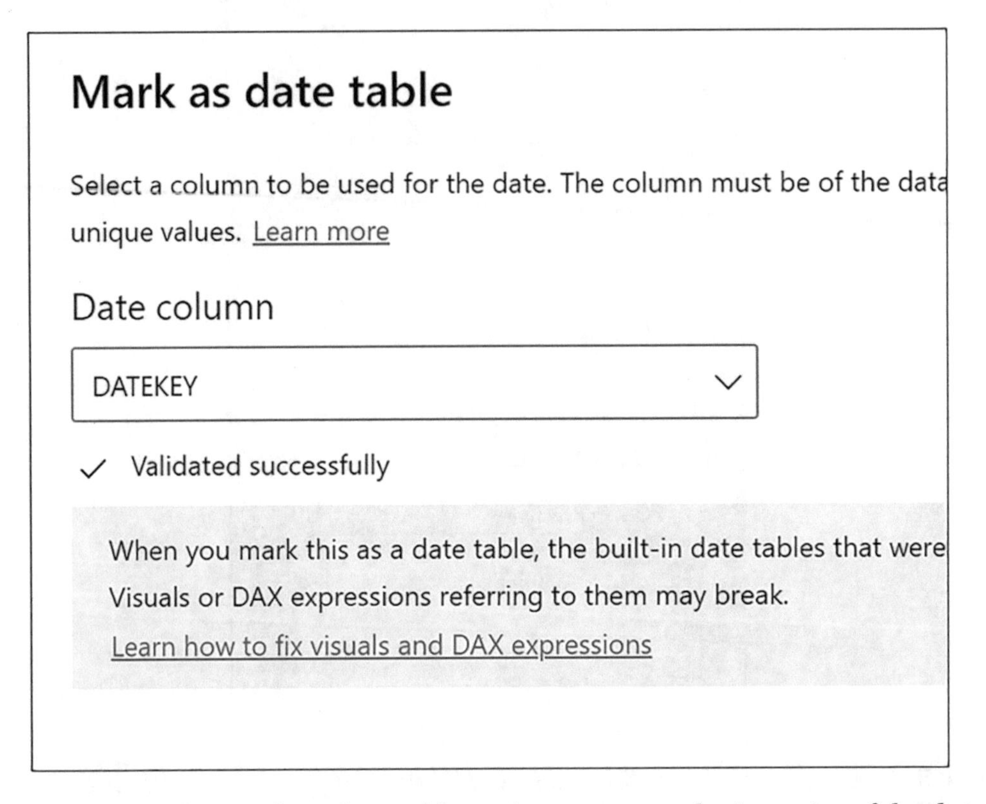

Figure 9-5. *Use the Mark as date table option to ensure the integrity of the date dimension*

Note You will find more information on the requirement to "Mark as date table" here: *https://www.sqlbi.com/articles/mark-as-date-table/*

The final step in the setup of the data dimension is to sort the month names correctly. You can see in Figure 9-6 that we've used the **Sort by column** button on the **Column Tools** tab to sort the Month by the Month No.

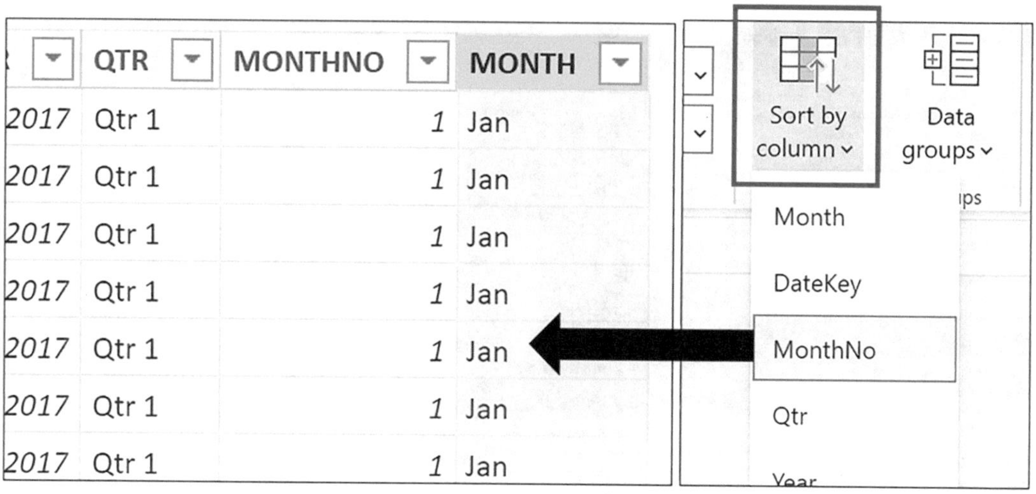

Figure 9-6. *Use the Sort by column option to sort the month names*

Now that we have generated our date dimension, we can reap the benefits of using the time intelligence functions inside DAX and analyze our data across years, quarters, months, and days in many insightful ways.

Using Time Intelligence Functions

Time intelligence functions use a *base date* from which to perform the required calculation. This base date is supplied by the current filter context. For example, the terms "previous month" and "same period last year" are relative terms, relative, that is, to the date that is in the current filter context. Therefore, with most of these functions, you must have a specific date filtered (a year, a quarter, a month, or a day) either by using slicers, by using the Filters pane, or by having dates in the visual. For example, if you want to find the previous month's sales, you must have a current month filtered in the visual or in a slicer; see Figure 9-7.

YEAR	MONTH	Total Cases	Previous Month's Total Cases
2017	Jan	6,657	
2017	Feb	5,705	6,657
2017	Mar	5,544	5,705
2017	Apr	5,364	5,544
2017	May	4,757	5,364
2017	Jun	3,011	4,757
2017	Jul	5,079	3,011
2017	Aug	3,182	5,079
2017	Sep	7,279	3,182
2017	Oct	5,602	7,279
2017	Nov	5,045	5,602
2017	Dec	7,521	5,045
2018	Jan	6,247	7,521
Total		421,281	

WINE	Total Cases	Previous Month's Total Cases
Bordeaux	998	478
Champagne	251	1,038
Chardonnay	430	
Chenin Blanc	483	586
Chianti	123	619
Grenache	916	893
Malbec	801	963
Piesporter		484
Pinot Grigio	338	487
Rioja	658	858
Sauvignon Blanc	1,465	1,721
Shiraz	512	98
Total	**6,975**	**8,225**

MONTH	YEAR
Jan	2017
Feb	2018
Mar	2019
Apr	2020
May	2021
Jun	
Jul	
Aug	
Sep	
Oct	
Nov	
Dec	

***Figure 9-7.** The "base date" is supplied by the filter context which can be through columns in the visual or by year and month slicers*

All time intelligence functions (except LASTNONBLANK and LASTNONBLANKVALUE) have an argument that requires specifying a column of dates to be used in the calculation. In most cases, in this argument, you supply the name of the column in your date table that holds the list of unique dates, for example, the DATEKEY column in our data; see Figure 9-8.

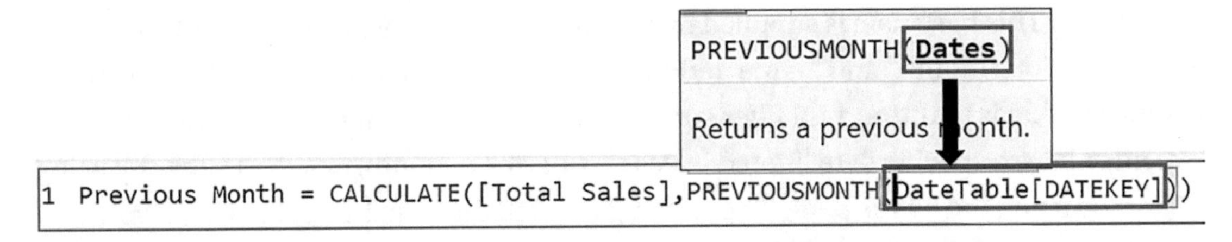

***Figure 9-8.** The "Dates" argument normally requires referencing the column that holds the list of unique dates*

Note There is an exception to this. In the LASTDATE and FIRSTDATE functions, you may need to reference the date column in your fact table.

For every DAX expression you construct using time intelligence functions, you could author an equivalent expression using standard DAX functions such as CALCULATE, FILTER, MAX, and MIN. However, if this were the case, there is one function you would also need, and that's the ALL function. For example, to find dates in May when the current filter context is filtering dates in June, you would have to use the ALL function to remove the current filter on June in the date table so that it could be refiltered for the dates in May. By using time intelligence functions and referencing the "Dates" column of the date table, the work of the ALL function is implicit. That's why when using time intelligence functions, you don't need to remove filters by using ALL and then reapply your specific filter.

The time intelligence functions we're going to explore in this chapter are outlined in Table 9-1. The return value is typically a virtual table containing a single column of dates. The dates returned into this column are also shown in Table 9-1.

Table 9-1. *Time intelligence functions and their return value*

Function	Dates Returned
PREVIOUSMONTH	The previous month from the month in the current filter context.
SAMEPERIODLASTYEAR	The same period last year from the month in the current filter context.
DATEADD	Prior (or future) years, quarters, months, or days from the current filter context.
DATESYTD	The year up to the date in the current filter context.
DATESBETWEEN	Between two dates.
DATESINPERIOD	Starting with a date and then going back (or forward) by any number of years, quarters, months, or days from the current filter context.
LASTDATE	The last date in the current filter context.
LASTNONBLANK	The last date in a column where the expression is nonblank in the current filter context.
LASTNONBLANKVALUE	The last value in a column where the expression is nonblank in the current filter context.

Typically, time intelligence functions generate a virtual one-column table containing filtered dates from the DATEKEY column in the date dimension (or whatever you've named this column). This virtual table is used as a table filter inside CALCULATE to filter the dates in the fact table.

However, DAX time intelligence functions either can be *table* functions that are nested inside CALCULATE as the filter argument or can return *scalar* values. The reason for this is that if a table function returns a one-column, one-row table, this virtual table is converted into a scalar value by the DAX engine; see Table 9-2.

Table 9-2. *Showing "Table" or "Scalar" functions, or both*

Table	Table or Scalar	Scalar
DATEADD	LASTDATE	LASTNONBLANKVALUE
DATESBETWEEN	LASTNONBLANK	
DATESINPERIOD		
DATESYTD		
PREVIOUSMONTH		
SAMEPERIODLASTYEAR		

For example, it would be possible to use LASTDATE as follows:

Used as a scalar

```
LastDate Example #1 =
LASTDATE(Winesales[SALE DATE])
```

Used as a table filter inside CALCULATE

```
LastDate Example #2 =
CALCULATE([Total Sales],LASTDATE(Winesales[SALE DATE]))
```

Used to return a scalar inside CALCULATE

```
LastDate Example #3 =
CALCULATE(LASTDATE(Winesales[SALE DATE]),DateTable[YEAR]=2020)
```

Let's now analyze our total cases values across different time frames. You can see the results of the following expressions in Figure 9-9. Note the use of slicers to filter the base date of **December 2021** from which the expressions are calculated.

Previous Month/Year – PREVIOUSMONTH/YEAR

These are the DAX expressions to calculate the previous month's or year's values, respectively:

```
Previous Month Total Cases =
CALCULATE ( [Total Cases],
    PREVIOUSMONTH ( DateTable[DATEKEY] )
)
```

```
Previous Year Total Cases =
CALCULATE ( [Total Cases],
    PREVIOUSYEAR ( DateTable[DATEKEY] )
)
```

The PREVIOUSYEAR function assumes that your financial year ends on December 31. If you use a different financial year, you can use the second argument of this function to define your year-end date. To avoid any date locale issues, use the date format "YYYY-MM-DD" (the function ignores the year, so use any year value); for example, if your year-end date is the March 31st, this would be your measure:

```
Year To Date Cases =
CALCULATE ( [Total Cases] ,
PREVIOUSYEAR ( DateTable[DATEKEY], "2021-03-31"
 )
)
```

Same Period Last Year – SAMEPERIODLASTYEAR

This is the DAX expression to calculate values in the same period in the previous year:

```
Same Period Last Year Cases =
CALCULATE ( [Total Cases],
    SAMEPERIODLASTYEAR ( DateTable[DATEKEY] )
)
```

Values for Any Time Ago – DATEADD

These are the DAX expressions that calculate values for 6 months ago and 30 days ago, respectively:

```
6 Months Ago Cases =
CALCULATE ( [Total Cases],
   DATEADD ( DateTable[DATEKEY], -6, MONTH )
)
```

```
30 Days Ago Cases =
CALCULATE ( [Total Cases] ,
     DATEADD ( DateTable[DATEKEY], -30, DAY )
)
```

Year to Date – DATESYTD

This expression will calculate year to date values for the year in the current filter context:

```
Year To Date Cases =
CALCULATE ( [Total Cases] ,
    DATESYTD ( DateTable[DATEKEY] )
)
```

The DATESYTD function, like PREVIOUSYEAR, assumes that your financial year ends in December, and just like PREVIOUSYEAR, you can use the second argument of this function to define your year-end date, using the format "YYYY-MM-DD" to avoid date locale issues, as follows:

```
Year To Date Cases =
CALCULATE ( [Total Cases] ,
    DATESYTD ( DateTable[DATEKEY], "2021-03-31")
)
```

WINE	Total Cases	Previous Month Total Cases	Same Period Last Year Cases	6 Months Ago Cases	Year To Date Cases
Bordeaux	998	951	563	139	18,514
Champagne	251	1,038	1,825	1,089	12,164
Chardonnay	430	218	887	940	12,671
Chenin Blanc	595	586	360	869	8,206
Chianti	123	619	1,279	1,996	8,837
Grenache	916	893	189	1,240	10,293
Malbec	801	963	914	769	11,082
Merlot		98	343	573	6,378
Piesporter		484	440	681	3,736
Pinot Grigio	338	487	902	752	6,400
Rioja	833	858	1,416	821	9,193
Sauvignon Blanc	1,465	1,721	929	335	12,247
Shiraz	616	98	301	516	4,675
Total	**7,366**	**9,014**	**10,348**	**10,720**	**124,396**

YEAR
- 2017
- 2018
- 2019
- 2020
- 2021 (selected)

MONTH
- Jan
- Feb
- Mar
- Apr
- May
- Jun
- Jul
- Aug
- Sep
- Oct
- Nov
- Dec (selected)

Figure 9-9. *Time intelligence calculations*

With the help of the time intelligence functions, these expressions have all been straightforward to write. Let's now move forward and explore some more complex calculations.

Total to Date or Cumulative Totals

The DAX measure for calculating total to date or a cumulative total for the "Total Sales" measure is as follows (see Figure 9-10):

```
Cumulative Total =
CALCULATE ( [Total Sales] ,
    DATESBETWEEN ( DateTable[DATEKEY], 0 ,
        LASTDATE ( DateTable[DATEKEY] )
    )
)
```

YEAR	MONTH	Total Sales	Cumulative Total
2017	Jan	$451,887	$451,887
2017	Feb	$385,299	$837,186
2017	Mar	$400,977	$1,238,163
2017	Apr	$327,070	$1,565,233
2017	May	$353,073	$1,918,306
2017	Jun	$241,419	$2,159,725
2017	Jul	$410,507	$2,570,232
2017	Aug	$194,755	$2,764,987
2017	Sep	$559,821	$3,324,808
2017	Oct	$438,513	$3,763,321
2017	Nov	$301,695	$4,065,016
2017	Dec	$584,269	$4,649,285
2018	Jan	$407,812	$5,057,097
2018	Feb	$299,495	$5,356,592
2018	Mar	$232,473	$5,589,065
Total		**$29,732,482**	**$29,732,482**

Figure 9-10. *The cumulative total sales*

This expression uses the DATESBETWEEN function that returns a table of dates that fall between a start date and an end date.

Notice that the start date for the DATESBETWEEN function is zero, which means the start date will be the earliest value in the dates column, or you could use the BLANK() function (we will look at this function in the following chapter). The end date is found by the LASTDATE function, which finds the last date in the current filter context. This will be the last date of the month sitting in any row of the Table visual or the last date of a month filtered in a slicer or Filters pane.

Rolling Annual Totals and Averages

To calculate rolling annual totals and averages, you must use two functions: DATESINPERIOD and LASTDATE. Let's do the rolling annual total first:

```
Rolling Annual Total Sales =
CALCULATE ( [Total Sales],
    DATESINPERIOD ( DateTable[DATEKEY],
```

```
        LASTDATE ( DateTable[DATEKEY] ) , -1 , YEAR ) )
```

The LASTDATE function in this measure finds the last date in the current filter context (i.e., the last date of the month sitting in any row of the Table visual, in a slicer, or in the Filters pane). The DATESINPERIOD function calculates the total sales, starting with this last date and going back by 1 year.

Now for the rolling annual average:

```
Rolling Annual Average Total Sales =
CALCULATE (
    [Total Sales] / COUNTROWS ( VALUES ( DateTable[MONTH] ) ),
    DATESINPERIOD (
        DateTable[DATEKEY],
        LASTDATE ( DateTable[DATEKEY] ),  -1,  YEAR
    )
)
```

The expression for the rolling annual average does much the same as the expression for the rolling annual total. However, we need to find the average monthly total for each rolling year. If we divided the "Total Sales" measure by 12, this would not be correct for the *first year* because in January, only one month is rolling; in February, only two months are rolling; in March, only three months; etc. This is why we need to use the COUNTROWS and VALUES functions to calculate the correct number of rolling months for the denominator and not simply divide by 12. The results of these measures are shown in Figure 9-11.

YEAR	MONTH	Total Sales	Rolling Annual Total Sales	Rolling Annual Average Total Sales
2017	Jan	$451,887	$451,887	$451,887
2017	Feb	$385,299	$837,186	$418,593
2017	Mar	$400,977	$1,238,163	$412,721
2017	Apr	$327,070	$1,565,233	$391,308
2017	May	$353,073	$1,918,306	$383,661
2017	Jun	$241,419	$2,159,725	$359,954
2017	Jul	$410,507	$2,570,232	$367,176
2017	Aug	$194,755	$2,764,987	$345,623
2017	Sep	$559,821	$3,324,808	$369,423
2017	Oct	$438,513	$3,763,321	$376,332
2017	Nov	$301,695	$4,065,016	$369,547
2017	Dec	$584,269	$4,649,285	$387,440
2018	Jan	$407,812	$4,605,210	$383,768
2018	Feb	$299,495	$4,519,406	$376,617
2018	Mar	$232,473	$4,350,902	$362,575
2018	Apr	$484,275	$4,508,107	$375,676
Total		**$29,732,482**	**$8,263,718**	**$688,643**

Figure 9-11. *The rolling annual and average sales*

We will meet the VALUES function later in this book, so at this stage, suffice to say that this function generates a virtual table containing only the values in the MONTH column of the date dimension that are visible in the filter context generated by the DATESINPERIOD expression. The COUNTROWS function counts the rows in the virtual table, giving us the correct number of cumulative months in the first year of our data.

Calculating the Last Transaction Date and the Last Transaction Value

If you want to find the first or last date for which there is data, for example, the last date for which there is a value for the "Total Sales" measure, you can use the functions FIRSTNONBLANK and LASTNONBLANK as follows:

```
Date of Last Transaction =
LASTNONBLANK ( DateTable[DATEKEY], [Total Sales] )
```

```
Date of First Transaction =
FIRSTNONBLANK ( DateTable[DATEKEY], [Total Sales] )
```

You could then find the value of the total sales on these dates by using LASTNONBLANKVALUE and FIRSTNONBLANKVALUE; see Figure 9-12.

```
Value of First Transaction =
FIRSTNONBLANKVALUE ( DateTable[DATEKEY], [Total Sales] )
```

WINE ▲	Date of First Transaction	Date of Last Transaction	Value of First Transaction	Value of Last Transaction
Bordeaux	18/01/2017	23/12/2021	$24,525	$21,750
Champagne	22/01/2017	11/12/2021	$52,050	$37,650
Chardonnay	07/01/2017	07/12/2021	$14,700	$22,500
Chenin Blanc	15/01/2017	22/12/2021	$7,350	$5,150
Chianti	09/01/2017	27/12/2021	$6,920	$4,920
Grenache	02/01/2017	30/12/2021	$2,100	$15,120
Malbec	01/01/2017	14/12/2021	$27,710	$22,865
Merlot	14/01/2017	25/10/2021	$4,680	$7,293
Piesporter	15/01/2017	19/11/2021	$9,450	$9,045
Pinot Grigio	19/01/2017	23/12/2021	$9,720	$5,730
Rioja	08/01/2017	26/12/2021	$6,975	$7,065
Sauvignon Blanc	01/01/2017	24/12/2021	$8,520	$13,240
Shiraz	10/01/2017	30/12/2021	$8,268	$15,912
Total	**01/01/2017**	**30/12/2021**	**$36,230**	**$31,032**

***Figure 9-12.** Calculating first and last transaction dates and values*

```
Value of Last Transaction =
LASTNONBLANKVALUE ( DateTable[DATEKEY], [Total Sales] )
```

The functions LASTNONBLANK and LASTNONBLANKVALUE can be used in more creative ways. Perhaps you need to calculate the date of the previous transaction, and perhaps you would like to find the difference in sales values between consecutive sales, as shown in Figure 9-13.

DATEKEY	Previous Sales Date	Previous Sales Value	Total Sales	Sales Difference
01 January 2017			$36,230	$36,230
02 January 2017	01 January 2017	$36,230	$2,100	($34,130)
03 January 2017	02 January 2017	$2,100	$10,560	$8,460
07 January 2017	03 January 2017	$10,560	$14,700	$4,140
08 January 2017	07 January 2017	$14,700	$6,975	($7,725)
09 January 2017	08 January 2017	$6,975	$6,920	($55)
10 January 2017	09 January 2017	$6,920	$8,268	$1,348
12 January 2017	10 January 2017	$8,268	$22,152	$13,884
13 January 2017	12 January 2017	$22,152	$22,800	$648
14 January 2017	13 January 2017	$22,800	$4,680	($18,120)
15 January 2017	14 January 2017	$4,680	$44,949	$40,269
17 January 2017	15 January 2017	$44,949	$7,410	($37,539)
18 January 2017	17 January 2017	$7,410	$24,525	$17,115
19 January 2017	18 January 2017	$24,525	$44,220	$19,695
20 January 2017	19 January 2017	$44,220	$34,480	($9,740)
22 January 2017	20 January 2017	$34,480	$52,050	$17,570

Figure 9-13. *Calculating the difference in values between consecutive transactions*

These are the expressions used to accomplish these tasks:

```
Previous Sales Date =
CALCULATE (
      LASTNONBLANK ( DateTable[DATEKEY],[Total Sales] ),
      DateTable[DATEKEY] < MAX (DateTable[DATEKEY] )
)

Previous Sales Value =
CALCULATE (
    LASTNONBLANKVALUE ( DateTable[DATEKEY], [Total Sales] ),
    DateTable[DATEKEY] < MAX ( DateTable[DATEKEY] )
)

Sales Difference =
[Total Sales] - [Previous Sales Value]
```

Because we are using the DATEKEY column from the date dimension in the Table visual in Figure 9-13, the expressions using LASTNONBLANK and LASTNONBLANKVALUE will be evaluated for *every* date in this column, regardless of whether each date has a transaction in the Winesales table. When you then populate the "Total Sales" measure into the Table visual, you will see blank values for dates where there are no transactions. To resolve this, use a visual-level filter and filter the "Total Sales" measure to exclude blank values.

The important factor in the evaluation of these expressions is the use of CALCULATE to modify the filter context in which the LASTNONBLANK and LASTNONBLANKVALUE are evaluated. The expression *"MAX (DateTable[DATEKEY])"* returns the date value in the current filter context, for example, **7 January 2017**; see Figure 9-14. The MAX function is used to return a scalar value. As there is only a single date in the current filter context, we could equally use MIN or SUM. The filter argument of CALCULATE therefore is saying "find the date in the DATEKEY column of the DateTable that is before the date returned by 'MAX (DateTable[DATEKEY])' but only if it has a sales value and is not blank." The LASTNONBLANK function returns this date, that is, **3 January 2017**. The LASTNONBLANKVALUE function returns the sales value associated with this date, **$10,560**.

DATEKEY	Previous Sales Date	Previous Sales Value	Total Sales	Sales Difference
01 January 2017			$36,230	$36,230
02 January 2017	01 January 2017	$36,230	$2,100	($34,130)
03 January 2017	02 January 2017	$2,100	$10,560	$8,460
07 January 2017	03 January 2017	$10,560	$14,700	$4,140
08 January 2017	07 January 2017	$14,700	$6,975	($7,725)

***Figure 9-14.** Focusing on an evaluation of the LASTNONBLANK and LASTNONBLANKVALUE expressions*

We can then simply subtract the "Previous Sales Value" measure from the "Total Sales" measure.

Finding the Difference Between Two Dates

Finding the difference in days between two dates in DAX can be done in a similar way to Excel; simply subtract one date from another. However, in DAX, you must nest the dates in the INT function to return a value in days as opposed to returning a date:

```
Days Difference =
INT ( [Date of Last Transaction] ) - INT ( [Date of First Transaction] )
```

DAX also has the same function DATEDIFF that we use in Excel to find the difference between weeks, months, years, etc. (see Figure 9-15).

```
Months Difference =
DATEDIFF ( [Date of First Transaction], [Date of Last Transaction], MONTH )
```

WINE	Date of First Transaction	Date of Last Transaction	Days Difference	Months Difference
Bordeaux	18/01/2017	23/12/2021	1,800	59
Champagne	22/01/2017	11/12/2021	1,784	59
Chardonnay	07/01/2017	07/12/2021	1,795	59
Chenin Blanc	15/01/2017	22/12/2021	1,802	59
Chianti	09/01/2017	27/12/2021	1,813	59
Grenache	02/01/2017	30/12/2021	1,823	59
Malbec	01/01/2017	14/12/2021	1,808	59
Merlot	14/01/2017	25/10/2021	1,745	57
Piesporter	15/01/2017	19/11/2021	1,769	58
Pinot Grigio	19/01/2017	23/12/2021	1,799	59
Rioja	08/01/2017	26/12/2021	1,813	59
Sauvignon Blanc	01/01/2017	24/12/2021	1,818	59
Shiraz	10/01/2017	30/12/2021	1,815	59
Total	**01/01/2017**	**30/12/2021**	**1,824**	**59**

Figure 9-15. *Calculating days between and months between two dates*

Hopefully, our foray into some of the more ubiquitous DAX time intelligence functions has whetted your appetite for performing calculations on dates. There are of course a number of other time intelligence functions that we haven't explored here but that you might find useful in the analysis of your data, so why not self-explore more of these valuable DAX functions. You will find them all here:

`https://docs.microsoft.com/en-us/dax/time-intelligence-functions-dax`

CHAPTER 10

Empty Values vs. Zero

In this chapter, we will look at a very specific DAX behavior, and that is how DAX treats empty, missing, and null values.[1]

Note We will be examining this behavior in the context of a *calculated column* and mostly creating expressions that would only be valid in this context. However, you must appreciate that the behavior of empty, missing, and null values is exactly the same in the context of DAX *measures*, and the examples at the end of this chapter will illustrate this.

The BLANK( ) Function

In DAX, there is a special way to identify null or empty values, and that's by using a value called "blank." To return blank values, we can use the BLANK() function as shown in a calculated column created in the Winesales table (Figure 10-1):

```
10 Percent =
IF ( Winesales[CASES SOLD] > 100,
Winesales[CASES SOLD] * 0.1, BLANK () )
```

[1] To follow along with the examples, use the Power BI Desktop file "2 DAX Blanks & Zeros.pbix".

© Alison Box 2022
A. Box, *Up and Running with DAX for Power BI*, https://doi.org/10.1007/978-1-4842-8188-8_10

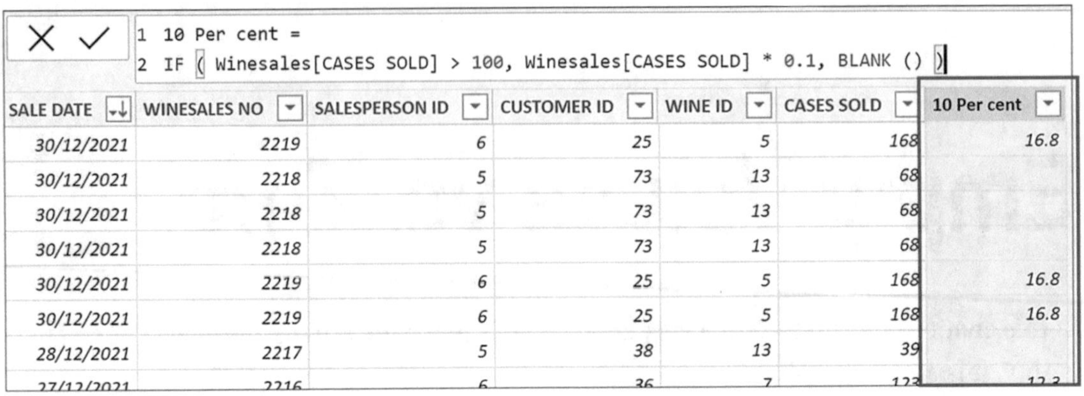

```
1 10 Per cent =
2 IF ( Winesales[CASES SOLD] > 100, Winesales[CASES SOLD] * 0.1, BLANK () )
```

SALE DATE	WINESALES NO	SALESPERSON ID	CUSTOMER ID	WINE ID	CASES SOLD	10 Per cent
30/12/2021	2219	6	25	5	168	16.8
30/12/2021	2218	5	73	13	68	
30/12/2021	2218	5	73	13	68	
30/12/2021	2218	5	73	13	68	
30/12/2021	2219	6	25	5	168	16.8
30/12/2021	2219	6	25	5	168	16.8
28/12/2021	2217	5	38	13	39	
27/12/2021	2216	6	36	7	123	12.3

Figure 10-1. *Use the BLANK() function to return blank values*

When constructing DAX expressions using IF, if you want to return BLANK() on the "Value if false" argument, you can just close off on the bracket because BLANK() is the default if no value is supplied in the argument. So we could rewrite the previous expression like this:

```
10 Percent =
IF ( Winesales[CASES SOLD] > 100,
Winesales[CASES SOLD] * 0.1 )
```

We can test for null or blank values as in the following calculated column:

Note In the sample .pbix file, sort the Winesales table by SALE DATE ascending to see the blanks and zeros in the CASES SOLD column.

```
Blank? =
IF ( Winesales[CASES SOLD] = BLANK(), "Blank", "Other")
```

Notice that testing for BLANK() includes **0** (zero), so we never get "Other" for zero (Figure 10-2).

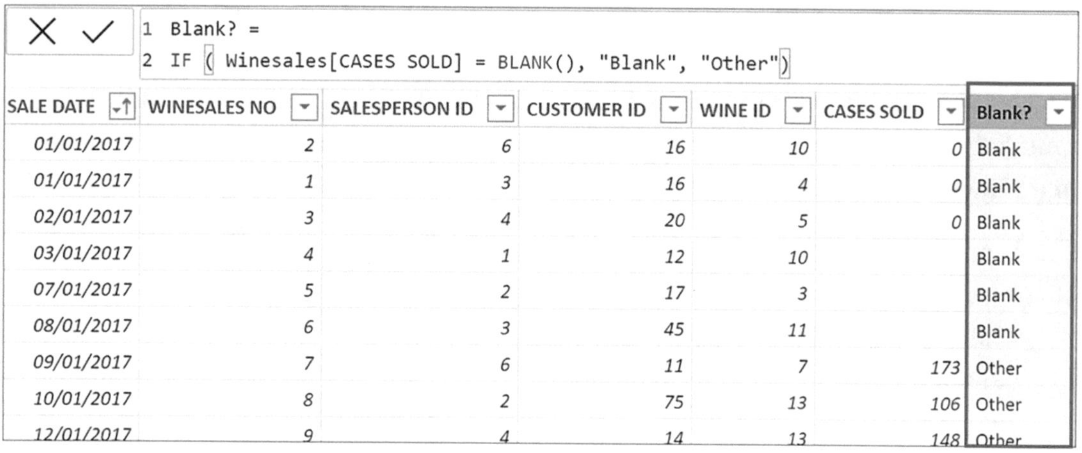

```
1 Blank? =
2 IF ( Winesales[CASES SOLD] = BLANK(), "Blank", "Other")
```

SALE DATE	WINESALES NO	SALESPERSON ID	CUSTOMER ID	WINE ID	CASES SOLD	Blank?
01/01/2017	2	6	16	10	0	Blank
01/01/2017	1	3	16	4	0	Blank
02/01/2017	3	4	20	5	0	Blank
03/01/2017	4	1	12	10		Blank
07/01/2017	5	2	17	3		Blank
08/01/2017	6	3	45	11		Blank
09/01/2017	7	6	11	7	173	Other
10/01/2017	8	2	75	13	106	Other
12/01/2017	9	4	14	13	148	Other

***Figure 10-2.** Testing for a blank includes zero values*

What's surprising, however, is that the reverse is true, so in the following calculated column, testing for **0** includes blank values, so again we don't get "Other" for blank values (see Figure 10-3):

```
Zero? =
IF ( Winesales[CASES SOLD] = 0, "Zero", "Other")
```

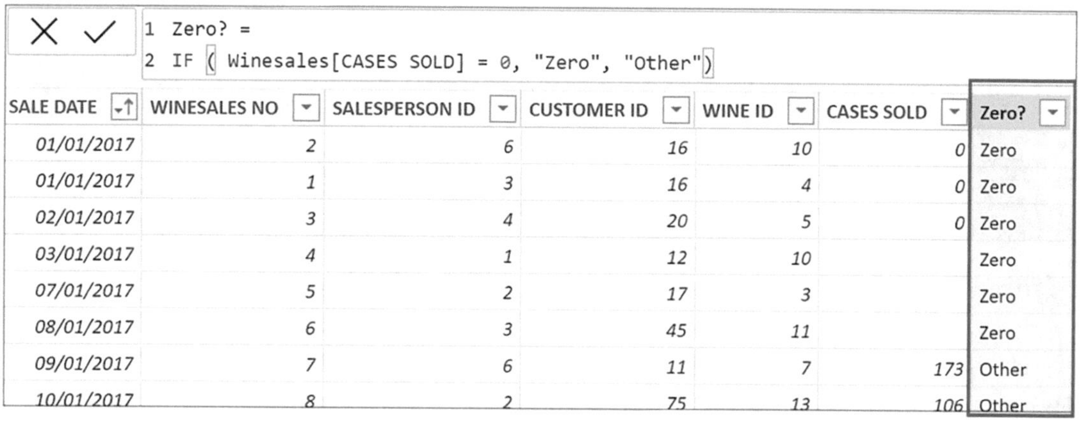

```
1 Zero? =
2 IF ( Winesales[CASES SOLD] = 0, "Zero", "Other")
```

SALE DATE	WINESALES NO	SALESPERSON ID	CUSTOMER ID	WINE ID	CASES SOLD	Zero?
01/01/2017	2	6	16	10	0	Zero
01/01/2017	1	3	16	4	0	Zero
02/01/2017	3	4	20	5	0	Zero
03/01/2017	4	1	12	10		Zero
07/01/2017	5	2	17	3		Zero
08/01/2017	6	3	45	11		Zero
09/01/2017	7	6	11	7	173	Other
10/01/2017	8	2	75	13	106	Other

***Figure 10-3.** Testing for zero includes blanks*

Therefore, we can see that DAX treats BLANK() and **0** (zero) as the same value when used in the predicate of the IF function, as in the previous two examples.

The ISBLANK Function

So what if you want to distinguish between **0** and blank values? You can use a DAX function that will "weed out" blanks as compared to **0**. That function is ISBLANK as used in this following calculated column (Figure 10-4):

```
Blank or Zero? =
IF (
    ISBLANK ( Winesales[CASES SOLD] ),
    "Blank",
    IF ( Winesales[CASES SOLD] = 0, "Zero", "Other" )
)
```

```
1 Blank or Zero? =
2 IF (
3     ISBLANK ( Winesales[CASES SOLD] ),
4     "Blank",
5     IF ( Winesales[CASES SOLD] = 0, "Zero", "Other" )
6 )
```

SALE DATE	WINESALES NO	SALESPERSON ID	CUSTOMER ID	WINE ID	CASES SOLD	Blank or Zero?
01/01/2017	2	6	16	10	0	Zero
01/01/2017	1	3	16	4	0	Zero
02/01/2017	3	4	20	5	0	Zero
03/01/2017	4	1	12	10		Blank
07/01/2017	5	2	17	3		Blank
08/01/2017	6	3	45	11		Blank
09/01/2017	7	6	11	7	173	Other
10/01/2017	8	2	75	13	106	Other
12/01/2017	9	4	14	13	148	Other

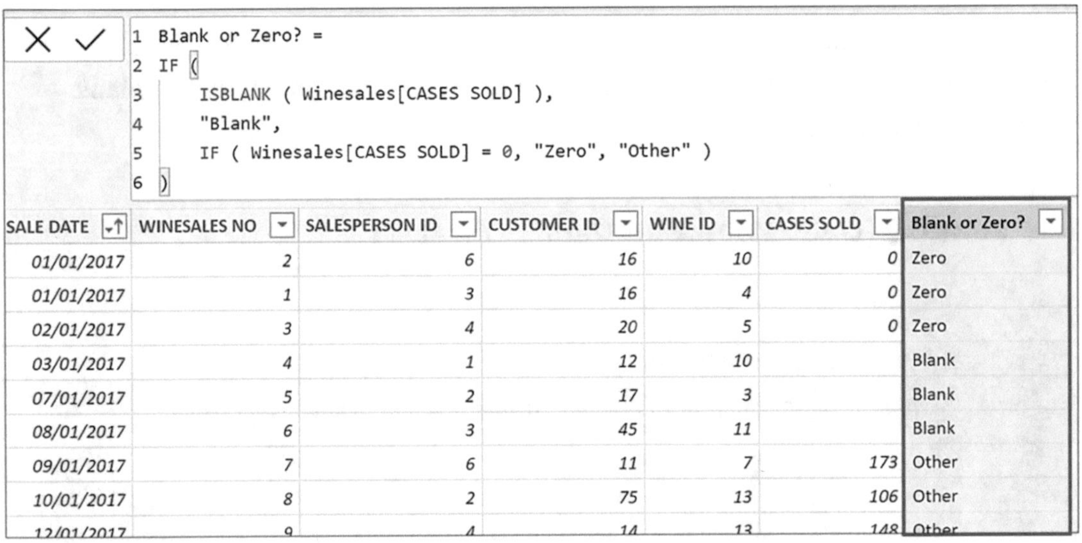

Figure 10-4. *Use the ISBLANK function to test for blanks and not zeros*

Using ISBLANK, we now have "Zero" returned for zero values and "Blank" returned for blank values, and any other values return "Other".

Testing for Zero

If you want to find just **0**, you can use this calculated column (Figure 10-5):

```
Zero? =
```

```
IF (
    NOT ( ISBLANK ( Winesales[CASES SOLD] ) )
        && Winesales[CASES SOLD] = 0,
    "Zero",
    "Other"
)
```

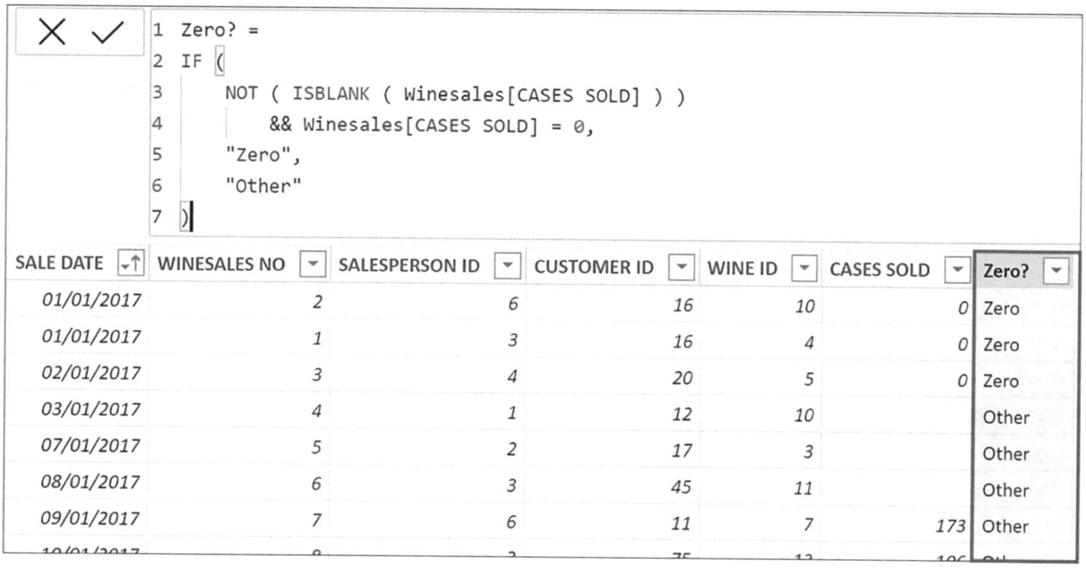

```
1 Zero? =
2 IF (
3     NOT ( ISBLANK ( Winesales[CASES SOLD] ) )
4         && Winesales[CASES SOLD] = 0,
5     "Zero",
6     "Other"
7 )
```

SALE DATE	WINESALES NO	SALESPERSON ID	CUSTOMER ID	WINE ID	CASES SOLD	Zero?
01/01/2017	2	6	16	10	0	Zero
01/01/2017	1	3	16	4	0	Zero
02/01/2017	3	4	20	5	0	Zero
03/01/2017	4	1	12	10		Other
07/01/2017	5	2	17	3		Other
08/01/2017	6	3	45	11		Other
09/01/2017	7	6	11	7	173	Other

***Figure 10-5.** Testing for zeros*

Now, we only see "Zero" where applicable.

Using Measures to Find Blanks and Zero

You can also use a measure inside ISBLANK. For example, to find how many customers have *no* sales, as opposed to **0** (zero) sales, this would be the DAX expression:

```
No. of Customers with No Sales =
COUNTROWS ( FILTER ( Customers, ISBLANK ( [Total Sales] ) ) )
```

Whereas this expression would find the number of customers who had *either* zero sales *or* no sales:

```
No. of Customers with Zero or No Sales =
```

```
COUNTROWS ( FILTER ( Customers,  [Total Sales] = 0 ) )
```

This expression would find the number of customers who had zero sales:

```
No. of Customers with Zero sales =
COUNTROWS (
    FILTER ( Customers, NOT ( ISBLANK ( [Total Sales] ) )
                                  && [Total Sales] = 0 )
)
```

You can see these measures used in Card visuals in Figure 10-6. To see the customers with no sales in the Table visual, use the “Show items with no data” option.

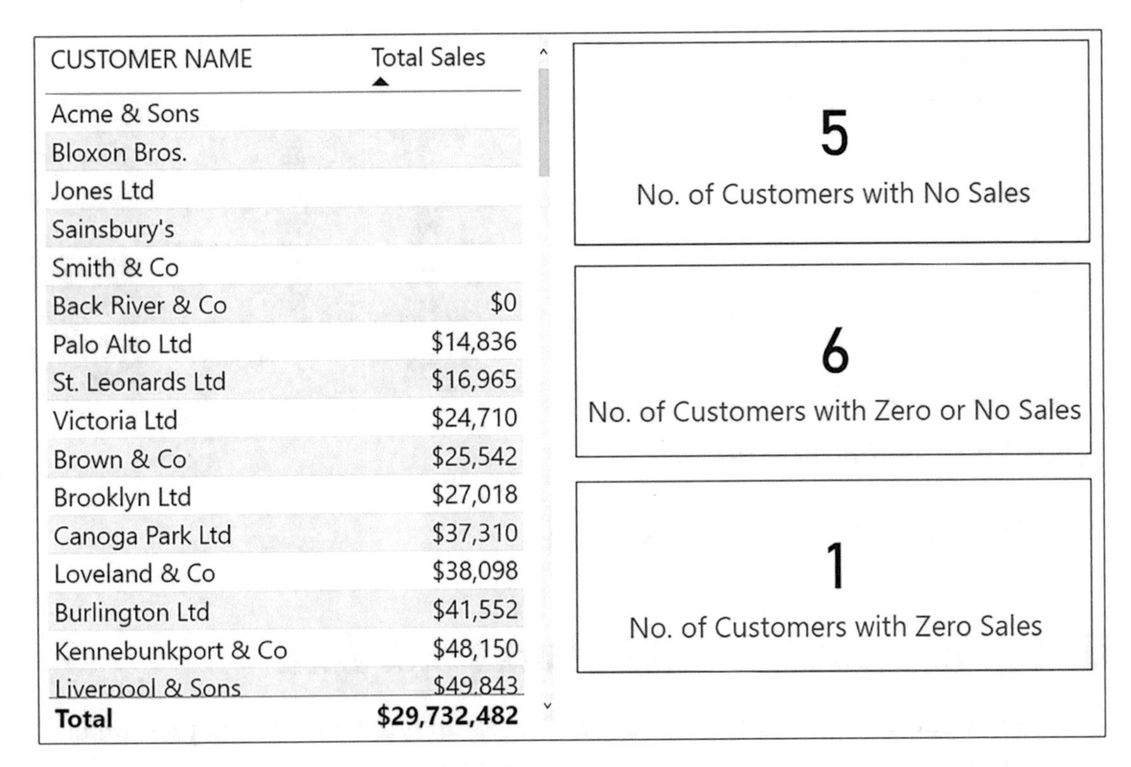

CUSTOMER NAME	Total Sales
Acme & Sons	
Bloxon Bros.	
Jones Ltd	
Sainsbury's	
Smith & Co	
Back River & Co	$0
Palo Alto Ltd	$14,836
St. Leonards Ltd	$16,965
Victoria Ltd	$24,710
Brown & Co	$25,542
Brooklyn Ltd	$27,018
Canoga Park Ltd	$37,310
Loveland & Co	$38,098
Burlington Ltd	$41,552
Kennebunkport & Co	$48,150
Liverpool & Sons	$49,843
Total	**$29,732,482**

Figure 10-6. *Customers with no sales and zero sales*

We can conclude, therefore, that we must be careful using the following expression:

“= IF ([expression] = 0)”

because it will include blank values as well as zero values.

Using the COALESCE Function

There is often a requirement to substitute a blank value for another value, such as zero. This would be the expression that would achieve this outcome:

```
If Blank Return Zero =
If ( ISBLANK ( [Total Sales] ), 0, [Total Sales] )
```

However, in March 2020, a new function was introduced into the DAX library, and that was the COALESCE function that provides us with a more succinct expression as in these two examples:

```
If Blank Return Zero =
COALESCE([Total Sales],0)
```

```
If Blank Return No Sales =
COALESCE([Total Sales],"No Sales")
```

The first argument of this function is the expression where you are looking for blank values, for example, the "Total Sales" measure. The second argument is the value you want returned if the expression is blank, for example, 0 or "No Sales", see Figure 10-7.

CUSTOMER NAME	Total Sales ▲	If Blank Return Zero	If Blank Return No Sales
Acme & Sons		$0	No Sales
Bloxon Bros.		$0	No Sales
Jones Ltd		$0	No Sales
Sainsbury's		$0	No Sales
Smith & Co		$0	No Sales
Back River & Co	$0	$0	$0
Palo Alto Ltd	$14,836	$14,836	$14,836
St. Leonards Ltd	$16,965	$16,965	$16,965
Victoria Ltd	$24,710	$24,710	$24,710
Brown & Co	$25,542	$25,542	$25,542
Brooklyn Ltd	$27,018	$27,018	$27,018
Canoga Park Ltd	$37,310	$37,310	$37,310
Loveland & Co	$38,098	$38,098	$38,098
Burlington Ltd	$41,552	$41,552	$41,552
Kennebunkport & Co	$48,150	$48,150	$48,150
Liverpool & Sons	$49,843	$49,843	$49,843
Total	**$29,732,482**	**$29,732,482**	**$29,732,482**

***Figure 10-7.** Use the COALESCE function to replace blanks with a value*

In this chapter, you have learned that DAX treats blanks and zeros as the same value unless you specifically use the ISBLANK function in your expression to distinguish between these two values. This chapter has also been a welcome transgression from the hard work of learning how to analyze your data by using some of the more difficult aspects of DAX such as using ALL to calculate percentages and using time intelligence to calculate rolling averages.

In the next chapter, we prepare ourselves for the more complex expressions to come. You must now learn how to use DAX variables in your code to facilitate authoring measures that require a more advanced knowledge of DAX.

CHAPTER 11

Using Variables: Making Our Code More Readable

We've managed very well so far without the use of variables in our DAX code. Indeed, variables haven't always been around in the DAX language. They came on board in 2015, five years after DAX was first developed. In this chapter, we will elaborate on why variables are so useful when writing DAX expressions, and once you've learned how to utilize them, we will be including them henceforth in our expression, where applicable.[1]

Using variables in your DAX expressions can help you write the more complex calculations that we will begin to tackle as we move forward in this book. There are three major advantages gained by using variables:

1. Improved performance
2. Improved readability
3. Reduced complexity

In this chapter, we will explore these three benefits of including variables when generating DAX code. We will also look at the immutable and constant nature of variables and when they may be a hindrance rather than a help.

To include variables in your code, use the keyword *VAR* followed by the name of the variable and then the definition of the variable. The keyword *RETURN* is then used at the end of the code to return the expression to be evaluated. For example:

```
Example Measure =
VAR MyVariable = SUM (Winesales[CASES SOLD])
RETURN
MyVariable * 1.1
```

[1] To follow along with the examples, use the Power BI Desktop file "1 DAX Sample Data.pbix".

© Alison Box 2022
A. Box, *Up and Running with DAX for Power BI*, https://doi.org/10.1007/978-1-4842-8188-8_11

Variable declarations are usually made at the beginning of the expression, and their value remains constant throughout the evaluation. However, you can declare variables within the expression to limit the scope.

Variables can be used in both measures and calculated columns to harvest the values generated by

- Expressions, for example, SUM (Winesales[CASES SOLD])
- Measures, for example, [Total Cases]
- Tables, for example, FILTER (Winesales, Winesales[CASES SOLD] >300)
- Values, for example, 0.1, 10, 20

When variables are used in calculated columns, they can also harvest values generated in columns.

The name of the variable must not contain spaces, and you can't use reserved words such as "date" or "min". Also, it makes sense if the name of the variable isn't the name of an existing table or column. Some people like to use the underscore to start the variable name.

Improved Performance

As an example of how variables can improve performance, let's look at a measure to calculate 10% or 5% of the CASES SOLD based on the CASES SOLD value being greater than 20,000 and 15,000, respectively. This would be the expression you might author:

```
10 PC or 5 PC =
IF (
    SUM ( Winesales[CASES SOLD] ) > 20000,
    SUM ( Winesales[CASES SOLD] ) * 0.1,
    IF (
        SUM ( Winesales[CASES SOLD] ) > 15000,
        SUM ( Winesales[CASES SOLD] ) * 0.5,
        SUM ( Winesales[CASES SOLD] )
    )
)
```

The problem with this expression, especially as far as performance goes, is that there are five repetitions of the SUM function, forcing the evaluation of these expressions five times. Also, the use of the nested IF is rather cumbersome. Using the SWITCH function in place of the nested IF is a small improvement:

```
10 PC or 5 PC #2 =
SWITCH (
    TRUE (),
    SUM ( Winesales[CASES SOLD] ) > 20000,
                   SUM ( Winesales[CASES SOLD] ) * 0.1,
    SUM ( Winesales[CASES SOLD] ) > 15000,
                    SUM ( Winesales[CASES SOLD] ) * 0.5,
                    SUM ( Winesales[CASES SOLD] )
)
```

This is the first time that we have met SWITCH, and its construct is as follows:

=SWITCH (expression, value1, result1, value2, result2 etc...else)

Notice that inside SWITCH, the function TRUE() is used as the expression to be evaluated and then Boolean statements are listed, followed by the value to be returned if the statements are true. The final argument is the "else" expression.

However, despite the fact that the measure using SWTICH is more compact to write, it doesn't offer any great improvement in performance as the SUM function is still being evaluated multiple times.

Therefore, let us now introduce the use of a variable by using the keyword *VAR* to define the variable and the keyword *RETURN* to return the expression to be evaluated, as follows:

```
10 PC or 5 PC #3 =
VAR TotalCasesValue =
    SUM ( Winesales[CASES SOLD] )
RETURN
    SWITCH (
        TRUE (),
        TotalCasesValue > 20000, TotalCasesValue * 0.1,
        TotalCasesValue > 15000, TotalCasesValue * 0.5,
        TotalCasesValue
    )
```

In this expression, not only do we avoid repeating the SUM function, but also the total cases calculation is performed only *once* when the variable is declared rather than being recalculated for every test.

Improved Readability

Variables can also help to clarify expressions that use nested measures or nested expressions where the readability of the expressions gets more convoluted. For example, consider the following expression that calculates growth percentage. Notice that the first variable defines a *measure* and the second variable defines an *expression.* The use of the variables and the RETURN statement result in the expression much simpler to understand:

```
Growth % =
VAR CurrentCases = [Total Cases]
VAR LastYrCases =
    CALCULATE ( [Total Cases], PREVIOUSYEAR (
                               DateTable[DateKey] ) )
RETURN
    DIVIDE ( CurrentCases - LastYrCases, LastYrCases )
```

Note Because this measure uses the PREVIOUSYEAR function, you must have a year filtered (e.g., by using a slicer) in the visual that uses the measure.

Not only can variables define measures and expressions, but they can also define tables. In Chapter 7, we calculated the number of high profit wines as follows:

```
High-profit Wines =
    CALCULATE ( [No Of Sales],
   FILTER ( Wines, Wines[PRICE PER CASE] >=
   Wines[COST PRICE] * 3 ))
```

However, we could use a variable to hold the table expression defined by the FILTER function and use that as the filter argument inside CALCULATE. Again, using the RETURN statement greatly streamlines the expression:

```
High-profit Wines #1 =
VAR TableOfWines =
    FILTER ( Wines, Wines[PRICE PER CASE] >=
         Wines[COST PRICE] * 3 )
RETURN
    CALCULATE ( [No Of Sales], TableOfWines )
```

We can use variables in calculated columns too, for instance, within the arguments of IF:

```
Cases Sold Increase =
VAR CasesSold = Winesales[CASES SOLD]
VAR MyValue1 = 1.1
VAR MyValue2 = 1.2
RETURN
IF(CasesSold > 100, CasesSold * MyValue1, CasesSold * MyValue2)
```

We will look at further examples of how variables can help you when used in the context of the calculated column when we delve into more complex DAX expressions in later chapters.

Reduced Complexity

Our next example of the benefit to be reaped by using a variable is by revisiting a calculation we built when exploring the FILTER function in Chapter 7. We calculated the number of sales where the value in the CASES SOLD column was above the average cases for all wines. This was the measure:

```
No. of Sales Where Cases is GT Avg All Wines =
     CALCULATE([No. of Sales],
     FILTER (Winesales,
     Winesales[CASES SOLD] >= [Avg Cases All Winesales] ) )
```

The problem with this code is that because it nests the measure "Avg Cases All Winesales" within the expression, this measure must already exist in our model, as would any measures we use in this context. We may be required to continually locate such measures in the Fields list in order to edit or debug them, leading to frustration and annoyance.

The preferred expression would use two variables as follows:

```
No. of Sales Where Cases is GT Avg All Wines #2 =
VAR AvgAllWines =
CALCULATE( AVERAGE ( Winesales[CASES SOLD] ) ,ALL ( Winesales ) )
VAR FilterAvgAll =
FILTER ( Winesales, Winesales[CASES SOLD] >= AvgAllWines )

RETURN
   CALCULATE ( [No. of Sales], FilterAvgAll )
```

Variables As Constants

There is one last important point to make regarding variables, and that is the term "variable" can be misleading. Perhaps if we called DAX variables "constants," this might be a more accurate description because that's what they really are. Consider the following expression:

```
Sales for Abel =
VAR MyAmount = [Total Sales]
RETURN
    CALCULATE ( MyAmount, SalesPeople[SALESPERSON] = "abel" )
```

We can see in Figure 11-1 that this expression does not return the sales amount for salesperson "Abel" but simply returns the total sales.

WINE	Total Sales	Sales for Abel
Bordeaux	$4,055,250	$4,055,250
Champagne	$7,373,700	$7,373,700
Chardonnay	$4,203,000	$4,203,000
Chenin Blanc	$1,236,950	$1,236,950
Chianti	$1,092,920	$1,092,920
Grenache	$1,078,950	$1,078,950
Malbec	$2,914,650	$2,914,650
Merlot	$900,276	$900,276
Piesporter	$1,384,155	$1,384,155
Pinot Grigio	$703,470	$703,470
Rioja	$1,527,795	$1,527,795
Sauvignon Blanc	$1,896,600	$1,896,600
Shiraz	$1,364,766	$1,364,766
Total	**$29,732,482**	**$29,732,482**

Figure 11-1. *Variables behave as constants and can't be modified by CALCULATE*

The reason for this is that the variable "MyAmount" is calculated where it is declared, in this case, before any other code. It then *does not* and *cannot* change by using CALCULATE to modify the filter. This is where we must use a measure such as "Total Sales" inside CALCULATE instead.

However, the immutable nature of variables is also their strength. For instance, consider the scenario where you want to identify the months where you've had exceptionally high sales. You've identified exceptionally high sales as those transactions where the sales value is greater than 5% of the total sales for that month.

This is the code you would probably write:

```
No of Sales GT 5% Wrong =
CALCULATE (
    [No of Sales],
    FILTER (
        Winesales,
        [Total Sales] >  [Total Sales] * 0.05
    )
)
```

However, this measure does not return the correct result. The value of the "Total Sales" measure when used inside an iterator such as the FILTER function calculates the total sales for *each* row in the Winesales table, not the total sales for each month. Therefore, the measure "Total Sales GT 5% Wrong" calculates the number of sales where the sales value is greater than 5% of the sales value on each row (i.e., each transaction) and so returns the number of sales; see Figure 11-2.

YEAR ▲	MONTH	Total Sales	No of Sales GT 5% Wrong	No of Sales
2018	May	$213,304	18	18
2018	Jun	$399,831	30	30
2018	Jul	$329,686	29	29
2018	Aug	$386,823	26	26
2018	Sep	$355,690	30	30
2018	Oct	$301,611	23	23
2018	Nov	$439,965	28	28
2018	Dec	$356,906	25	25
2019	Jan	$148,855	14	14
2019	Feb	$167,738	14	14
2019	Mar	$213,333	12	12
2019	Apr	$170,815	15	15
2019	May	$517,246	31	31
2019	Jun	$281,813	24	24

Figure 11-2. *The "No of Sales GT 5% Wrong" measure returns the number of sales*

This expression uses the concept of context transition that we will meet in a later chapter, but nevertheless, it's intuitive to understand that if FILTER is iterating the Winesales table, it must be scanning the table row by row.

The correct expression must calculate the total sales in the current filter context, which is the total sales for each month, that has been lost by the iteration of FILTER. To reapply this filter, CALCULATE can use the filter that is placed on the Winesales table, the code for which would be a challenge even to experienced DAX users:

```
No of Sales GT 5% Difficult =
CALCULATE (
    [No of Sales],
```

```
    FILTER (
        Winesales,
        [Total Sales] > CALCULATE ( [Total Sales], Winesales ) * 0.05
    )
)
```

This measure has been labelled as the "difficult" expression because it uses two challenging DAX concepts that we've yet to meet: context transition and table expansion. However, you may be relieved to know that you don't need this advanced knowledge to arrive at the correct calculation. You can use variables instead, and this will render the expression very easy:

```
No of Sales GT 5% Easy =
VAR PerCentToFind = [Total Sales] * 0.05
RETURN
    CALCULATE ( [No of Sales],
FILTER ( Winesales, [Total Sales] > PerCentToFind ) )
```

The "easy" expression uses a variable to calculate 5% of the "Total Sales" measure, and this is evaluated first and remains constant. This variable is then used to calculate the number of sales in each month that have a total sales value that is greater than the value stored by the variable.

The moral of this story? Let's just be grateful for variables!

CHAPTER 12

Returning Values in the Current Filter

There is often a requirement when designing reports to display the value or values selected in slicers or in the Filters pane. This might be to show these values in the title of a visual using conditional formatting or to show them in Card visuals, as shown in Figure 12-1.

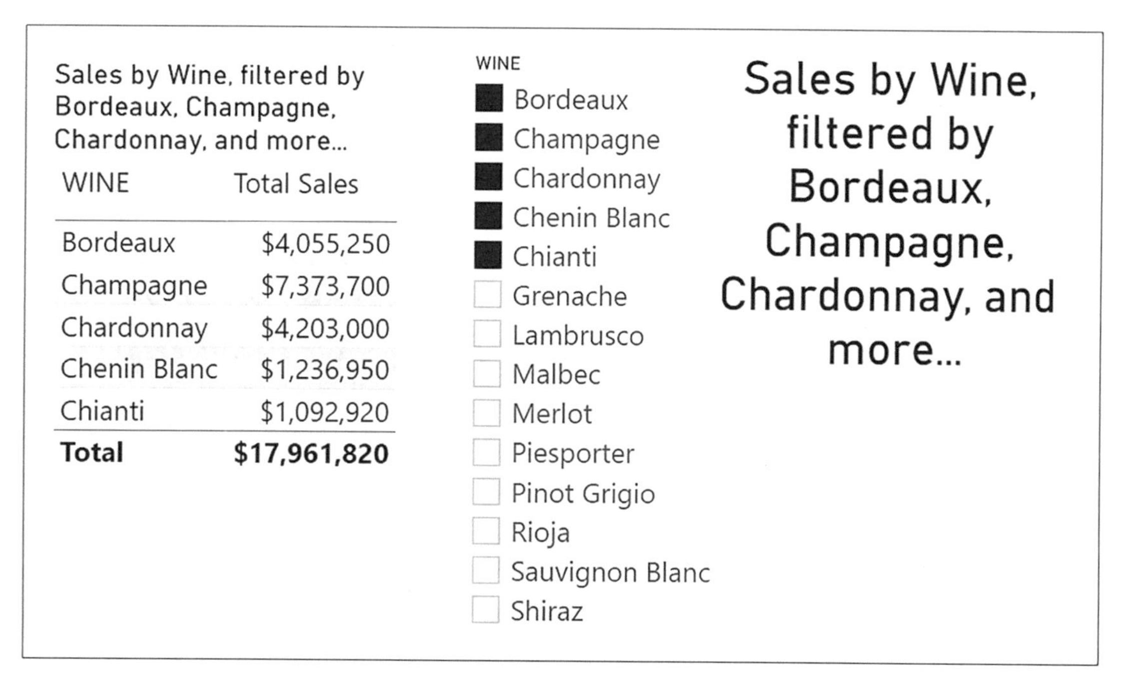

Figure 12-1. *Displaying the values in the current filter context*

If this is your goal, we have three DAX functions that do this job: SELECTEDVALUE, CONCATENATEX, and VALUES. In this chapter, we will be exploring the use of these functions to return filter selections. You will learn how to generate dynamic titles for

© Alison Box 2022
A. Box, *Up and Running with DAX for Power BI*, https://doi.org/10.1007/978-1-4842-8188-8_12

your visuals that label the data filtered within them. However, this chapter will also introduce the concept of the *parameter table*, a table that is unrelated to other tables in the model and used to capture values selected by the user. Such values can then be used dynamically within your calculations.

The SELECTEDVALUE and CONCATENATEX functions fall into the category of functions that return scalar values and can return either a numeric or a text value. This is why it's not a verity to say the measures only return scalar values, as that would imply that they can only return numeric values. Measures using either of these functions will often return a text value. The VALUES function is unusual in that it can return either a scalar value or a table, and therefore, we will hold off looking at this function until the end of the chapter.

You've learned that a measure must return a *single* value whether numeric or text and SELECTEDVALUE and CONCATENATEX are no exception. SELECTEDVALUE will return the value in the current filter context but only if there is *one* value to return. However, sometimes, the filter context holds more than one value, when we make multiple selections in slicers for instance, so how can we return values in this scenario?

If the requirement is to return multiple values that are in the filter context, we must use another function: CONCATENATEX. This function falls into the "X" group of iterating functions that you learned about in Chapter 5. In order that a single value is returned, measures using CONCATENATEX will concatenate multiple values in the current filter context and so return a single text string.

Therefore, we have two functions SELECTEDVALUE and CONCATENATEX, one of them being an iterator, that are very different from each other. However, they are used for the same purpose, and that is flagging up items that have been filtered out by slicer or filter selections. Let's now look at the first of these: SELECTEDVALUE.

The SELECTEDVALUE Function

The SELECTEDVALUE function returns the value in the filter context when there's only one value in the specified column, otherwise, it returns the alternate result. It has the following syntax:

= SELECTEDVALUE(column name, alternate result)

where:

column name is the column from which you want to find the value.

alternate result (optional) is the value returned when the column has been filtered to more than one distinct value or no value. When not provided, the default value is BLANK().

Here is an example of the SELECTEDVALUE syntax:

= SELECTEDVALUE (Wines[TYPE], "Many")

Before we look more closely at this function, it's important that we recap on what we mean by "the current filter context" by considering the following measure:

```
Total Cases =
SUM ( Winesales[CASES SOLD] )
```

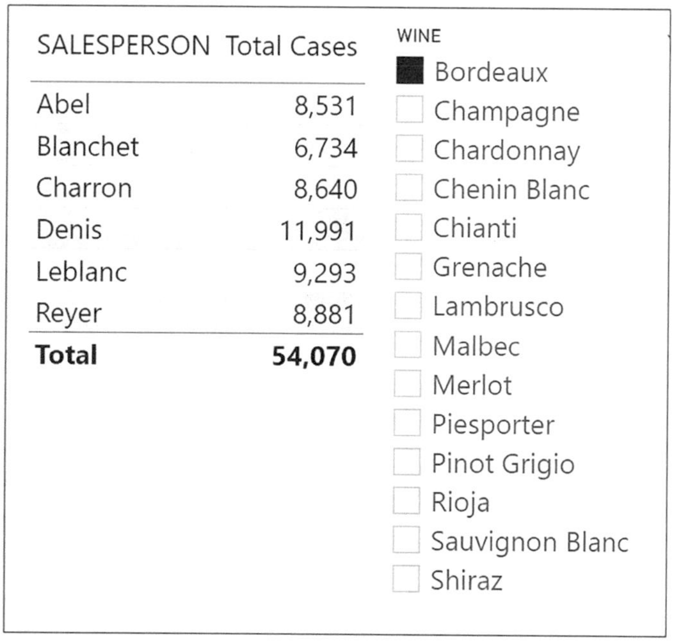

SALESPERSON	Total Cases
Abel	8,531
Blanchet	6,734
Charron	8,640
Denis	11,991
Leblanc	9,293
Reyer	8,881
Total	**54,070**

WINE

- [x] Bordeaux
- [ ] Champagne
- [ ] Chardonnay
- [ ] Chenin Blanc
- [ ] Chianti
- [ ] Grenache
- [ ] Lambrusco
- [ ] Malbec
- [ ] Merlot
- [ ] Piesporter
- [ ] Pinot Grigio
- [ ] Rioja
- [ ] Sauvignon Blanc
- [ ] Shiraz

Figure 12-2. *The filters for the evaluation of the "Total Cases" measure are placed on both the SALESPERSON and WINE columns*

This visual in Figure 12-2 contains the "Total Cases" measure filtered by the SALESPERSON and WINE columns. For the first evaluation of **8,531** cases, there is a filter on salesperson "Abel" and "Bordeaux" wine. However, it's the filter on the WINE column from the slicer on which we will focus. If we could see the filter on the Wines dimension, it would look something like Figure 12-3 where the table has been filtered to one row.

WINE ID	WINE	SUPPLIER	TYPE	WINE COUNTRY	PRICE PER CASE	COST PRICE
1	Bordeaux	Laithwaites	Red	France	$75.00	$25.00

***Figure 12-3.** The slicer filters just one row in the Wines dimension*

We know that this filter is then propagated to the fact table along with the filter on the SalesPeople dimension, both these filters making up the current filter context.

Often, we have many slicers on the report canvas, and it's not always apparent to users of the report which slicers they have clicked on. It would be beneficial if we could provide them with this information as in Figure 12-4.

You have selected Bordeaux

SALESPERSON	Total Cases
Abel	8,531
Blanchet	6,734
Charron	8,640
Denis	11,991
Leblanc	9,293
Reyer	8,881
Total	**54,070**

WINE
- Bordeaux
- Champagne
- Chardonnay
- Chenin Blanc
- Chianti
- Grenache
- Lambrusco
- Malbec
- Merlot
- Piesporter
- Pinot Grigio
- Rioja
- Sauvignon Blanc
- Shiraz

You have selected Bordeaux
Wine Selected

***Figure 12-4.** Informing users of slicer selections*

This is where the SELECTEDVALUE function can help us. You can see in Figure 12-4 that the wine selected in the slicer is shown in both the title of the Table visual using conditional formatting and in the Card visual. This is the measure that we used in these examples:

```
Wine Selected =
"You have selected " & SELECTEDVALUE ( Wines[WINE] )
```

This example uses the SELECTEDVALUE function to return the value from the WINE column sitting in the current filter context. This is also the first time that we've used the ampersand (&) in a DAX expression. Just like Excel, the ampersand is the DAX concatenate operator and is used to string parts of a DAX expression together.

Note If you need help in using conditional formatting in the Title of a visuals, follow this link: `https://docs.microsoft.com/en-us/power-bi/create-reports/desktop-conditional-format-visual-titles`

But what if there's more than one value selected in the slicer? As we will see in the following, one option is to use CONCATENATEX, but there is another, much easier solution because the SELECTEDVALUE function allows you to supply an alternative result when multiple items have been selected, as shown here:

```
Wine Selected #2 =
"You have selected " &
SELECTEDVALUE ( Wines[WINE],"multiple wines" )
```

However, because the "alternate result" argument of SELECTEDVALUE kicks in whether there are *multiple* selections or *no* selection, we have a problem. You'll notice that if you have nothing selected in the slicer, the Table visual title and Card visual will still tell you that you have multiple wines selected (Figure 12-5)!

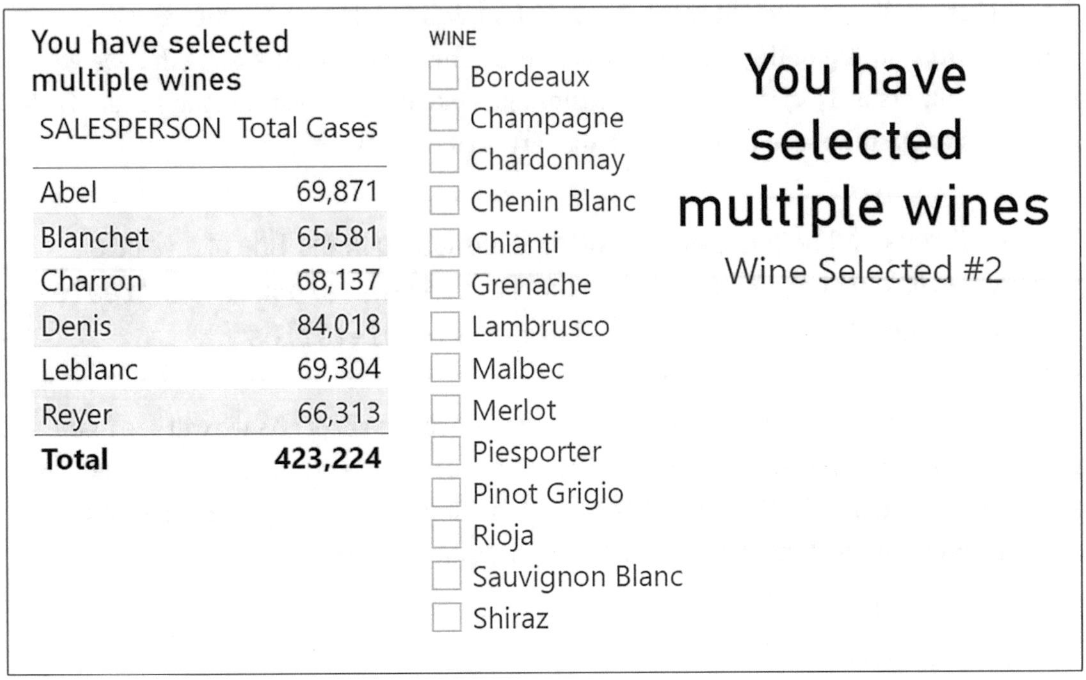

You have selected multiple wines

SALESPERSON	Total Cases
Abel	69,871
Blanchet	65,581
Charron	68,137
Denis	84,018
Leblanc	69,304
Reyer	66,313
Total	**423,224**

WINE

- Bordeaux
- Champagne
- Chardonnay
- Chenin Blanc
- Chianti
- Grenache
- Lambrusco
- Malbec
- Merlot
- Piesporter
- Pinot Grigio
- Rioja
- Sauvignon Blanc
- Shiraz

You have selected multiple wines

Wine Selected #2

Figure 12-5. *The "alternate result" shows for no selection as well as for many selected*

One way to avoid this problem is to ensure users can't make multiple selections or no selection by turning on "Single select" on the Slicer settings formatting card. The other way is to use CONTCATENATEX as we will be discovering later in this chapter.

The SELECTEDVALUE function also allows you to test for specific values in the current filter. In Figure 12-6, the Card visual[1] shows "Expensive Wine" if the PRICE PER CASE value of the wine selected in the slicer is greater than $75.00; otherwise, it shows "Cheap Wine".

[1] For information on formatting the Card visual, visit `https://docs.microsoft.com/en-us/power-bi/visuals/power-bi-visualization-card`

Figure 12-6. *Using SELECTEDVALUE to test for values in the current filter*

This is the expression used in Figure 12-6:

```
High Price =
IF (
    SELECTEDVALUE ( Wines[PRICE PER CASE] ) > 75,
    "Expensive Wine",
    "Cheap Wine"
)
```

It's important to note here that when using SELECTEDVALUE, you can select any value sitting in *any* column of the row that has been filtered, not just the column used in the slicer.

However, we still have a problem when a user selects multiple values in a slicer. You may not want to use "single select" in the slicer but instead be able to select multiple items and list the items in a Table or Card visual. We've also seen that the "alternate result" of SELECTEDVALUE displays when there is no selection as well as when there are many selected. Let's now see how we can solve this problem by using the CONCATENATEX function.

The CONCATENATEX Function

We know that any function that ends in an "X" is an iterating function, and CONCATENATEX is no exception. It iterates the table referenced in its first argument and then concatenates the values in the column referenced in its second argument. Specifically, the CONCATENATEX function has the following arguments:

= CONCATENATEX(table, expression, delimiter, order by, order)

where:

table is the table to be iterated.

expression is the column (or expression) whose values you want concatenating for every row in **table**.

delimiter is the character you want to separate the values, for example, a comma or an ampersand.

order by (optional) is usually a column by which you want to sort the values.

order (optional) is ASC or DESC.

Now let's look at an example of an expression using CONCATENATEX:

```
Types of Wine =
CONCATENATEX ( Wines, Wines[WINE] , ", " , Wines[WINE ID], ASC )
```

In this measure, CONCATENATEX iterates the Wines table and, for every row in the table, returns a concatenated list of values from the WINE column, separated with a comma and sorted ascending by WINE ID. In Figure 12-7, you can see the values that this expression returns when the TYPE column from the Wines tables has been placed in the Table visual. The "Types of Wine" measure displays all the wines beside their type (i.e., Red or White), separated by a comma and sorted by the WINE ID column ascending.

TYPE	Types of Wine
Red	Bordeaux, Malbec, Grenache, Chianti, Merlot, Rioja, Shiraz
White	Champagne, Chardonnay, Piesporter, Pinot Grigio, Sauvignon Blanc, Chenin Blanc, Lambrusco
Total	**Bordeaux, Champagne, Chardonnay, Malbec, Grenache, Piesporter, Chianti, Pinot Grigio, Merlot, Sauvignon Blanc, Rioja, Chenin Blanc, Shiraz, Lambrusco**

Figure 12-7. *The values returned by the "Types of Wine" measure*

We can use just the first three arguments and place this measure in a Card visual, using a slicer to filter by the WINE column:

```
Types of Wine #1 =
        CONCATENATEX (
                Wines, Wines[WINE] ,
                   ", " )
```

Here, CONCATENATEX will simply return all the wine names in the current filter; see Figure 12-8. At last, we've been able to solve the problem of displaying slicer selections when multiple items have been selected.

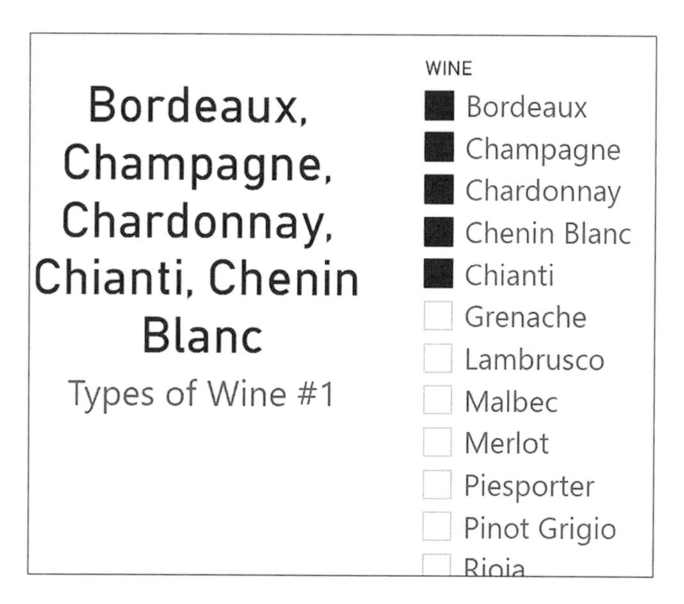

Figure 12-8. *The "Types of Wine #1" measure in a Card visual sliced by WINE*

However, we're not quite there yet. If there is no selection in the slicer, the Card visual returns all the wine names which probably isn't what you want. To resolve this, we need to take our "Types of Wine #1" expression a little further.

In Figure 12-9, we have used two similar measures in the title of a Table visual using conditional formatting: "Types of Wine #2" and "Types of Wine #3". Both measures return the phrase "Sales by Wine, filtered by", and the list of wines will grow as the selection grows. To avoid cluttering the visual with many wine names, the "Types of Wine #3" shows "and More" when more than three wines have been selected. When there is *no* selection in the slicer, the title of the visuals shows "Sales by Wine", rather than "you have selected multiple wines", as in the case of the measures using SELECTEDVALUE.

"Types of Wine #2

Sales by Wine, filtered by Bordeaux, Champagne, Chardonnay, Chianti, Chenin Blanc

SALESPERSON	Total Cases
Abel	34,091
Blanchet	33,554
Charron	34,785
Denis	35,555
Leblanc	32,183
Reyer	27,152
Total	**197,320**

"Types of Wine #3

Sales by Wine, filtered by Bordeaux, Champagne, Chardonnay, and more...

SALESPERSON	Total Cases
Abel	34,091
Blanchet	33,554
Charron	34,785
Denis	35,555
Leblanc	32,183
Reyer	27,152
Total	**197,320**

WINE

- ■ Bordeaux
- ■ Champagne
- ■ Chardonnay
- ■ Chenin Blanc
- ■ Chianti
- ☐ Grenache
- ☐ Lambrusco
- ☐ Malbec
- ☐ Merlot
- ☐ Piesporter
- ☐ Pinot Grigio
- ☐ Rioja
- ☐ Sauvignon Blanc

"Types of Wine #2

Sales by Wine

SALESPERSON	Total Cases
Abel	69,871
Blanchet	65,581
Charron	68,137
Denis	84,018
Leblanc	69,304
Reyer	66,313
Total	**423,224**

"Types of Wine #3

Sales by Wine

SALESPERSON	Total Cases
Abel	69,871
Blanchet	65,581
Charron	68,137
Denis	84,018
Leblanc	69,304
Reyer	66,313
Total	**423,224**

WINE

- ☐ Bordeaux
- ☐ Champagne
- ☐ Chardonnay
- ☐ Chenin Blanc
- ☐ Chianti
- ☐ Grenache
- ☐ Lambrusco
- ☐ Malbec
- ☐ Merlot
- ☐ Piesporter
- ☐ Pinot Grigio

Figure 12-9. *Using CONCATENATEX to solve the problem of multiple selections and no selection*

Therefore, using CONCATENATEX, we have solutions for all four problem scenarios:

1. No selection in the slicer
2. Selections in the slicer
3. Three or fewer wines selected
4. More than three wines selected

The measure required that solves problem scenarios #1 and #2 is relatively straightforward. However, we need to extend this expression to accommodate scenarios #3 and #4, and this is where the expression will become a little more ambitious.

Therefore, let's tackle the situation where users make selections in the slicer or there is no selection.

To resolve this scenario, the measure we build must return either

1. "Sales of Wines" if there are no selections in the slicer

or

2. "Sales of Wines filtered by" followed by a list of wines selected in the slicer

Therefore, we need a way to find out whether the filter on the WINES column has reduced the number of rows in the Wines dimension. If it has, there must be selections in the slicer. If it hasn't, there must be no selection in the slicer. What we can do here is use the function named VALUES that generates a virtual one-column table that lists the values in the WINE column in the current filter context. We can then use the ALL function to return another virtual one-column table containing all the wine names. If these tables have the same number of rows in them, then there must be no selections in the slicer.

Note We deep dive into the VALUES function later in this chapter.

Here is the expression that we can build. Note the use of variables to harvest the values returned by COUNTROWS:

```
Types of Wine #2 =
VAR NoFilteredWines =
         COUNTROWS (  VALUES ( Wines[WINE] ) )
VAR  NoAllWines=
            COUNTROWS ( ALL( Wines[WINE] ))
RETURN
     IF ( NoFilteredWines = NoAllWines ,
               "Sales by Wine",
               "Sales by Wine, filtered by "
               &
         CONCATENATEX (
                    Wines, Wines[WINE] ,
                        ", " ) )
```

Let's now turn our attention to resolving the scenario of users selecting more than three wines in the slicer. If they select three or fewer wines or no wines, then the measure will return the same as "Types of Wine #2". However, if they select four or more wines, we want the measure to return a list of the first three wines selected followed by "and more...". Therefore, we need to generate a list of just the top three wine names selected in the slicer. We can use a table function named TOPN to do this job. As its name suggests, TOPN will build a virtual table containing only the top N (e.g., **3**) values as in the following expression:

TOPN (3, VALUES (Wines[WINE]))

Notice again how the VALUES function is used to generate a one-column table listing the wine names in the current filter context. The TOPN function will extract the top three of these wine names into its own table that can then be used by CONCATENATEX to concatenate these values. We can then concatenate "and more..." using the ampersand.

You can see the following expression will solve our final scenario. All we need to do is add the IF function to execute the TOPN expression, followed by the TOPN expression itself, added to the bottom of the code (highlighted in gray):

```
Types of Wine #3 =
VAR NoFilteredWines =
          COUNTROWS (  VALUES ( Wines[WINE] ) )
VAR  NoAllWines=
             COUNTROWS (ALL ( Wines[WINE]))
RETURN
     IF ( NoFilteredWines = NoAllWines ,
               "Sales by Wine",
               "Sales by Wine, filtered by "
               &
        IF ( NoFilteredWines <=3,

         CONCATENATEX (
                    Wines ,
                        Wines[WINE] ,
                        ", ") ,
```

```
            CONCATENATEX (
             TOPN ( 3, VALUES ( Wines[WINE] )),
                          Wines[WINE] ,
                          ", ")
                                   & " and more..."
))
```

In building these measures, you have learned how CONCATENATEX can be used to string together slicer selections. However, it has also been a valuable exercise in the use of table functions and table expressions to generate virtual in-memory tables that are then used within the expression. This concept lies at the heart of DAX, building temporary tables that contain the values used by scalar functions. It might also be worth noting here that the measure "Types of Wine #3" is an order of magnitude more advanced than anything you have tackled so far in this book, but you now have the skills to author such complex code.

We have also covered the details of the SELECTEDVALUE function on which we are now going to refocus. This is because we can put it to better use than alerting users to whatever has been chosen in a slicer. We understand that SELECTEDVALUE will return a single value, and in this way, we can use this function to harvest ad hoc values in columns of *unrelated tables*. These unrelated tables have a name, *parameter tables* whose use we are now going to explore.

Using Parameter Tables

You can use SELECTEDVALUE to return a user-selected parameter. This chosen parameter can then be used as a value inside a measure.

Consider the Table visual in Figure 12-10. Here, we have a slicer that allows us to select a sales projection scenario for our "Total Sales" measure as follows:

- "Best case" (increase by 20%)
- "Probable" (increase by 10%)
- "Worst case" (decrease by 10%)

The total sales is then calculated accordingly in the "What If Scenario" measure.

WINE	Total Sales	What If Scenario
Bordeaux	$4,055,250	$4,866,300
Champagne	$7,373,700	$8,848,440
Chardonnay	$4,203,000	$5,043,600
Chenin Blanc	$1,236,950	$1,484,340
Chianti	$1,092,920	$1,311,504
Grenache	$1,078,950	$1,294,740
Malbec	$2,914,650	$3,497,580
Merlot	$900,276	$1,080,331
Piesporter	$1,384,155	$1,660,986
Pinot Grigio	$703,470	$844,164
Rioja	$1,527,795	$1,833,354
Sauvignon Blanc	$1,896,600	$2,275,920
Shiraz	$1,364,766	$1,637,719
Total	**$29,732,482**	**$35,678,978**

Scenario
Best
Probable
Worst

Figure 12-10. *Using a parameter table to analyze sales projection scenarios*

To create these scenarios, we've used the **Enter data** button on the Home tab and created this table, called "What If", as shown Figure 12-11.

Scenario	Value
Best	1.2
Worst	0.9
Probable	1.1

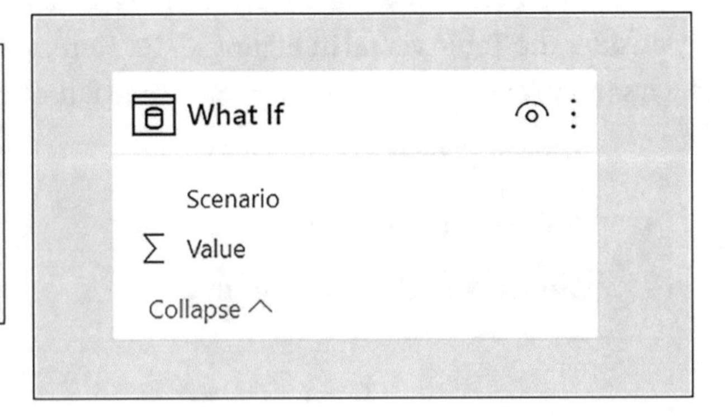

Figure 12-11. *The "What If" parameter table*

Notice in Figure 12-11 that this table is not related to any other tables in the data model. We now need to place a slicer on the canvas populated with the "Scenario" column from the "What If" table, and we're ready to create this measure:

```
What If Scenario =
[Total Sales] * SELECTEDVALUE ( 'What If'[Value] )
```

When we select a value from the Scenario slicer, for example, "Probable", this value is filtered in the "What If" table. There is only one row in the "What If" table, and the value sitting in the Value column is then used to multiply the value of the "Total Sales" measure.

You have learned that you can build parameter tables and by using SELECTEDVALUE can construct expressions that test for specific values selected from the parameter table. Once you know you can do this, you can use the values selected to drive specific calculations. Let's look at an example of this. You may have found that one of the frustrations of working in Power BI is that you can only populate column values into slicers. However, this question often arises: Can I put measures into slicers? The answer is yes, you can! Consider Figure 12-12.

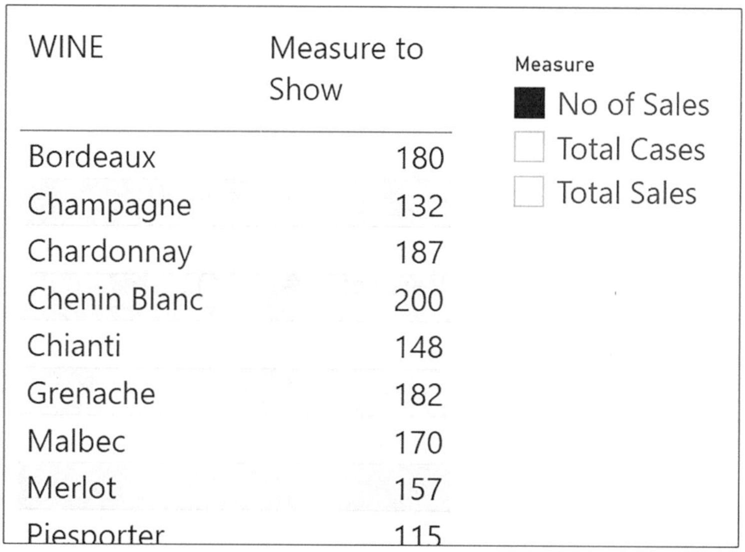

WINE	Measure to Show
Bordeaux	180
Champagne	132
Chardonnay	187
Chenin Blanc	200
Chianti	148
Grenache	182
Malbec	170
Merlot	157
Piesporter	115

Measure
- No of Sales
- Total Cases
- Total Sales

Figure 12-12. *Creating slicers for measures*

Here, we have a slicer that lists three measures. On selecting a measure in the slicer, the "Measure to Show" measure in the Table visual calculates the selected measure.

To build this example, again, we started with creating the parameter table and named it "Select Measures". This table has two columns. The column named "Measure" lists the measures, but appreciate that these names are arbitrary; you don't have to use the exact measure names. The second column named "Value" assigns a value to the "Measure" name. As with all parameter tables, this table is unrelated to any other tables in the model; see Figure 12-13.

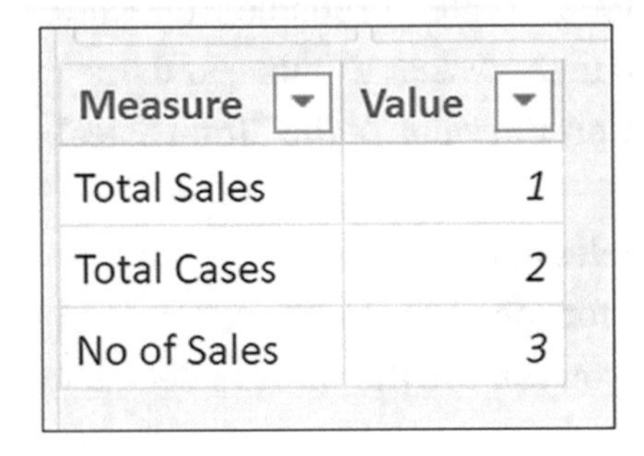

Measure	Value
Total Sales	1
Total Cases	2
No of Sales	3

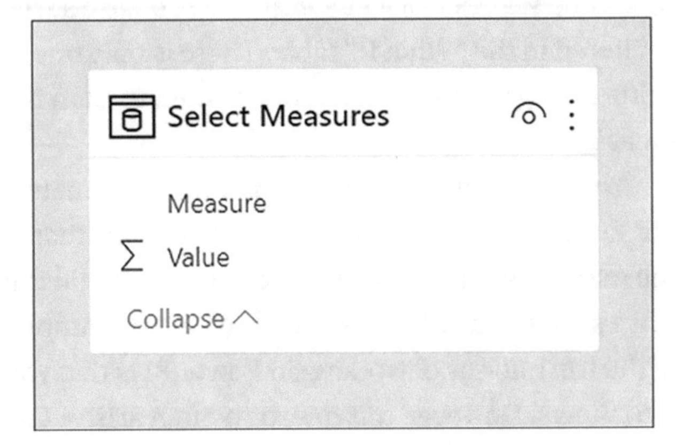

Figure 12-13. *The "Select Measures" parameter table*

A slicer was then placed on the canvas containing the "Measure" column from the "Select Measures" table.

This is the expression for "Measure to Show":

```
MEASURE toShow =
    SWITCH (
        SELECTEDVALUE ( 'Select Measures'[Value] ),
        1, [Total Sales],
        2, [Total Cases],
        3, [No. of Sales]
    )
```

Note the use of the SWITCH function in place of using IF, but either does the job. This measure was then placed in the Table visual alongside the WINE column from the Wines table.

The Values Function

It was debatable whether I would include the VALUES function in this book because in recent years, its requirement has largely been replaced by the SELECTEDVALUE function. However, the reason I changed my mind is that if you're a DAX user, you would know and understand the VALUES function, even if you were rarely required to use it.

Before the arrival of the SELECTEDVALUE function in 2017, the VALUES function was one of the major DAX functions. For this reason, you will meet VALUES when you browse other people's code, and therefore, it would be a good idea if you knew the purpose of the function within an expression. Also, these two functions, SELECTEDVALUE and VALUES, are not interchangeable; sometimes, only VALUES will do. Indeed, we've already had cause to use the VALUES function when we were exploring CONCATENATEX.

VALUES is particularly useful when you want to convert a *column* reference into a *table* reference or when you want to reapply "lost" filters.

This function has a very simple syntax. Inside the function, you either reference a table or a column:

=VALUES (table name or column name)

Here are two examples of VALUES syntax; the first references a table and the second, a column:

= VALUES (Wines)

= VALUES (Wines[WINE])

This function is a *table function* and returns a virtual table as follows:

- When the input parameter is a column name, it returns a one-column table that contains the distinct values from the specified column *using the current filter context.* Duplicate values are removed, and only unique values are returned.

- When the input parameter is a table name, it returns a table containing the rows from the specified table *using the current filter context*, and duplicate rows are preserved.

Although SELECTEDVALUE has largely replaced VALUES, they are two quite different functions. SELECTEDVALUE is a scalar function that will return a single value. Therefore, inside SELECTEDVALUE, you can only reference the column name where the scalar value you require is located. The VALUES function, on the other hand, is described as being a table function, and inside VALUES, you can reference either a column name or a table name. If you reference a column name inside VALUES, that column is converted to a table and so allows you to use columns as table expressions. Because this is one of the benefits of using this function, VALUES is more commonly used with a column reference, and it's this behavior of VALUES on which we will concentrate in this section.

A Table or a Scalar Function?

However, if SELECTEDVALUE returns a scalar value and VALUES returns a table, how can VALUES be replaced by SELECTEDVALUE? This is where the VALUES function gets interesting because although it's described as a table function, VALUES can return either a table *or a scalar value.*

The reason for this is that when a DAX table expression returns a *one-column, one-row table,* it's converted by the DAX engine from a table to a scalar value (remember that the LASTNONBLANK function also exhibited this behavior; see Chapter 9). This is when VALUES changes its nature and switches from returning a table to returning a scalar value.

We can now explore an example of this behavior.

Note The following examples of DAX measures using the VALUES and SELECTEDVALUE functions are for explanation purposes only. We write measures that return the wine names that we've already put into a visual, and clearly, there's no purpose to these calculations. The reason we're using these particular expressions is to explain more readily how the VALUES function works. We later put the VALUES function to more realistic and beneficial use.

Consider the following expression that will return a one-column table containing the name of the wine sitting in the current filter context.

```
Values Wine = VALUES ( Wines[Wine] )
```

Before we put this measure into a Table visual, we must turn off the Total row of the visual (for reasons we will explain presently).[2] When the measure is placed into the visual, it returns the values in the WINE column in the current filter; see Figure 12-14.

WINE	Values Wine
Bordeaux	Bordeaux
Champagne	Champagne
Chardonnay	Chardonnay
Chenin Blanc	Chenin Blanc
Chianti	Chianti
Grenache	Grenache
Lambrusco	Lambrusco
Malbec	Malbec
Merlot	Merlot
Piesporter	Piesporter
Pinot Grigio	Pinot Grigio
Rioja	Rioja
Sauvignon Blanc	Sauvignon Blanc
Shiraz	Shiraz

Figure 12-14. *The VALUES function returns the value in the current filter context*

We get no error on the evaluation of the "Values Wine" measure, so it would appear that VALUES is behaving like a scalar function (remember that all measures must return scalars). We can see how this is possible. In the first evaluation for "Bordeaux" wine, the VALUES expression creates a virtual table containing a list of unique values in the WINE column that are in the current filter. It therefore generates a *one-column, one-row table* containing the value "Bordeaux". If we could see this table, it may well look like the table containing a single value as shown in Figure 12-15.

Figure 12-15. *The one-column, one-row table generated by VALUES*

[2] For information on removing the Total row, visit `https://community.powerbi.com/t5/Desktop/How-to-remove-the-quot-Total`

This table contains a single value that can be converted to a scalar, and this is why it can be used successfully in the measure "Values Wine".

However, let's now replace the Total row in the Table visual. When we do this, the measure will now return an error as shown in Figure 12-16.

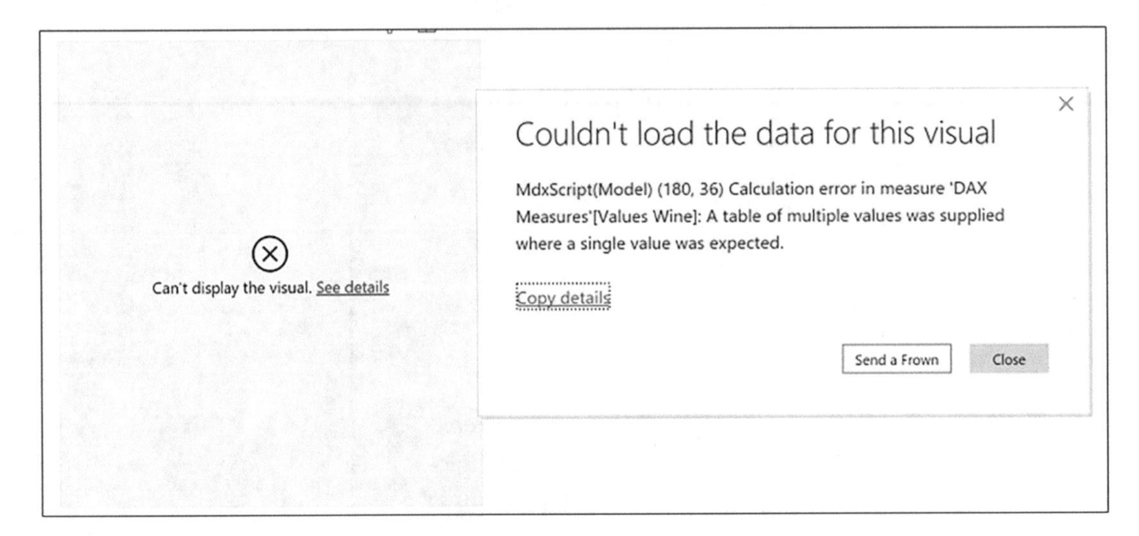

Figure 12-16. *An error is returned when the VALUES function evaluates the Total row*

The error message reads:

"A table of multiple values was supplied where a single value was expected."

Why do we get this error when the Total row shows but not when it's absent? When a DAX expression is evaluated for the Total row, there is no longer a single value being returned by VALUES, but now *all* the wine names are in the filter context. Therefore, the VALUES function will return a virtual table containing all the values in the WINE column. This is the "table of multiple values" that the error message is referring to (Figure 12-17).

Figure 12-17. *VALUES returns a "table of multiple values" when evaluating the Total row*

Therefore, we can deduce that it's the evaluation of the Total row that's the problem because you can't put multiple values into a "cell" in the Total row. This is why in the Table visual in Figure 12-14, we must remove the Total row for our expression to work. However, you might think this is a bit of a workaround and at some point want to show the Total row value for your measure.

To remedy this, rather than removing the Total row from the visual, instead, we can get DAX to distinguish between the evaluation for each wine and the evaluation for the Total row. For this, we must use a DAX function that returns TRUE if there is just one value in the current filter context. Its name is unsurprisingly HASONEVALUE. Here is the expression we need:

```
Values Wine =
IF ( HASONEVALUE ( Wines[WINE] ),
    VALUES ( Wines[WINE] ),
   "All Wines" )
```

But doesn't the preceding expression return the same values as this one?

```
Selected Value Wine =
SELECTEDVALUE ( Wines[WINE], "All Wines" )
```

WINE	Values Wine	Selected Value Wine
Bordeaux	Bordeaux	Bordeaux
Champagne	Champagne	Champagne
Chardonnay	Chardonnay	Chardonnay
Chenin Blanc	Chenin Blanc	Chenin Blanc
Chianti	Chianti	Chianti
Grenache	Grenache	Grenache
Lambrusco	Lambrusco	Lambrusco
Malbec	Malbec	Malbec
Merlot	Merlot	Merlot
Piesporter	Piesporter	Piesporter
Pinot Grigio	Pinot Grigio	Pinot Grigio
Rioja	Rioja	Rioja
Sauvignon Blanc	Sauvignon Blanc	Sauvignon Blanc
Shiraz	Shiraz	Shiraz
Total	**All Wines**	**All Wines**

Figure 12-18. *The VALUES function returns the same values as the SELECTEDVALUE function*

Well, yes, it does (Figure 12-18), and because the VALUES expression is more complex, you would probably prefer to use SELECTEDVALUE. Whenever you use VALUES to return a scalar value, you could use SELECTEDVALUE instead. What's more, with SELECTEDVALUE, you don't have to account for only one value in the filter context as it's implicit in the "alternate result" argument.

You may be wondering why you would want to return the wine names anyway, using either SELECTEDVALUE or using VALUES, when you've already got them as the first column in the visual!

Replacing "Lost Filters"

You may feel that these examples, although explaining how VALUES works, are not "real-world" calculations. However, you have now learned how the VALUES function operates, that it can return either a table or a scalar value. We need to find a better use for VALUES and also find a situation where we can't substitute SELECTEDVALUE. A better example of the VALUES function is when we use VALUES as a table function, rather than returning a scalar. So let's look at this next scenario.

One of the problems with filtering using slicers is that you lose the original unfiltered value. One way to overcome this problem is to use two visuals and then use "Edit Interactions"[3] so that a slicer filters one of the visuals but not the other (Figure 12-19).

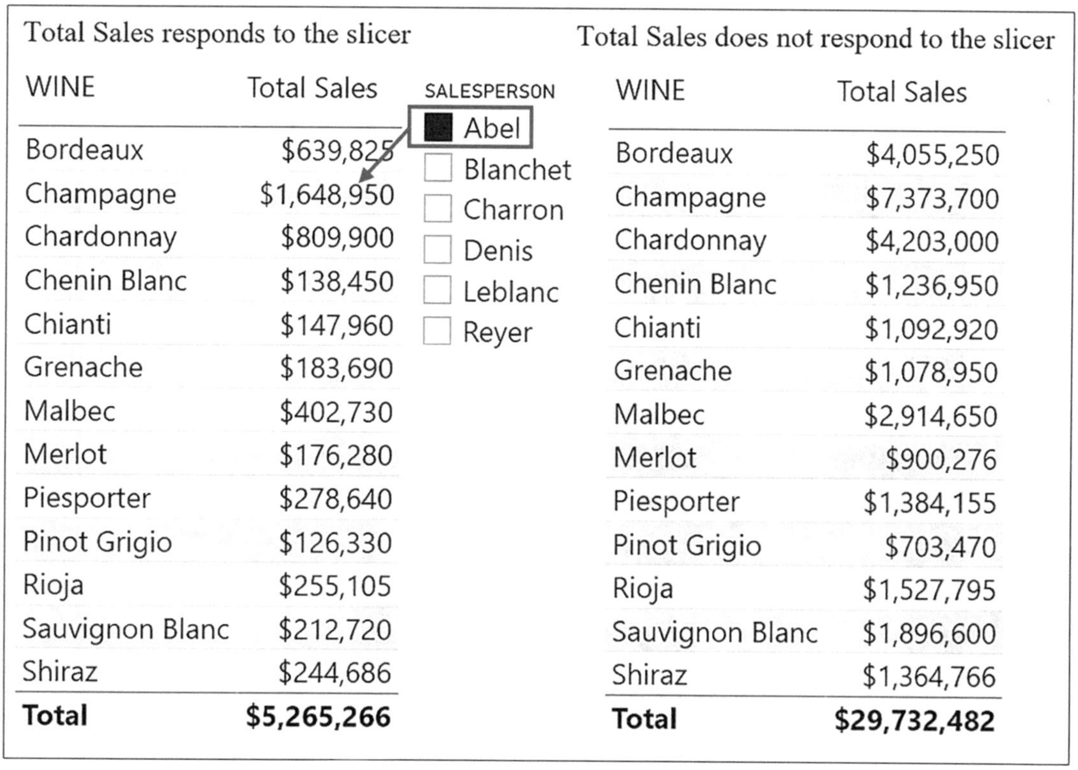

Total Sales responds to the slicer

WINE	Total Sales
Bordeaux	$639,825
Champagne	$1,648,950
Chardonnay	$809,900
Chenin Blanc	$138,450
Chianti	$147,960
Grenache	$183,690
Malbec	$402,730
Merlot	$176,280
Piesporter	$278,640
Pinot Grigio	$126,330
Rioja	$255,105
Sauvignon Blanc	$212,720
Shiraz	$244,686
Total	**$5,265,266**

SALESPERSON

- ■ Abel
- ☐ Blanchet
- ☐ Charron
- ☐ Denis
- ☐ Leblanc
- ☐ Reyer

Total Sales does not respond to the slicer

WINE	Total Sales
Bordeaux	$4,055,250
Champagne	$7,373,700
Chardonnay	$4,203,000
Chenin Blanc	$1,236,950
Chianti	$1,092,920
Grenache	$1,078,950
Malbec	$2,914,650
Merlot	$900,276
Piesporter	$1,384,155
Pinot Grigio	$703,470
Rioja	$1,527,795
Sauvignon Blanc	$1,896,600
Shiraz	$1,364,766
Total	**$29,732,482**

Figure 12-19. *Using "Edit Interactions", you can prevent slicers from filtering a visual*

[3] For information on how to edit the interactions of visuals, visit `https://docs.microsoft.com/en-us/power-bi/create-reports/service-reports-visual-interactions`

However, we want a single visual that retains the unfiltered values alongside the filtered ones, as in Figure 12-20. This is the DAX expression for the "Total Sales Not Filtered" measure:

```
Total Sales Not Filtered =
CALCULATE ( [Total Sales],
    ALL ( Winesales ),
    VALUES ( Wines[WINE] )
)
```

WINE	Total Sales	Total Sales Not Filtered
Bordeaux	$639,825	$4,055,250
Champagne	$1,648,950	$7,373,700
Chardonnay	$809,900	$4,203,000
Chenin Blanc	$138,450	$1,236,950
Chianti	$147,960	$1,092,920
Grenache	$183,690	$1,078,950
Malbec	$402,730	$2,914,650
Merlot	$176,280	$900,276
Piesporter	$278,640	$1,384,155
Pinot Grigio	$126,330	$703,470
Rioja	$255,105	$1,527,795
Sauvignon Blanc	$212,720	$1,896,600
Shiraz	$244,686	$1,364,766
Total	**$5,265,266**	**$29,732,482**

SALESPERSON
- ■ Abel
- ☐ Blanchet
- ☐ Charron
- ☐ Denis
- ☐ Leblanc
- ☐ Reyer

Figure 12-20. *A table visual where the "Total Sales Not Filtered" measure ignores the slicer filter*

Now let's examine the "Total Sales Not Filtered" measure in more detail. The first filter argument to CALCULATE is the ALL function that acts as a modifier and removes any cross-filters on the Winesales fact table coming from both the WINE column and the

SALESPERSON column. In the second filter argument, VALUES is used to build a virtual one-column, one-row table containing the wine name in the current filter context, that is, "Bordeaux" in the first evaluation. This is equivalent to "Wines[WINE] = "Bordeaux". CALCULATE then applies this new filter to the Winesales table that is then refiltered accordingly. The end result is that there is a filter on the WINE column but no longer a filter on the SALESPERSON column, and therefore, we see sales for all salespeople for each wine. When the measure calculates the Total row, it constructs a virtual one-column table containing all the wine names to be used as the filter.

However, the following expression is an alternative way of achieving the same result:

```
Total Sales Not Filtered #2 =
CALCULATE ( [Total Sales], ALL ( Winesales ), Wines )
```

In this measure, we've referenced the entire Wines table as the filter instead of using VALUES to generate a virtual one-column table. We can do this because the Wines table has been filtered down to one row (or all rows for the evaluation of the Total row) and the entire table can be used as a table expression.

This is the first time we have referenced a table in the filter argument to CALCULATE rather than a table *expression*, and we're going to do this again later on. Remember that the Wines table will contain a single row containing the wine in the current filter context, or all the rows of the Wines table when evaluating the Total row. This expression is perhaps a better one because we don't need to nest yet another function.

Converting Columns to Tables

We've already established that the VALUES function is a useful function to add to your DAX "toolbox" even though you can normally use SELECTEDVALUE instead. What you will discover as you work with DAX is that VALUES is more commonly used with a column reference because one of its major uses is to convert columns into tables. For example, this expression:

"= Wines[WINE]" is a *column*,

but this expression:

"= VALUES (Wines[WINE])" is a *table*.

We will look later at using VALUES in this way when we look at the TREATAS function later in this book.

With its dual personality of returning either a table or a scalar value, and particularly how it can convert a column to a table, VALUES is a function well worth getting to know.

In this chapter, we have explored three functions, SELECTEDVALUE, CONCATENATEX, and VALUES, that allow you to use the value or values sitting in the current filter. You have learned that by creating parameter tables, you can harvest these values to use within your DAX expressions. But more than this, when working with CONCATENATEX, you have understood how, by using variables, you can hold the values returned by these functions so they can be referenced later within the expression. You have also successfully generated a number of temporary in-memory tables to control filters placed on the data model. All these techniques are ubiquitous to writing DAX expressions and will hold you in good stead as you move forward and author more complex code.

CHAPTER 13

Controlling the Direction of Filter Propagation

Up to now, you have understood that filters *only* flow from the one side of the relationship to the many, from dimensions into the fact table, as indicated by the arrows in the linking lines in Model view; see Figure 13-1.

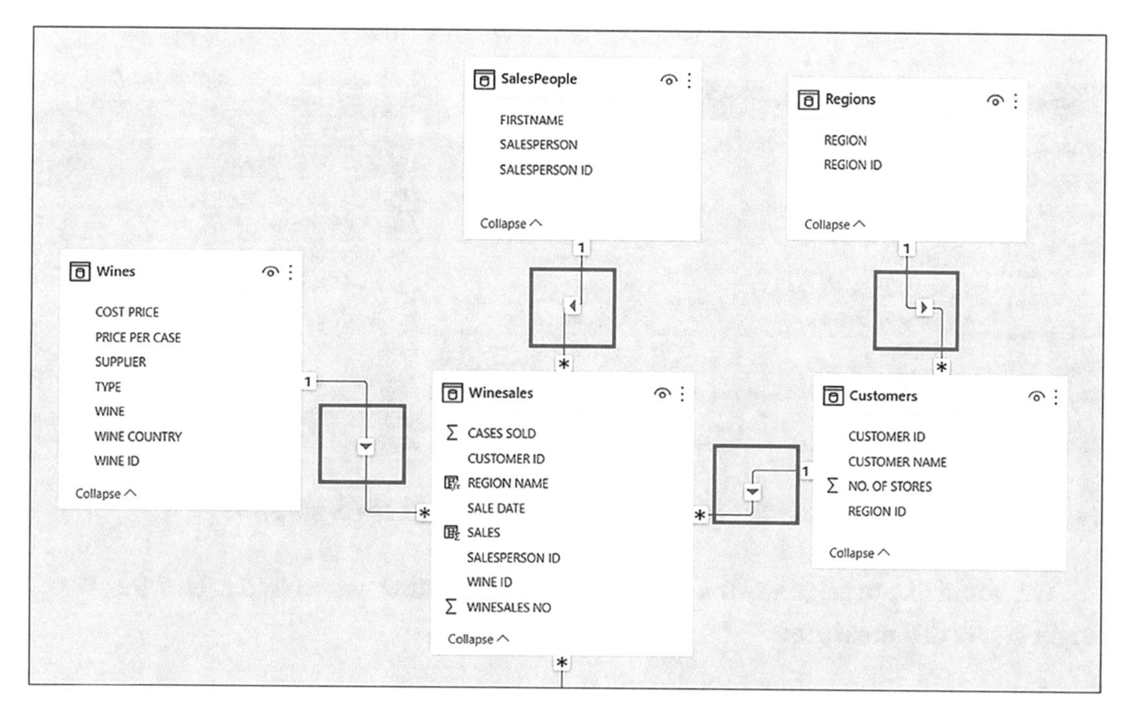

***Figure 13-1.** Filters only flow from dimensions into fact tables*

However, there will be situations when you will want to author measures that require filters to propagate in the opposite direction. In this chapter, we explore these situations and learn how to reverse the direction of the filters using two methods:

© Alison Box 2022
A. Box, *Up and Running with DAX for Power BI*, https://doi.org/10.1007/978-1-4842-8188-8_13

1. The CROSSFILTER function to programmatically reverse the filters
2. Editing the data model to make filter propagation flow both to and from the fact table

However, regarding method #2, we will be warning you of the downside if you change the structure of your data model. In fact, it's important to understand that if you want filters to flow in the opposite direction, this will always be problematic whichever way you choose to work it.

Programming Bidirectional Filters

For example, let's look at a problem we explored in Chapter 4 when you were learning about the filter context and which at that time, you were not able to resolve. In the Customers dimension, we have the column NO. OF STORES; see Figure 13-2.

CUSTOMER ID	CUSTOMER NAME	REGION ID	County	Area	Country	NO. OF STORES
1	Landstuhl Ltd	1800	West Midlands	England	United Kingdom	21
2	Erlangen & Co	800	Greater London	England	United Kingdom	21
4	Black Ltd	500	Central Bedfordshire	England	United Kingdom	19
5	Snoqualmie & Sons	1200	Greater Manchester	England	United Kingdom	13
6	Leeds & Co	1000	Merseyside	England	United Kingdom	11
7	Newcastle upon Tyne & Sons	1900	County Durham	England	United Kingdom	18
8	Charlottesville & Co	300	Greater London	England	United Kingdom	13
9	Brown & Co	1400	Derbyshire	England	United Kingdom	24
10	Lavender Bay Ltd	500	Leicestershire	England	United Kingdom	10
12	El Cajon & Sons	400	Greater London	England	United Kingdom	1
13	Sedro Woolley Ltd	1800	West Midlands	England	United Kingdom	23

***Figure 13-2.** The Customers table and the NO. OF STORES column*

We would like to calculate the number of stores in which we've sold each wine. We might create this measure:

```
Total Stores =
SUM ( Customers[ NO. OF STORES] )
```

However, as you can see in Figure 13-3, this measure does not work.

WINE	Total Stores
Bordeaux	1,181
Champagne	1,181
Chardonnay	1,181
Chenin Blanc	1,181
Chianti	1,181
Grenache	1,181
Lambrusco	1,181
Malbec	1,181
Merlot	1,181

***Figure 13-3.** The "Total Stores" measure does not return the correct results*

In Chapter 4, we established the reason for the incorrect values. The filter on the Wines dimension only propagates to the fact table and does *not* propagate onward to the Customers dimension; see Figure 13-4.

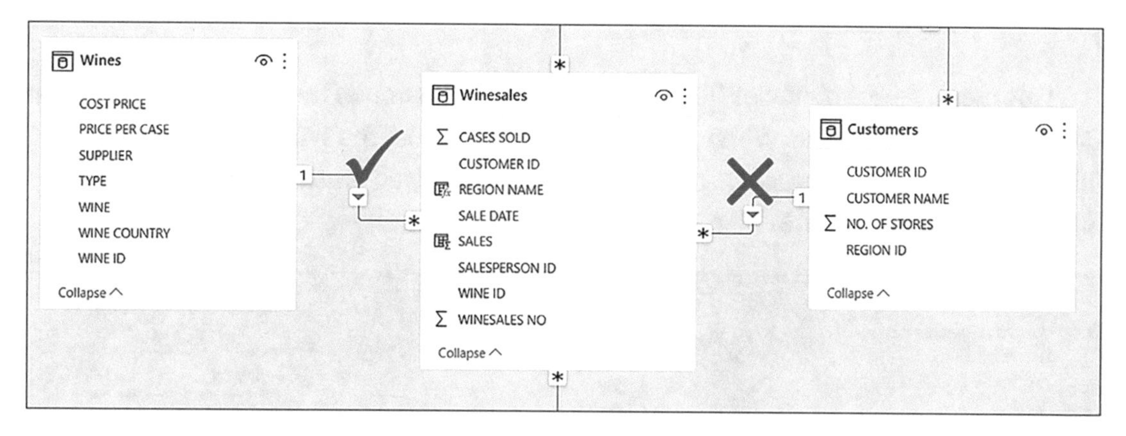

***Figure 13-4.** The filter does not propagate from Winesales to Customers*

So how do we find the number of stores in which we've sold our wines? The answer lies in using a function called CROSSFILTER.

The CROSSFILTER function returns no value but is used as a modifier to the CALCULATE function. It *programmatically* sets the direction of the filter propagation in the execution of the measure in which it is used. It has the following syntax:

= CROSSFILTER (column1, column2, direction)

where:

column1 is the column name that represents the many side of the relationship to be used.

column2 is the column name that represents the one side of the relationship to be used.

direction is the cross-filter direction to be used in the measure and can be set to "both" to generate bidirection filters.

Here is an example of CROSSFITLER syntax:

= CROSSFILTER (Winesales[CUSTOMERID], Customers[CUSTOMERID], both)

The CROSSFILTER function specifies the cross-filtering direction to be used by a measure, so we can now, in memory, *change* the direction in which the filters propagate.

We can rewrite our original "Total Stores" measure like this:

```
Total Stores =
CALCULATE (
    SUM ( Customers[NO. OF STORES] ),
    CROSSFILTER ( Winesales[CUSTOMER ID], Customers[CUSTOMER ID], BOTH )
)
```

This measure uses CROSSFILTER to change the direction of the relationship between Customers and Winesales. When this measure is evaluated, the Winesales table is cross-filtered by the Wines dimension, and this filter is propagated *onward* to the Customers dimension; see Figure 13-5.

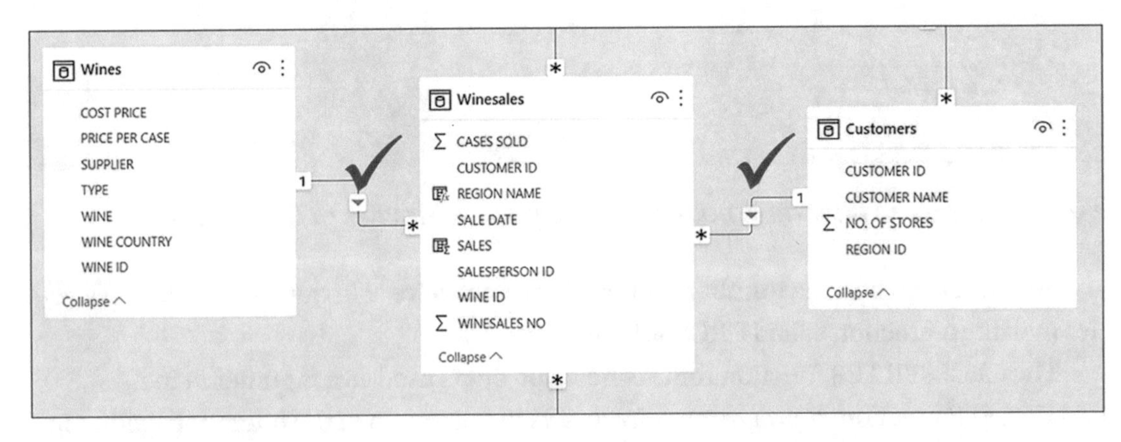

***Figure 13-5.** The CROSSFILTER function can change the direction of the filter programmatically*

So in the first instance for "Bordeaux" wine, the Customers table becomes cross-filtered to contain only customers who bought this wine, and we can see that there were **728** stores in which we've sold "Bordeaux"; see Figure 13-6.

WINE	Total Stores
Bordeaux	728
Champagne	709
Chardonnay	805
Chenin Blanc	757
Chianti	626
Grenache	685
Malbec	736
Merlot	749
Piesporter	563
Pinot Grigio	696
Rioja	832
Sauvignon Blanc	777
Shiraz	727
Total	**1,181**

Figure 13-6. *The "Total Stores" measure is now calculated correctly*

However, note the value in the Total row, **1,181**. It is not the total of the values for all the wines in the Table visual. Changing the filter propagation to bidirectional has a side effect. Many of the same customers have bought each wine, and so their total number of stores is included in multiple evaluations. However, the Total row sums the number of stores for all customers for all wines.

Why You Should Never Use Bidirectional Relationships

The CROSSFILTER function allows you to *programmatically* change the direction of filter propagation in the execution of a specific measure. However, you may know that there's an easier way to change the filter direction, and that's to change the structure of the data model. To do this, you can double-click on the linking line between two tables in Model view to edit the relationship, setting the "Cross filter direction" to "Both"; see Figure 13-7.

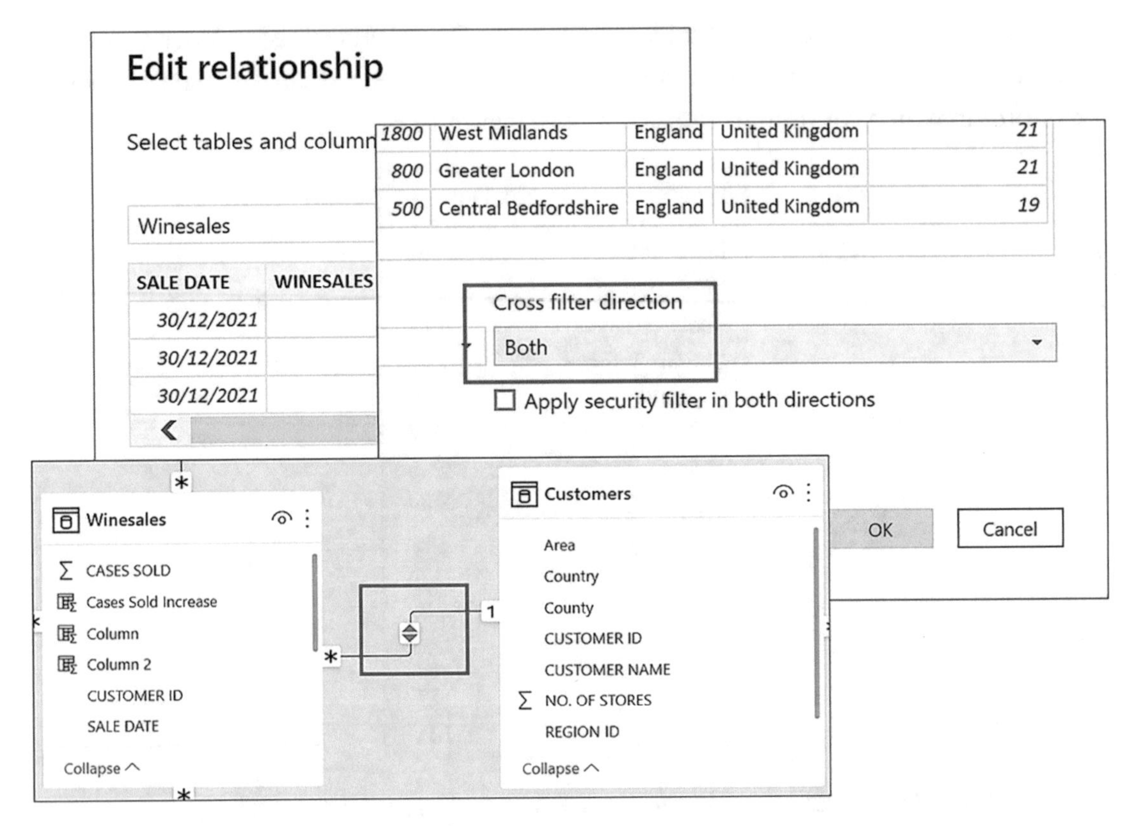

Figure 13-7. *You can edit the relationship and set the cross-filter to both*

However, a quick fix as this is, we would never recommend that you do this for two reasons. Firstly, bidirectional relationships are much less efficient and can hinder the performance of the data model, but perhaps, more importantly, they introduce *ambiguity* into the data model. It's beyond the scope of this book to elaborate on the concept of ambiguity, but for more information on these issues, check out this link:

`www.sqlbi.com/articles/bidirectional-relationships-and-ambiguity-in-dax/`

However, even at a more basic level, you will find that creating many bidirectional relationships in your model will render the data model unpredictable when filters are propagated, and you will start to lose control of what filters what. You will find it much easier if your model abides by the rule of single directional relationships, and if you must change the filter direction, use CROSSFILTER.

There are usually three reasons why people edit a relationship to bidirectional filtering, all of which are not valid reasons:

1. There is a lack of understanding of the subtleties of the Power BI data model and filter propagation.

2. People don't know enough DAX to be able to programmatically change the filter direction using CROSSFILTER.

3. People want to cross-filter slicers when the slicers use columns from different dimensions.

Let's take a look at the last of these reasons: wanting to cross-filter slicers when using columns from different dimensions. If this is your objective, you don't need to use bidirectional filtering. You can do this by using a *measure* in a visual-level filter on the slicer you want cross-filtered.

For example, in Figure 13-8, you can see we have two slicers: one using the CUSTOMER NAME column from the Customers dimension and one using the WINE column from the Wines dimension. If we select from the CUSTOMER NAME slicer, for example, "Ballard & Sons", the WINE slicer won't change to reflect the wines that "Ballard & Sons" has bought. We always see all the wines regardless of selections made in the CUSTOMER NAME slicer.

CUSTOMER NAME	WINE
Acme & Sons	Bordeaux
Back River & Co	Champagne
Ballard & Sons	Chardonnay
Barstow Ltd	Chenin Blanc
Beaverton & Co	Chianti
Black Ltd	Grenache
Bloxon Bros.	Lambrusco
Bluffton Bros	Malbec
Branch Ltd	Merlot
Brooklyn & Co	Piesporter
Brooklyn Ltd	Pinot Grigio
Brown & Co	Rioja
Burlington Ltd	Sauvignon Blanc
Burningsuit Ltd	Shiraz
Busan & Co	
Canoga Park Ltd	

Figure 13-8. *Slicers don't cross-filter from one dimension to another*

You already know why this is. If the Customers table is filtered, the filter is propagated to the Winesales table but not filtered onward to the Wines table because filters don't flow from the many side of the relationship to the one side. However, we can force the Wines table to cross-filter accordingly. We can do this by placing a visual filter on the WINE slicer using a *measure*, such as "Total Sales", and filter only Wines that have a "Total Sales" value. In fact, we can use any measure that does a calculation on the Winesales table and then set this filter to "Show items when the value is not blank" as shown in the visual filter in Figure 13-9.

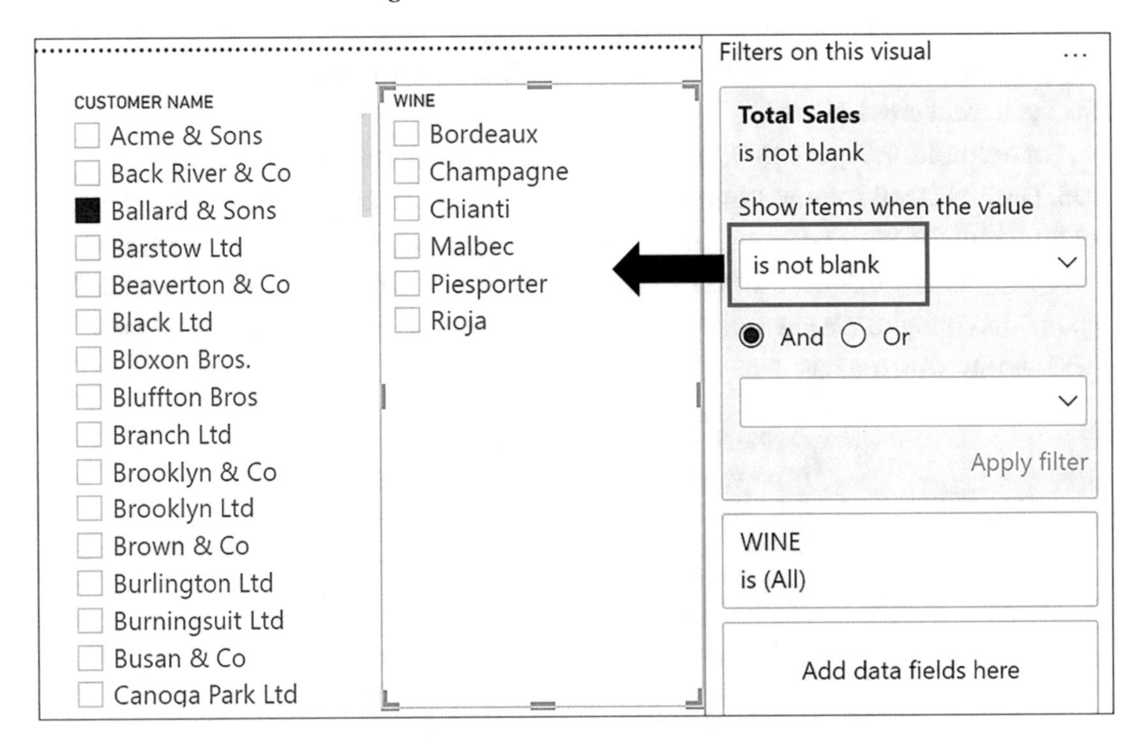

Figure 13-9. *Use a visual filter populated with a measure and set to "is not blank" to cross-filter slicers*

So there really is no excuse for editing relationships to bidirectional! Always design measures using the CROSSFILTER function to do this. However, as we have seen, the problem of measures that use bidirectional filters, whether using CROSSFILTER or editing the relationship, is that the Total row shows a misleading value. There is no real solution to this outcome; the total is correct but may not be the total you want to show. You can, of course, always turn off the display of the total row in the Table or Matrix visual.

CHAPTER 14

Working with Multiple Relationships Between Tables

In our data model, all our tables have *single* relationships between other tables. Indeed, it's only possible to have *one active* relationship between any two tables, but you can have as many *inactive* relationships as you want. In this chapter, you will learn how to use multiple relationships between tables and activate inactive relationships. This may be because you require multiple links from a dimension table into the fact table. However, there is another less obvious use of inactive relationships that we will discover in this chapter, and that is using comparison dimension tables. Here, we can use measures to force filter propagation through the comparison dimension table, therefore being able to compare a column from a default dimension with its counterpart in a comparison dimension.

If you attempt to build a second relationship or subsequent relationships between any two tables, all but the first relationship will be inactive, indicated by a dotted relationship line. Consider the tables in Figure 14-1. We now have two date columns in our Winesales table: SALE DATE and ORDER DATE.[1] The first relationship was established between the DATEKEY column in the DateTable and the SALE DATE column in the Winesales table. When we attempt to create a second relationship between the DateTable and the Winesales table by using ORDER DATE, we get a dotted line indicating that this relationship is inactive.

[1] To follow along with the examples, use the Power BI Desktop file "3 DAX USERELATIONSHIP.pbix".

© Alison Box 2022
A. Box, *Up and Running with DAX for Power BI*, https://doi.org/10.1007/978-1-4842-8188-8_14

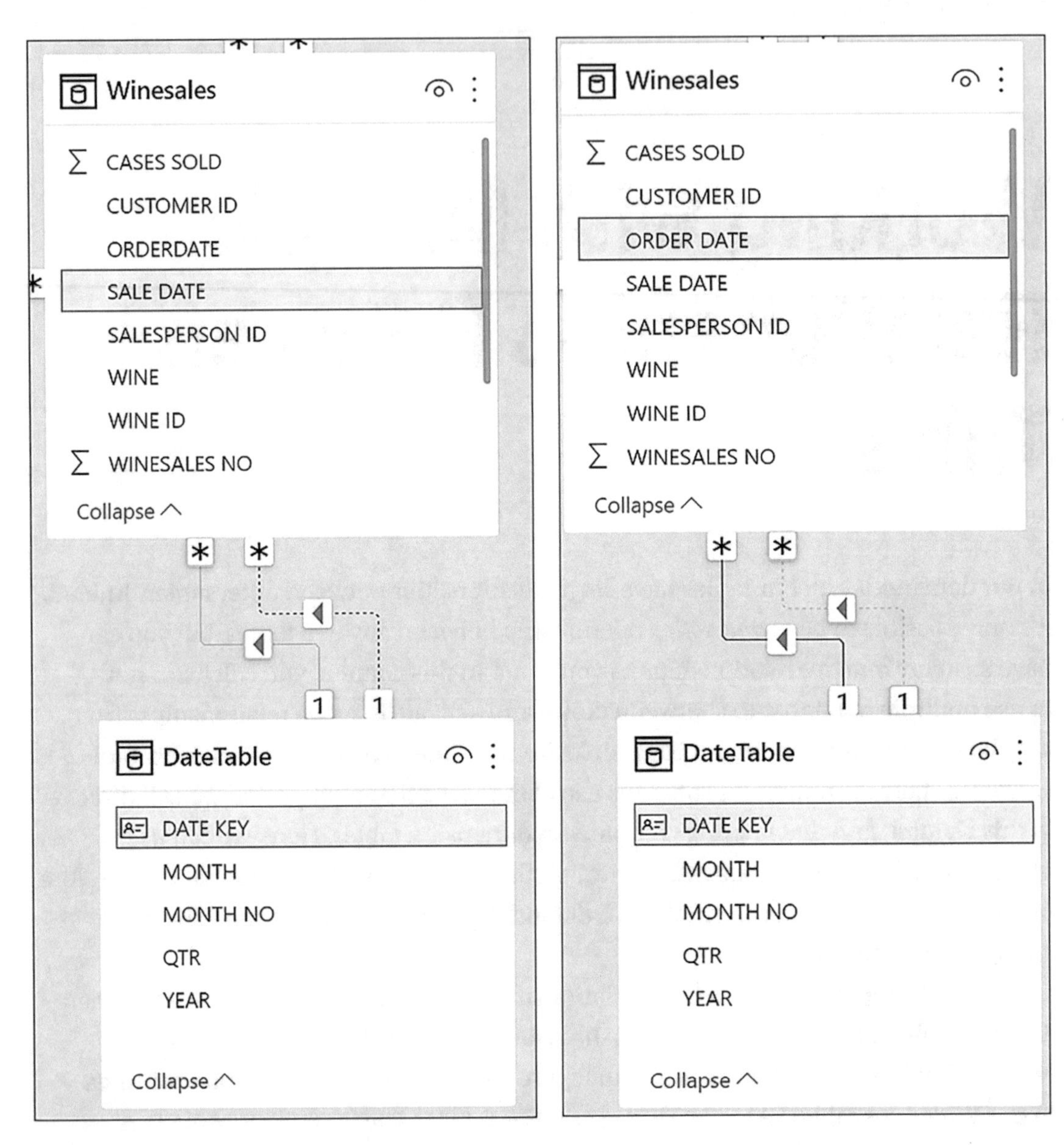

Figure 14-1. *Active and inactive relationships*

All measures will use the active relationship by default, so how do you use the inactive relationship? For example, if we build a Table visual containing the YEAR and MONTH columns from the DateTable (Figure 14-2), we can find the number of sales in each month using this measure:

```
No. of Sales =
COUNTROWS ( Winesales )
```

YEAR	MONTH	No. of Sales
2017	Jan	32
2017	Feb	29
2017	Mar	27
2017	Apr	28
2017	May	27
2017	Jun	14
2017	Jul	23
2017	Aug	16
2017	Sep	36
Total		**2,197**

Figure 14-2. *Using YEAR and MONTH from the DateTable filters the SALE DATE column in the Winesales table*

In this visual, the "No. of Sales" measure filters the YEAR and MONTH columns in the DateTable, which is propagated to the Winesales table using the *active* relationship and therefore filters the SALE DATE column for that year and month. However, to calculate the number of orders, we will need to use the *inactive* relationship so that the ORDER DATE column is filtered for that year and month instead. To do this, we can use the USERELATIONSHIP function.

Activating Inactive Relationships

The USERELATIONSHIP function, like the CROSSFILTER function, returns no value but is used as a modifier to the CALCULATE function. It programmatically uses an inactive relationship to propagation filters in the execution of the measure in which it is used. It has the following syntax:

= USERELATIONSHIP (column1, column2)

where:

column1 is the column name that represents the many side of the relationship to be used.

column2 is the column name that represents the one side of the relationship to be used.

Here is an example of the USERELATIONSHIP syntax:

= **USERELATIONSHIP (Winesales[SALE DATE], DateTable[ORDER DATE])**

You must have an *inactive* relationship in place in order to use the USERELATIONSHIP function.

This is the measure to calculate the number of orders in each month shown in Figure 14-3:

```
No. of Orders =
CALCULATE (
    COUNTROWS ( Winesales ),
    USERELATIONSHIP ( Winesales[ORDER DATE], DateTable[DATEKEY] ))
```

•

YEAR	MONTH	No. of Sales	No. of Orders
2017	Jan	32	43
2017	Feb	29	26
2017	Mar	27	24
2017	Apr	28	31
2017	May	27	22
2017	Jun	14	13
2017	Jul	23	24
2017	Aug	16	20
2017	Sep	36	35
2017	Oct	25	24
2017	Nov	28	31
Total		**2,197**	**2,197**

Figure 14-3. *Calculating the number of sales and number of orders*

When this measure is evaluated, the year and month filtered in the DateTable are propagated to the Winesales table to cross-filter the ORDER DATE column to find the orders in that month.

Comparing Values in the Same Column

The USERELATIONSHIP function can be used for another purpose: dynamic comparisons between values from the same column in a dimension. In other words, being able to compare a column from a default dimension with its counterpart in a comparison dimension.

Consider the example shown in Figure 14-4. Here, we are comparing 2020 sales (the "Total Sales" measure) to 2021 sales (the "Compare Year" measure), but the benefit here is that we are making the comparison *in the same Table visual*, rather than using separate visuals for each year.

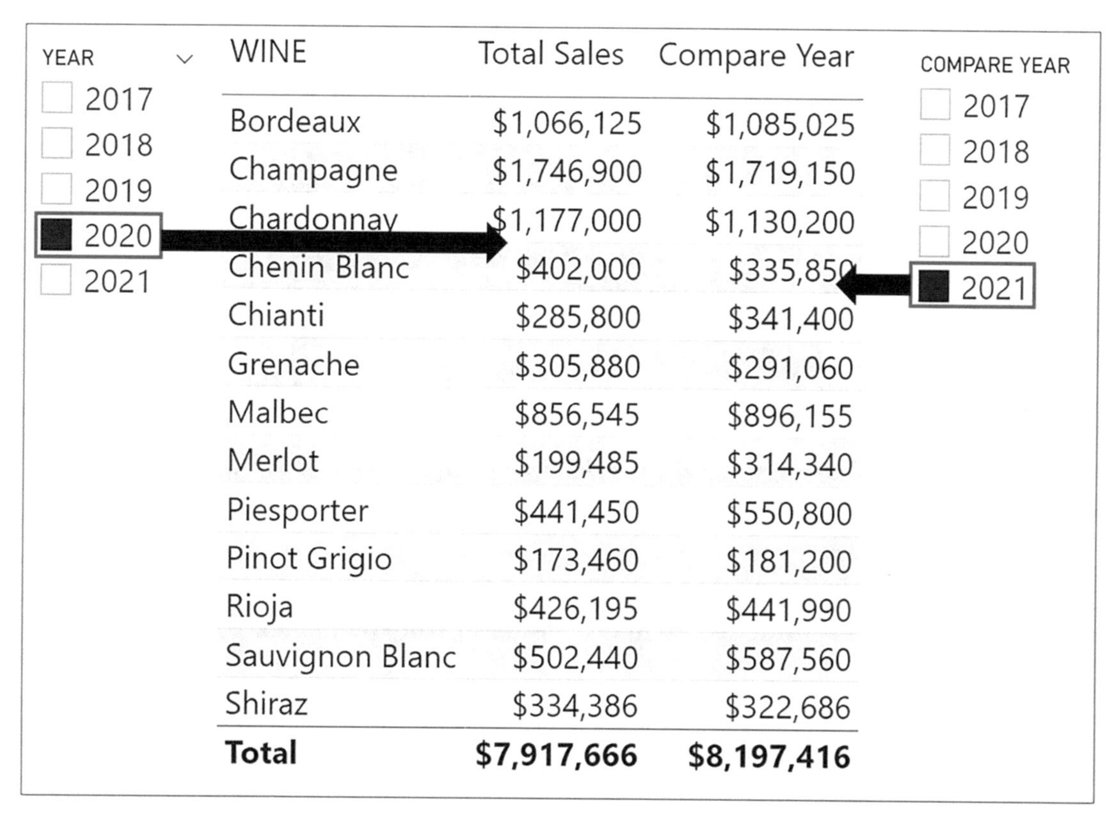

WINE	Total Sales	Compare Year
Bordeaux	$1,066,125	$1,085,025
Champagne	$1,746,900	$1,719,150
Chardonnay	$1,177,000	$1,130,200
Chenin Blanc	$402,000	$335,850
Chianti	$285,800	$341,400
Grenache	$305,880	$291,060
Malbec	$856,545	$896,155
Merlot	$199,485	$314,340
Piesporter	$441,450	$550,800
Pinot Grigio	$173,460	$181,200
Rioja	$426,195	$441,990
Sauvignon Blanc	$502,440	$587,560
Shiraz	$334,386	$322,686
Total	**$7,917,666**	**$8,197,416**

Figure 14-4. *Comparing sales for two different years selected in slicers*

The starting point for the "Compare Year" measure is to create a comparison DateTable in the data model by duplicating the original DateTable. We've named the duplicate DateTable "DateTable Compare". This table is then related to the Winesales table using the DATEKEY column from the "DateTable Compare" table and the SALE DATE column from the Winesales table; see Figure 14-5.

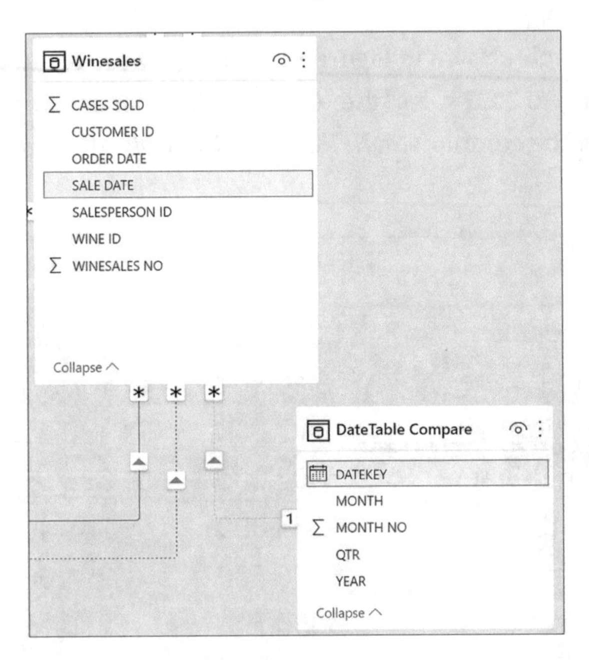

Figure 14-5. *Relate the comparison table to the fact table but set the relationship to inactive*

You must then edit this relationship to ensure that it's marked as *inactive* by checking off "Make this relationship active" in the **Edit Relationship** dialog; see Figure 14-6.

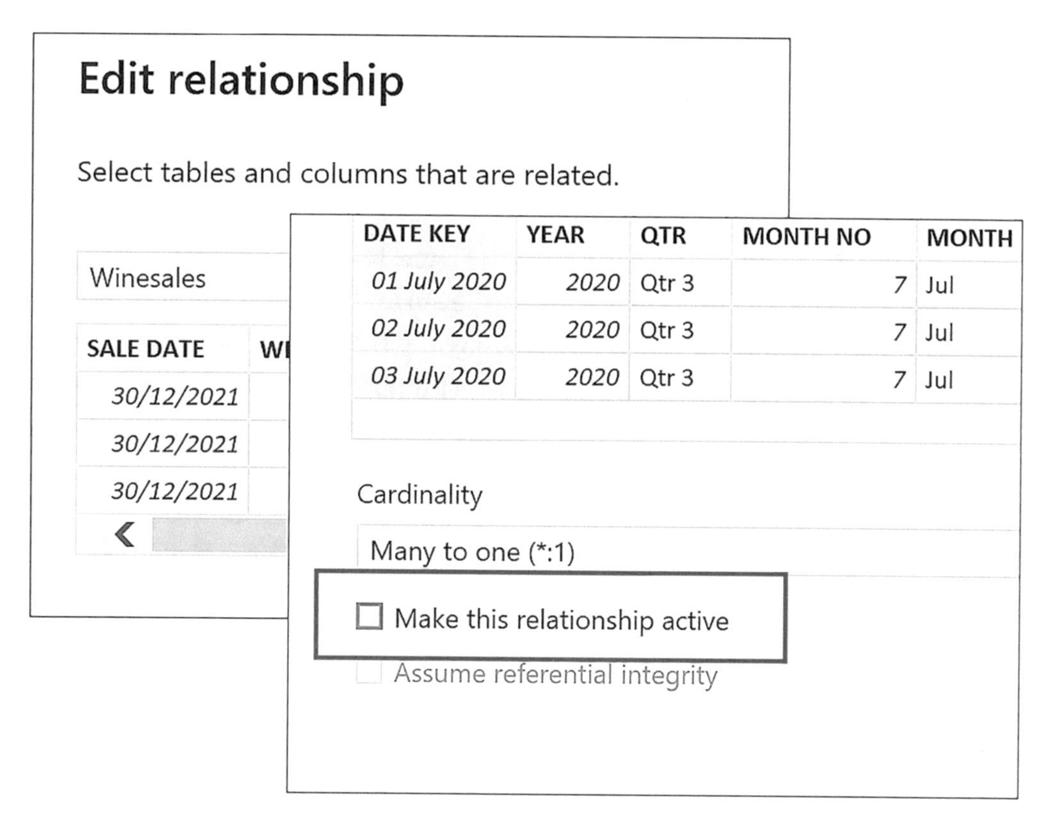

Figure 14-6. *Making a relationship inactive*

The next step is to create the two slicers as shown in Figure 14-4. The slicer on the left, named "YEAR", is created using the YEAR column from the DateTable. Selecting a year from this slicer filters the "Total Sales" measure. The slicer on the right, named "COMPARE YEAR", uses the YEAR column from the "DateTable Compare" table. Selecting a year from this slicer filters the "Compare Year" measure as follows:

```
Compare Year =
CALCULATE (
    [Total Sales],
    ALL ( DateTable ),
    USERELATIONSHIP ( Winesales[SALE DATE], 'DateTable Compare'[DATEKEY] )
)
```

Notice the use of the ALL function to remove the filter on the YEAR column of the DateTable coming through from the YEAR slicer that is used by the active relationship.

Using USERELATIONSHIP to make comparisons between your data is a simple strategy and doesn't require complex DAX, so let's take this idea a step further. Let's see if we can answer this question: Of the customers who bought wine X, who also bought wine Y? For example, of the customers who bought "Champagne", who also bought "Pinot Grigio"?

We've set out the solution to this question in Figure 14-7. The "No. of Sales" measure is being filtered by the WINE slicer on the left and shows the number of sales of "Champagne" for each customer. The "Compare Wine" measure is being filtered by the COMPARE WINE slicer on the right and shows the number of sales of "Pinot Grigio" for each customer. Finally, we've created a "Both Wines" measure that shows the customers who bought both wines, showing the combined number of sales for both wines.

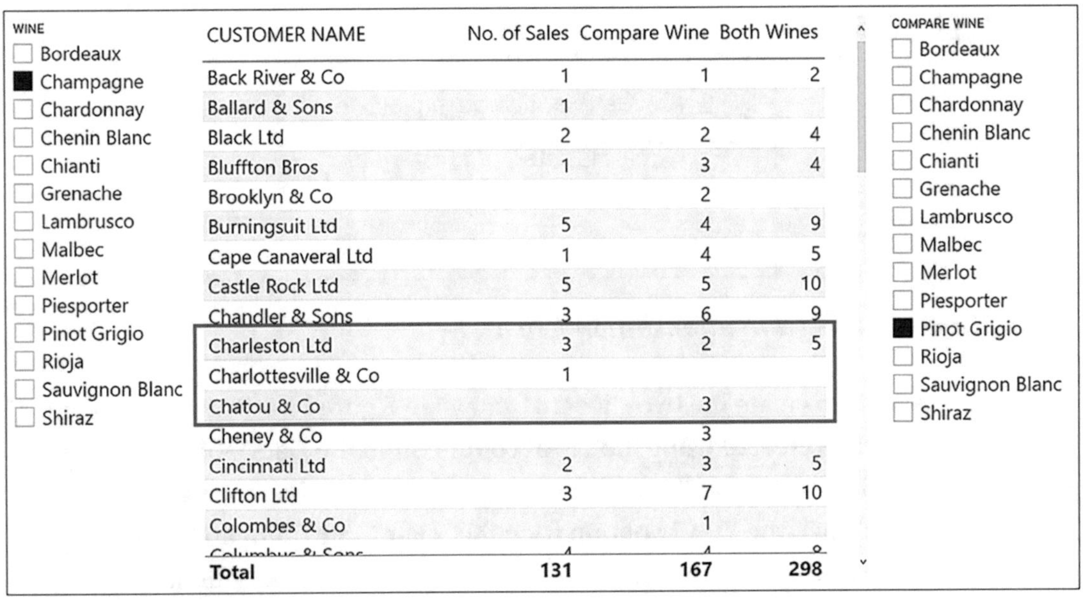

WINE

- ☐ Bordeaux
- ■ Champagne
- ☐ Chardonnay
- ☐ Chenin Blanc
- ☐ Chianti
- ☐ Grenache
- ☐ Lambrusco
- ☐ Malbec
- ☐ Merlot
- ☐ Piesporter
- ☐ Pinot Grigio
- ☐ Rioja
- ☐ Sauvignon Blanc
- ☐ Shiraz

CUSTOMER NAME	No. of Sales	Compare Wine	Both Wines
Back River & Co	1	1	2
Ballard & Sons	1		
Black Ltd	2	2	4
Bluffton Bros	1	3	4
Brooklyn & Co		2	
Burningsuit Ltd	5	4	9
Cape Canaveral Ltd	1	4	5
Castle Rock Ltd	5	5	10
Chandler & Sons	3	6	9
Charleston Ltd	3	2	5
Charlottesville & Co	1		
Chatou & Co		3	
Cheney & Co		3	
Cincinnati Ltd	2	3	5
Clifton Ltd	3	7	10
Colombes & Co		1	
Columbus & Sons	4	4	8
Total	**131**	**167**	**298**

COMPARE WINE

- ☐ Bordeaux
- ☐ Champagne
- ☐ Chardonnay
- ☐ Chenin Blanc
- ☐ Chianti
- ☐ Grenache
- ☐ Lambrusco
- ☐ Malbec
- ☐ Merlot
- ☐ Piesporter
- ■ Pinot Grigio
- ☐ Rioja
- ☐ Sauvignon Blanc
- ☐ Shiraz

Figure 14-7. *Customers who bought either wines or both wines*

You can see in Figure 14-7 that

- "Charleston Ltd" bought both "Champagne" and "Pinot Grigio".
- "Charlottesville & Co" bought "Champagne" but not "Pinot Grigio".
- "Chatou & Co" bought "Pinot Grigio" but not "Champagne".

If we put the "Both Wines" measure into a Table visual of its own, we see only customers who bought both wines (Figure 14-8).

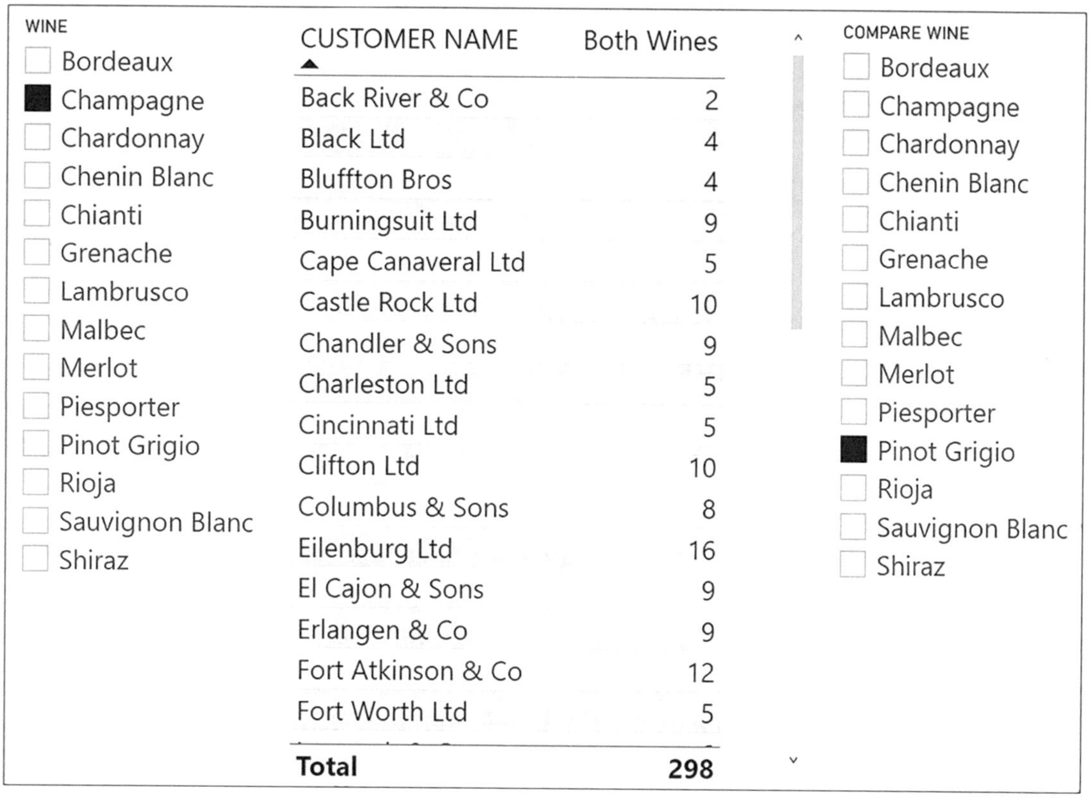

WINE
- Bordeaux
- Champagne (selected)
- Chardonnay
- Chenin Blanc
- Chianti
- Grenache
- Lambrusco
- Malbec
- Merlot
- Piesporter
- Pinot Grigio
- Rioja
- Sauvignon Blanc
- Shiraz

CUSTOMER NAME	Both Wines
Back River & Co	2
Black Ltd	4
Bluffton Bros	4
Burningsuit Ltd	9
Cape Canaveral Ltd	5
Castle Rock Ltd	10
Chandler & Sons	9
Charleston Ltd	5
Cincinnati Ltd	5
Clifton Ltd	10
Columbus & Sons	8
Eilenburg Ltd	16
El Cajon & Sons	9
Erlangen & Co	9
Fort Atkinson & Co	12
Fort Worth Ltd	5
Total	**298**

COMPARE WINE
- Bordeaux
- Champagne
- Chardonnay
- Chenin Blanc
- Chianti
- Grenache
- Lambrusco
- Malbec
- Merlot
- Piesporter
- Pinot Grigio (selected)
- Rioja
- Sauvignon Blanc
- Shiraz

***Figure 14-8.** Customers who bought both wines*

The expressions for the measures in Figure 14-7 are almost the same as those we used when we were comparing years in Figure 14-4. First, you need to duplicate the Wines dimension. We've called this duplicate table "Wines Compare" and then related this duplicate table to the fact table, remembering to set the relationship as "inactive."

These are the three measures we used in Figure 14-7:

```
No. of Sales =
COUNTROWS ( Winesales)

Compare Wine =
CALCULATE (
    [No. of Sales],
    ALL ( Wines ),
    USERELATIONSHIP ( Winesales[WINE ID], 'Wines Compare'[WINE ID] )
)
```

```
Both Wines =
IF (
SELECTEDVALUE ( Wines[WINE] ) = SELECTEDVALUE ( 'Wines Compare'[WINE] ),
[No. of Sales],
--If the same wine is selected in both slicers, don't add the number of
sales together
    IF (
        [No. of Sales] && [Compare Wine],
        [No. of Sales] + [Compare Wine]
    )
--If customers have sales for both wines, add the number of sales together
)
```

However, those of you that are observant may notice that the value in the Total row of the "Both Wines" measure in Figure 14-8 **(298)** is not correct. It totals *all rows* for the selected wines not just those rows for customers who have bought both wines. To calculate the correct total if "Champagne" and "Pinot Grigio" are selected **(258),** you can use SUMX (iterating the Customers table) and edit the "Both Wines" measure as follows:

```
Both Wines =
SUMX (
    Customers,
    IF (
        SELECTEDVALUE ( Wines[WINE] ) = SELECTEDVALUE ( 'Wines
        Compare'[WINE] ),
        [No. of Sales],
        IF ( [No. of Sales] && [Compare Wine], [No. of Sales] +
        [Compare Wine] )
    )
)
```

We hope you feel inspired by these examples to create comparisons in your own data by using the USERELATIONSHIP function. And of course, you now know how to activate inactive relationships.

CHAPTER 15

Understanding Context Transition

Nothing in life that's worth anything is easy.

–Barack Obama

You could also say that nothing in DAX that's worth anything is easy. Certainly, the concept of context transition is one of the more challenging theories to get to grips with in DAX. It can't be explained in a few short paragraphs, and therefore, we dedicate this entire chapter to teaching you the details of what context transition is and how it is used within DAX expressions. It's only then can you move forward in the following chapter to explore some practical applications of this concept. Once you understand the purpose of context transition in your code, a whole range of challenging calculations becomes possible. In fact, most DAX expressions you meet will probably be using context transition, and indeed, there will come a time when most DAX expressions you write will use it.[1]

To explain context transition in its simplest terms, it allows you to programmatically perform aggregations at the dimension, or group granularity, rather than the row granularity. For example, the expression *"AVERAGE (Winesales[CASES SOLD])"* calculates the average cases sold across transactions. This expression, using context transition, *"AVERAGEX (Wines, [Total Cases])"*, will calculate the average of the *aggregated values*, in this case, the average of the values returned by the "Total Cases" measure. Mostly, context transition happens in memory when an expression is being evaluated, and therefore, we can't see it happening.

[1] To follow along with the examples, use the Power BI Desktop file "1 DAX Sample Data.pbix".

© Alison Box 2022
A. Box, *Up and Running with DAX for Power BI*, https://doi.org/10.1007/978-1-4842-8188-8_15

Overview of DAX Evaluations Contexts

To understand context transition, you must have a firm handle on how DAX expressions are evaluated. You must clearly understand the difference between filter context and row context and be able to use these concepts correctly in your code. Therefore, our starting point in this chapter will be to remind ourselves of the difference between these two conditions in which our expressions are evaluated.

Row Context Revisited

When using the row context, a DAX expression iterates every row in a table. The values used in the expression are *the values sitting in the current row*, which may be different for every row. For example, the CASES SOLD value is mostly different for each row of the Winesales table, and so this calculated column

```
10 Percent of Cases Sold =
Winesales[CASES SOLD] * 0.1
```

will iterate all the rows in the Winesales table, finding a different value for CASES SOLD on each row and multiplying it by 0.1.

We can categorically state therefore that all calculated columns are evaluated in the row context. But measures will also use the row context if they *iterate a table*. For example, this measure (that we met when looking at the SUMX function in Chapter 5)

```
Total Sales =
SUMX ( Winesales,
    Winesales[CASES SOLD] * RELATED ( Wines[PRICE PER CASE] )
)
```

is evaluated first in a filter context, for example, filtered for “Bordeaux” wine, but then the SUMX function iterates the Winesales table and using the row context multiplies the CASES SOLD value sitting in each row by the PRICE PER CASE value from the Wines table. This will be the price of the wine in the current row of the Winesales table. The SUMX function then sums the result of all these row-level calculations, for example, for “Bordeaux” wine.

Therefore, what we can also state is that any DAX expression *that iterates a table*, whether in a calculated column *or* inside a measure, uses the row context.

Filter Context Revisited

All DAX measures are evaluated in a filter context. There are no exceptions to this rule. We understand that the filter context is typically generated from the current state of the Power BI report when the measure is evaluated, be it the structure of the visual, any slicers affecting the visual, or any filters in the Filters pane. But there is another way that the filter context can be generated, and this is what we're now going to investigate.

How Row Context Becomes Filter Context

There is a specific situation when a DAX expression is evaluated that will turn the row context into a filter context. This is what we know as *context transition*. To understand this specific situation, let's consider these five DAX expressions, two calculated columns and three measures (you don't need to know at this point what the expressions are calculating):

1. **Column 1 =**
   CALCULATE (SUM (Winesales[CASES SOLD))

2. **Column 2 =**
   [Total Cases]

3. **Measure 1 =**
   AVERAGEX (Wines, [Total Cases])

4. **Measure 2 =**
   AVERAGEX (Wines, CALCULATE (SUM (Winesales[CASES SOLD])))

5. **Measure 3 =**
   CALCULATE ([No. of Sales], FILTER (Winesales, [Total Cases] > 350))

Question: What is common to all these expressions?
The answer is that all five expressions share the same three attributes as follows:

1. They all use the *CALCULATE* function.
   But surely **Column 2** and **Measure 1** don't? At this point, there's something more we need to teach you regarding measures. *All measures implicitly invoke CALCULATE* even if they don't call the function explicitly. Therefore, **Column 2** and **Measure 1**, which both reference the measure "Total Cases", are both calling CALCULATE implicitly. The other expressions are using CALCULATE explicitly.

2. They all *iterate tables* creating a row context.
   Column 1 and **Column 2** are calculated columns, and all calculated columns iterate tables. We know that the functions AVERAGEX and FILTER are iterators too, so **Measure 1**, **Measure 2**, and **Measure 3** all iterate tables, creating a row context. **Measure 1** and **Measure 2** iterate the Wines table, and **Measure 3** iterates the Winesales table.

3. They all invoke *context transition*.
   This is where the row context, generated by an iteration of either a calculated column or inside a measure, is turned into a filter context.

Therefore, the specific situation to which we alluded is this: context transition occurs whenever

- The expression uses CALCULATE either explicitly or implicitly (because you're using a measure)

AND

- The expression (either in a column or in a measure) iterates a table using the row context

You now understand *when* context transition happens, but what exactly *is* "context transition"? To answer this question, let's first take this expression and use it in a calculated column:

```
Total Cases Column =
SUM ( Winesales[CASES SOLD] )
```

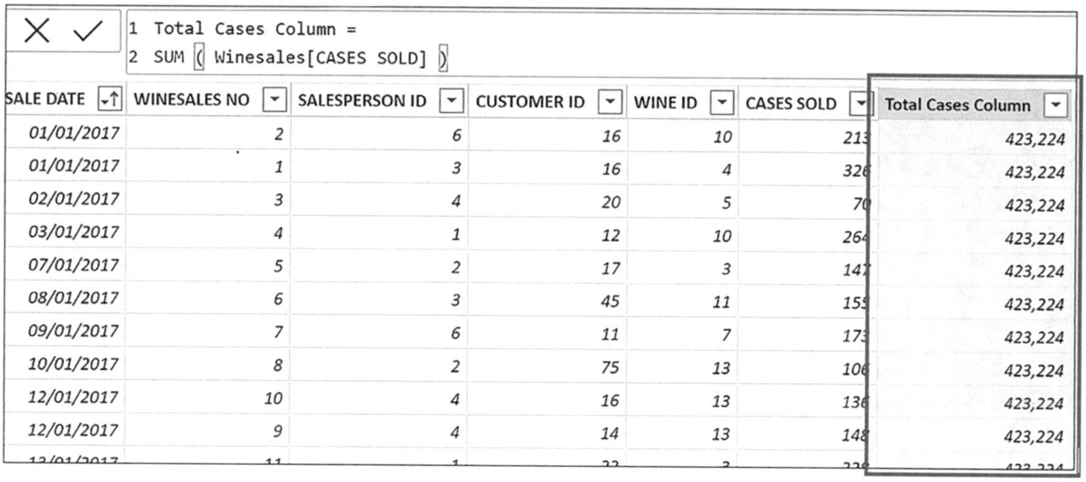

```
1 Total Cases Column =
2 SUM ( Winesales[CASES SOLD] )
```

SALE DATE	WINESALES NO	SALESPERSON ID	CUSTOMER ID	WINE ID	CASES SOLD	Total Cases Column
01/01/2017	2	6	16	10	21	423,224
01/01/2017	1	3	16	4	32	423,224
02/01/2017	3	4	20	5	7	423,224
03/01/2017	4	1	12	10	26	423,224
07/01/2017	5	2	17	3	14	423,224
08/01/2017	6	3	45	11	15	423,224
09/01/2017	7	6	11	7	17	423,224
10/01/2017	8	2	75	13	10	423,224
12/01/2017	10	4	16	13	13	423,224
12/01/2017	9	4	14	13	14	423,224

Figure 15-1. *The calculated column returns the grand total on every row*

You can see in Figure 15-1 that in every row, the expression returns the same value, the grand total of CASES SOLD. As a calculated column, the expression iterates the table using the row context, and therefore, there is no filter present. Aggregate functions such as SUM, by definition, require the rows to be aggregated to first be filtered. Because there is no filter on the table, this expression can only use the values from the entire table and so sums all the values for CASES SOLD.

We have just learned that context transition happens when there's an iteration, *and* we use CALCULATE. We can therefore now take our first look at context transition in action in a calculated column by editing our expression and wrapping CALCULATE around it:

```
Total Cases Column =
CALCULATE (
    SUM ( Winesales[CASES SOLD] )
)
```

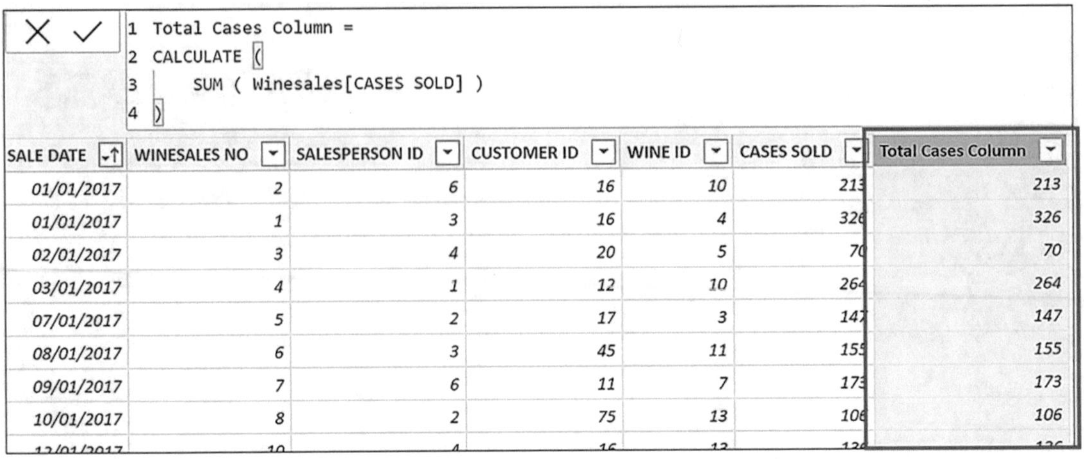

```
1 Total Cases Column =
2 CALCULATE (
3     SUM ( Winesales[CASES SOLD] )
4 )
```

SALE DATE	WINESALES NO	SALESPERSON ID	CUSTOMER ID	WINE ID	CASES SOLD	Total Cases Column
01/01/2017	2	6	16	10	213	213
01/01/2017	1	3	16	4	326	326
02/01/2017	3	4	20	5	70	70
03/01/2017	4	1	12	10	264	264
07/01/2017	5	2	17	3	147	147
08/01/2017	6	3	45	11	155	155
09/01/2017	7	6	11	7	173	173
10/01/2017	8	2	75	13	106	106

Figure 15-2. *Using CALCULATE evokes context transition in the calculated column*

Figure 15-2 shows that the result of this expression returns the CASES SOLD value of each row. What has happened here is the expression iterates the table generating a row context, as do all calculated columns. But we're also using CALCULATE in the iteration, and by doing so, the expression ignores the row context and replaces it with a filter context. Notice that although the expression uses CALCULATE, there are no filter arguments inside CALCULATE. Therefore, what is the filter being used by CALCULATE? The answer is rather a strange one (at least to new DAX users). A filter is placed on each value in each of the columns sitting in the current row. For example, in the first row of the table where the calculation returns **213**, the filter is this:

```
SALE DATE = 01/01/2017
WINESALES NO = 2
SALESPERSON ID = 6
CUSTOMER ID = 16
WINE ID = 10
CASES SOLD  = 213
```

The calculated column, "Total Cases Column", iterates the Winesales table, and because of the presence of CALCULATE, context transition occurs. All rows that share the same set of filters (as described before) are grouped and become filtered in their own right. The CASES SOLD values summed are the cases sold values sitting in each group. Because our rows are unique, each group comprises a single row, and therefore, the expression returns the same value as CASES SOLD. This is why, were you to have a duplicate first row in our example, you would see **426** (213 x 2) in "Total Cases Column" because the duplicate rows would be grouped before CASES SOLD was summed.[2] However, each of our rows is unique, so each filter generated by the context transition returns one row, which is the current row. This is an example of using CALCULATE in a calculated column where we have an iteration (and therefore a row context) and so CALCULATE evokes the context transition.

However, context transition also happens whenever you use a *measure* where there is a row context, for example, if you put a measure into a calculated column.

Note It is recommended that you are in Data view to create the calculated columns as described in the following.

This is because all measures call CALCULATE implicitly, and so context transition will also occur. For example, let's take this measure:

```
Total Cases =
SUM ( Winesales[CASES SOLD] )
```

Now let's edit our calculated column, "Total Cases Column", to perform the same calculation (i.e., summing the CASES SOLD column) but this time expressed as the "Total Cases" measure:

```
Total Cases Column =
[Total Cases]
```

[2] For information on removing duplicate rows, visit `www.excelnaccess.com/removing-duplicate-rows-in-power-bi/`

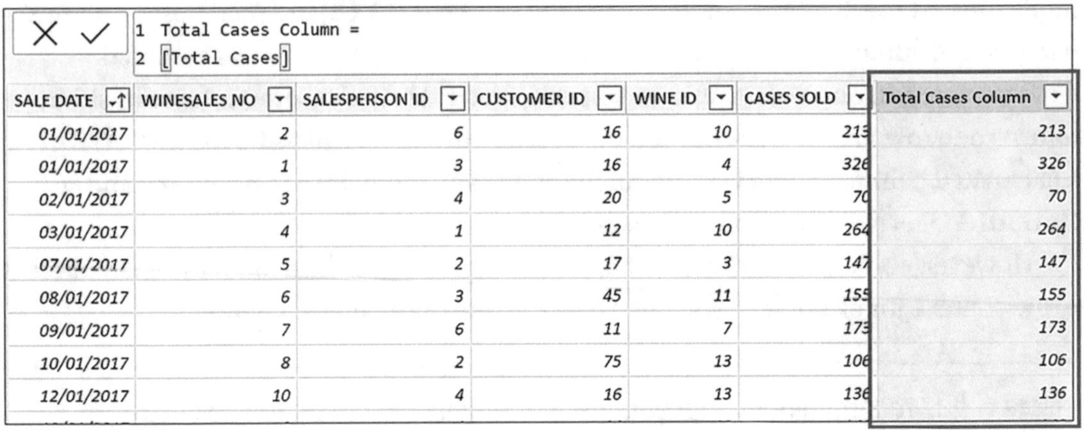

```
1 Total Cases Column =
2 [Total Cases]
```

SALE DATE	WINESALES NO	SALESPERSON ID	CUSTOMER ID	WINE ID	CASES SOLD	Total Cases Column
01/01/2017	2	6	16	10	213	213
01/01/2017	1	3	16	4	326	326
02/01/2017	3	4	20	5	70	70
03/01/2017	4	1	12	10	264	264
07/01/2017	5	2	17	3	147	147
08/01/2017	6	3	45	11	155	155
09/01/2017	7	6	11	7	173	173
10/01/2017	8	2	75	13	106	106
12/01/2017	10	4	16	13	136	136

Figure 15-3. *Using a measure in a calculated column evokes context transition*

You will notice in Figure 15-3 that the results of this expression are the same as when we used CALCULATE explicitly. Therefore, these two expressions

```
Total Cases Column =
CALCULATE (
    SUM ( Winesales[CASES SOLD] ) )
```

and

```
Total Cases Column =
[Total Cases]
```

are the *same* expressions.

At this stage in understanding context transition, I appreciate you're thinking: Why would I want to create a calculated column that returns the same value as the value sitting in the current row? Also, our Winesales table, being the fact table, could potentially contain millions of rows, so any context transition occurring in a calculated column would be very slow. In short, what is the purpose of context transition?

To answer this question, let's see how context transition performs when invoked in dimension tables, rather than in the fact table. Let's now repeat the same expressions we've been working with, but rather than placing them in the fact table, this time we will put them in the Wines dimension.

These are the calculated columns that we can create in the Wines dimension:

```
Wine Total Cases 1=
SUM ( Winesales[CASES SOLD] )

Wine Total Cases 2 =
CALCULATE (
    SUM ( Winesales[CASES SOLD] ) )

Wine Total Cases 3 =
[Total Cases]
```

Observing the behavior of these calculated columns in Figure 15-4, let's look more closely at the evaluation of each of these expressions.

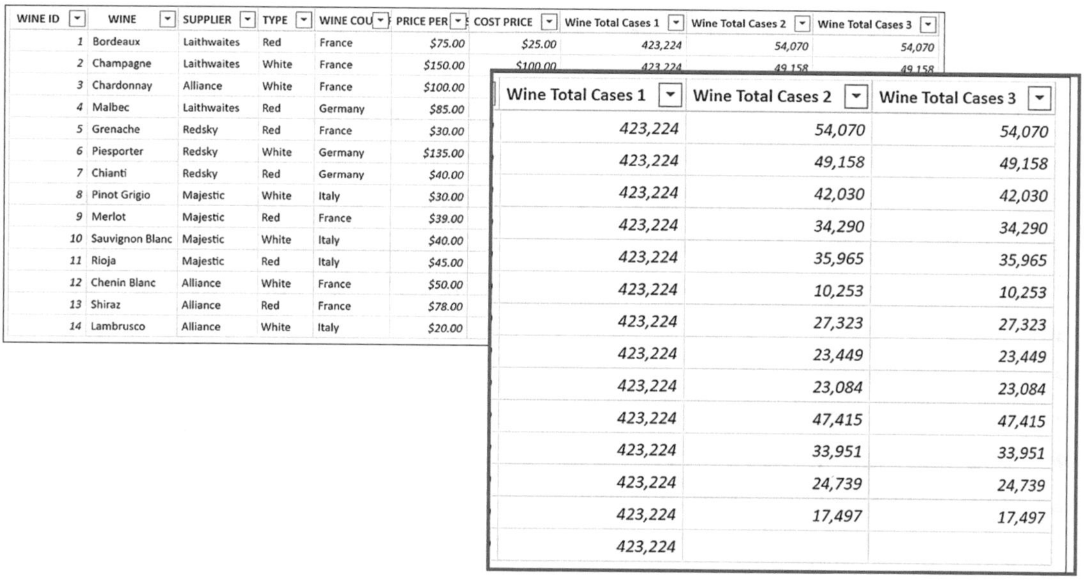

WINE ID	WINE	SUPPLIER	TYPE	WINE COU	PRICE PER	COST PRICE	Wine Total Cases 1	Wine Total Cases 2	Wine Total Cases 3
1	Bordeaux	Laithwaites	Red	France	$75.00	$25.00	423,224	54,070	54,070
2	Champagne	Laithwaites	White	France	$150.00	$100.00	423,224	49,158	49,158
3	Chardonnay	Alliance	White	France	$100.00				
4	Malbec	Laithwaites	Red	Germany	$85.00				
5	Grenache	Redsky	Red	France	$30.00				
6	Piesporter	Redsky	White	Germany	$135.00				
7	Chianti	Redsky	Red	Germany	$40.00				
8	Pinot Grigio	Majestic	White	Italy	$30.00				
9	Merlot	Majestic	Red	France	$39.00				
10	Sauvignon Blanc	Majestic	White	Italy	$40.00				
11	Rioja	Majestic	Red	Italy	$45.00				
12	Chenin Blanc	Alliance	White	France	$50.00				
13	Shiraz	Alliance	Red	France	$78.00				
14	Lambrusco	Alliance	White	Italy	$20.00				

Wine Total Cases 1	Wine Total Cases 2	Wine Total Cases 3
423,224	54,070	54,070
423,224	49,158	49,158
423,224	42,030	42,030
423,224	34,290	34,290
423,224	35,965	35,965
423,224	10,253	10,253
423,224	27,323	27,323
423,224	23,449	23,449
423,224	23,084	23,084
423,224	47,415	47,415
423,224	33,951	33,951
423,224	24,739	24,739
423,224	17,497	17,497
423,224		

***Figure 15-4.** The three calculated columns in the Wines dimension*

The first of these calculated columns, "Wine Total Cases 1", uses the *"SUM (Winesales[CASES SOLD])"* expression. There is no measure in this expression, and it's not using CALCULATE, either implicitly or explicitly. The expression uses the SUM function that requires a filter context. In the absence of any filter, it sums the CASES SOLD values in all the rows of the Winesales table giving us the grand total of CASES SOLD.

The second calculated column, "Wine Total Cases 2", is using CALCULATE that converts the row context invoked by the iteration of the calculated column into a filter context. At this point, we need to remind ourselves that the filter context always propagates through the entire data model. The filter coming through from context transition behaves no differently from a filter coming through from a visual or a slicer on the report canvas. When the expression in the calculated column, "Wine Total Cases 2", evaluates the first row of the Wines dimension, it turns the entire row into a filter and filters "Bordeaux" wine. We could imagine that in memory on the evaluation of the first row, our Wines dimension looks something like the table in Figure 15-5.

WINE ID	WINE	SUPPLIER	TYPE	WINE COUNTRY	PRICE PER CASE	COST PRICE
1	Bordeaux	Laithwaites	Red	France	$75.00	$25.00

Figure 15-5. *The Wines dimension is filtered by context transition*

Does Figure 15-5 look familiar? The filter on the Wines dimension for "Bordeaux" is the same filter that would be applied if we had used a slicer or any other means by which we could filter "Bordeaux" in the report. We know that because the Wines dimension is related to the Winesales fact table in a many-to-one relationship, this filter, *generated by context transition*, is propagated onward to the Winesales table. Therefore, our Winesales table is now cross-filtered to contain only "Bordeaux" wines, and the CASES SOLD values are summed accordingly.

What we can conclude, therefore, is that a calculated column that uses CALCULATE where context transition occurs behaves just like a measure in a visual on the report canvas, in that it filters and then aggregates.

Looking at the third calculated column, "Wine Total Cases 3", here, we are using the "Total Cases" measure that defines the same expression as in "Wine Total Cases 2". Because all measures implicitly call CALCULATE, "Wine Total Cases 2" and "Wine Total Cases 3" are the same expressions. Whenever you see a measure, even if it doesn't use CALCULATE explicitly, you should always imagine that it's wrapped inside CALCULATE.

To summarize the outputs of the three calculated columns, "Wine Total Cases 2" and "Wine Total Cases 3" both use context transition in their evaluation, but "Wine Total Cases 1" does not.

How Context Transition Can Return "Surprising Results"

In our investigation of context transition, we've been using calculated columns to see context transition in action. However, we don't need to see context transition to understand that it happens, and besides which, you're probably not going to be creating these types of calculated columns in reality.

Mostly, context transition happens behind the scenes, in memory, when you construct iterating measures *that reference another measure* (because all measures implicitly call CALCULATE).

Let's, at this point, remind ourselves of the specific situation where context transition occurs:

- When the expression uses CALCULATE either explicitly or implicitly via a measure

AND

- When the expression iterates a table using the row context

Typically, this is when we nest measures inside the iterating "X" aggregate functions like AVERAGEX or MAXX or we use measures inside the FILTER function. Because we can't see context transition happening, being oblivious of its existence means we'll struggle to understand how DAX works. Marco Russo and Alberto Ferrari in their *The Definitive Guide to DAX* explain understanding context transition as follows:

"Being ignorant of certain behaviors can ensure surprising results. Nevertheless, once you master the behavior, you start leveraging it as you see fit. The only difference between a strange behavior and a useful feature - at least in DAX - is your level of knowledge."[3]

Marco and Alberto talk about "strange behaviors" and "surprising results." The only reason these behaviors would seem strange or surprising to you is that you don't understand the behavior of context transition, the fact that in the evaluation of measures, there's a world of difference between iterations referencing *measures* that call CALCULATE and iterations referencing *expressions* that do not. To illustrate this, we're going to take a look at authoring expressions where getting it right, which is whether

[3] Marco Russo and Alberto Ferrari (2020), *The Definitive Guide to DAX*, 2nd ed, p.154 [Microsoft Press]

you nest a measure or whether you nest an expression, is key. In these examples, we're going to see how DAX expressions can return "surprising results" unless, of course, you understand the behavior of context transition.

Filters Using AVERAGE

In the first example, we must reference an *expression* in our measure to get the correct calculation; nesting the measure that defines the same expression won't work.

Consider the calculation to find the number of sales for each wine where cases sold is greater than the average cases sold for that wine. For example, the average number of cases sold for "Bordeaux" is **300** and we want to calculate how many sales of "Bordeaux" have cases sold greater than this value (this is purely an intellectual exercise and not a particularly useful calculation).

We've already created these two measures:

```
Avg Cases =
AVERAGE ( Winesales[CASES SOLD] )

No. of Sales =
COUNTROWS ( Winesales )
```

Now to calculate the number of sales where the CASES SOLD value is greater than the average cases, we could author this measure:

```
No. Of Sales GT Avg #1=
VAR AvgCasesTable =
    FILTER ( Winesales, Winesales[CASES SOLD] > [Avg Cases] )
RETURN
    CALCULATE ( [No. Of Sales], AvgCasesTable )
```

Note the use of the "Avg Cases" *measure* (highlighted) nested in the FILTER expression that iterates the Winesales table. We know that in the presence of a nested measure inside an iteration, context transition is invoked.

Unfortunately, the "No. Of Sales GT Avg #1" measure does not return the correct results; it returns blanks. This is a surprising result, I think you'll agree; see Figure 15-6.

WINE	Avg Cases	No. Of Sales GT Avg #1
Bordeaux	300	
Champagne	372	
Chardonnay	225	
Chenin Blanc	124	
Chianti	185	
Grenache	198	
Malbec	202	
Merlot	147	
Piesporter	89	
Pinot Grigio	140	
Rioja	172	
Sauvignon Blanc	282	
Shiraz	86	
Total	**192**	

Figure 15-6. *The "No. Of Sales GT Avg #1" does not return a value*

Clearly, we must take a closer look at what's happening here. This measure, "No. Of Sales GT Avg #1", uses the FILTER function that iterates the Winesales table to filter rows where the CASES SOLD value is greater than the value calculated by the "Avg Cases" measure. But what is the value calculated by "Avg Cases"? If we put this measure, that is, *"[Avg Cases]"*, into the Winesales table as a calculated column, we can see what the FILTER function is testing the CASES SOLD value against; see Figure 15-7.

```
1 Average Cases Column = [Avg Cases]
```

SALE DATE	WINESALES NO	SALESPERSON ID	CUSTOMER ID	WINE ID	CASES SOLD	Average Cases Column
23/12/2021	2207	3	12	1	290	290
14/12/2021	2184	1	34	1	190	190
13/12/2021	2181	4	3	1	330	330
06/12/2021	2169	5	11	1	188	188
20/11/2021	2145	4	44	1	149	149
20/11/2021	2195	4	2	1	473	473
14/11/2021	2134	3	37	1	329	329
15/10/2021	2083	3	16	1	197	197
07/10/2021	2065	5	39	1	451	451
05/10/2021	2060	3	18	1	304	304
21/09/2021	2037	3	25	1	240	240
19/09/2021	2033	1	17	1	382	382

Figure 15-7. *The "Avg Cases" measure evaluated in a calculated column filters the Winesales table to the single row that's being evaluated*

Note Remember that in the first evaluation in the Table visual in Figure 15-6, the Winesales table will be cross-filtered in memory for "Bordeaux" wine, which is WINE ID 1.

What we find is that the values returned by "Avg Cases" are the same as the CASES SOLD values. This is because "Avg Cases" is a measure, and therefore, it evokes context transition as FILTER iterates the Winesales table. This creates a filter on each row of the Winesales table that's being evaluated in memory. Because each row is unique, it calculates the average of the CASES SOLD value only for the current row, which is the same as the CASES SOLD value. Therefore, the "Avg Cases" value is never greater than the CASES SOLD value. You could test this by changing ">" to ">=" where instead of blanks being returned, you would get the same values as "No. of Sales".

Let's now replace the measure inside the FILTER function with an expression (highlighted) that calculates the average:

```
No. Of Sales GT Avg #2 =
VAR AvgCasesTable =
    FILTER ( Winesales, Winesales[CASES SOLD] >
```

```
                        AVERAGE ( Winesales[CASES SOLD] ) )
RETURN
    CALCULATE ( [No. Of Sales], AvgCasesTable )
```

This time we get the correct results; see Figure 15-8.

WINE	Avg Cases	No. Of Sales GT Avg #2
Bordeaux	300	89
Champagne	372	68
Chardonnay	225	107
Chenin Blanc	124	100
Chianti	185	73
Grenache	198	87
Malbec	202	83
Merlot	147	81
Piesporter	89	50
Pinot Grigio	140	86
Rioja	172	106
Sauvignon Blanc	282	88
Shiraz	86	103
Total	**192**	**932**

Figure 15-8. *Using a nested expression inside FILTER and not a nested measure returns the correct results*

To understand why the second version of the measure using the expression works, we can again put the expression, *"AVERAGE (Winesales[CASES SOLD])"*, into a calculated column in the Winesales table, filtered for "Bordeaux" wines (as this is the in-memory cross-filter on the Winesales table in the first evaluation).

We can see in Figure 15-9 that the expression returns the average of the cases sold for the wine in the current filter context.

Note In Figure 15-9, we are simulating the rows in the Winesales table that would be visible in the current filter *in memory*, which you will not be able to see in Data view. Therefore, when you put *"AVERAGE (Winesales[CASES SOLD])"* into a calculated column, you will see the average for all transactions (**192**), not just those for "Bordeaux" **(300)**.

```
1 Avg Cases Column = AVERAGE ( Winesales[CASES SOLD] )
```

SALE DATE	WINESALES NO	SALESPERSON ID	CUSTOMER ID	WINE ID	CASES SOLD	Avg Cases Column
18/01/2017	18	1	12	1	327	300
19/01/2017	22	1	19	1	386	300
27/01/2017	29	1	26	1	401	300
07/02/2017	43	1	30	1	266	300
19/02/2017	58	4	2	1	168	300
08/04/2017	94	1	19	1	284	300
11/04/2017	98	5	11	1	347	300
26/06/2017	156	2	18	1	394	300
13/08/2017	187	1	36	1	297	300
03/10/2017	233	6	10	1	376	300
04/10/2017	235	1	26	1	232	300
28/10/2017	253	6	44	1	232	300

***Figure 15-9.** The "AVERAGE (Winesales[CASESSOLD])" expression evaluated in a calculated column returns the average cases for each wine*

We know that the average number of cases sold for "Bordeaux" is **300**. So in the first evaluation for "Bordeaux", there are **89** transactions where cases sold is greater than **300**.

We can understand therefore that despite the fact that the expression and the measure both calculate the same average, we must nest the average *expression* inside the measure being evaluated, not nest the *measure* that calculates the average.

Filters Using MAX

In our second example of how DAX can return "surprising results," we will calculate cumulative totals, as shown in Figure 15-10.

YEAR	MONTH	Total Sales	Cumulative Total
2017	Jan	$451,887	$451,887
2017	Feb	$385,299	$837,186
2017	Mar	$400,977	$1,238,163
2017	Apr	$327,070	$1,565,233
2017	May	$353,073	$1,918,306
2017	Jun	$241,419	$2,159,725
2017	Jul	$410,507	$2,570,232
2017	Aug	$194,755	$2,764,987
2017	Sep	$559,821	$3,324,808
2017	Oct	$438,513	$3,763,321
2017	Nov	$301,695	$4,065,016
2017	Dec	$584,269	$4,649,285
2018	Jan	$407,812	$5,057,097
2018	Feb	$299,495	$5,356,592
2018	Mar	$232,473	$5,589,065
Total		**$29,732,482**	**$29,732,482**

Figure 15-10. *Calculating cumulative totals in the "Cumulative Total" measure*

To generate the "Cumulative Total" measure, we must again use a nested expression in the parent measure, not a nested measure, and this time we'll be using the aggregate function, MAX.

Note We've calculated cumulative totals before using the time intelligence function, DATESBETWEEN. However, in this section, we explore an alternative method of achieving the same result.

To calculate cumulative totals in the Table visual in Figure 15-10, for any given date in the current filter context, we must sum a value (in our example, the total sales value) up to the latest date in the current filter context. For example, if "May 2017" is the current filter, we must sum values up to and including **31 May 2017**. We might think, therefore, that we need to first construct a measure that finds the latest date in the current filter context using the MAX function like so:

```
Max Date =
```

```
MAX ( DateTable[DATEKEY] )
```

We could then use this "Max Date" measure (highlighted) in the following expression (note the use of ALL to remove the filter on the DateTable that is currently filtering each month):

```
Cumulative Total Wrong =
VAR FilteredDatesTable =
    FILTER (
        ALL ( DateTable ),
        DateTable[DATEKEY] <= [Max Date]
    )
RETURN
    CALCULATE ( [Total Sales], FilteredDatesTable )
```

Looking at the result of this expression in the visual in Figure 15-11, clearly, this hasn't worked.

YEAR	MONTH	Total Sales	Cumulative Total Wrong
2017	Jan	$451,887	$29,732,482
2017	Feb	$385,299	$29,732,482
2017	Mar	$400,977	$29,732,482
2017	Apr	$327,070	$29,732,482
2017	May	$353,073	$29,732,482
2017	Jun	$241,419	$29,732,482
2017	Jul	$410,507	$29,732,482
2017	Aug	$194,755	$29,732,482
2017	Sep	$559,821	$29,732,482
2017	Oct	$438,513	$29,732,482
2017	Nov	$301,695	$29,732,482
2017	Dec	$584,269	$29,732,482
2018	Jan	$407,812	$29,732,482
2018	Feb	$299,495	$29,732,482
Total		**$29,732,482**	**$29,732,482**

Figure 15-11. *The "Cumulative Total Wrong" measure returns incorrect results*

Let's take a look at what's going wrong with "Cumulative Total Wrong". Here, we're using the measure "Max Date", which defines the maximum date. The FILTER function with ALL generates a virtual DateTable containing all the rows in the DateTable. It then iterates this virtual table to compare each date in the DATEKEY column to the date calculated by "Max Date". What is the value of "Max Date"? The "Max Date" measure evokes context transition and so filters each row to a single row. Therefore, when iterating the DateTable, it will always return the same date that is sitting in the current row of the DateTable. To understand this, we can put the "Max Date" measure into a calculated column in the DateTable as shown in Figure 15-12.

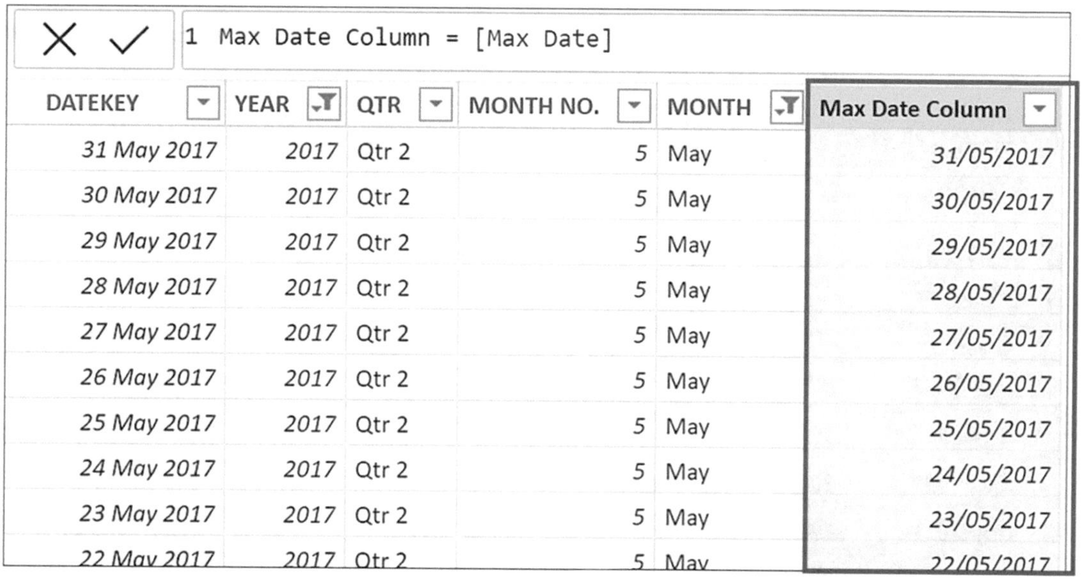

```
1 Max Date Column = [Max Date]
```

DATEKEY	YEAR	QTR	MONTH NO.	MONTH	Max Date Column
31 May 2017	2017	Qtr 2	5	May	31/05/2017
30 May 2017	2017	Qtr 2	5	May	30/05/2017
29 May 2017	2017	Qtr 2	5	May	29/05/2017
28 May 2017	2017	Qtr 2	5	May	28/05/2017
27 May 2017	2017	Qtr 2	5	May	27/05/2017
26 May 2017	2017	Qtr 2	5	May	26/05/2017
25 May 2017	2017	Qtr 2	5	May	25/05/2017
24 May 2017	2017	Qtr 2	5	May	24/05/2017
23 May 2017	2017	Qtr 2	5	May	23/05/2017
22 May 2017	2017	Qtr 2	5	May	22/05/2017

Figure 15-12. *The "Max Date" measure evaluated in a calculated column filters the DateTable to the single row that's being evaluated*

Because DATEKEY is always *equal to* "Max Date", all the dates are filtered by the FILTER function, and so CALCULATE calculates the total cases for all dates (to see this in another way, try replacing the "<=" with "<" where you will now get blanks returned).

Therefore, to remedy this, we must use an expression (highlighted) to calculate the latest date in the current filter context, and this is the correct measure:

```
Cumulative Total =
VAR FilteredDatesTable =
    FILTER (
        ALL ( DateTable ),
```

```
        DateTable[DATEKEY] <= MAX ( DateTable[DATEKEY] )
    )
RETURN
    CALCULATE ( [Total Sales], FilteredDatesTable )
```

In this measure, the FILTER function with ALL iterates the virtual DateTable table to compare each date in the DATEKEY column to the date calculated by the expression *"MAX (DateTable[DATEKEY]".* This expression will find the latest date in the current filter context; for example, it will return **31 May 2017** when evaluating "May 2017"; see Figure 15-13.

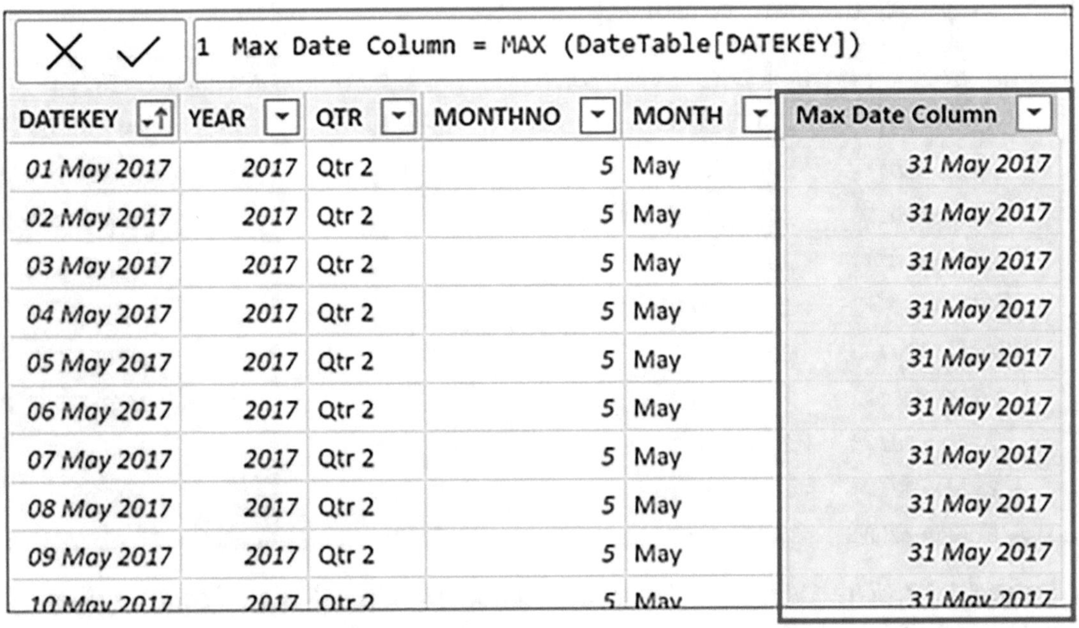

```
1 Max Date Column = MAX (DateTable[DATEKEY])
```

DATEKEY	YEAR	QTR	MONTHNO	MONTH	Max Date Column
01 May 2017	2017	Qtr 2	5	May	31 May 2017
02 May 2017	2017	Qtr 2	5	May	31 May 2017
03 May 2017	2017	Qtr 2	5	May	31 May 2017
04 May 2017	2017	Qtr 2	5	May	31 May 2017
05 May 2017	2017	Qtr 2	5	May	31 May 2017
06 May 2017	2017	Qtr 2	5	May	31 May 2017
07 May 2017	2017	Qtr 2	5	May	31 May 2017
08 May 2017	2017	Qtr 2	5	May	31 May 2017
09 May 2017	2017	Qtr 2	5	May	31 May 2017
10 May 2017	2017	Qtr 2	5	May	31 May 2017

Figure 15-13. *The MAX expression evaluated in a calculated column returns the maximum date for the month in the current filter*

Note In Figure 15-13, we're again simulating the rows in the DateTable that would be visible in memory in the current filter. You cannot see in-memory filters in Data view. Therefore, if you put the *"MAX (DateTable[DATEKEY]"* expression into a calculated column in the DateTable, you will see the last date for *all dates*, that is, **31 December 2021**.

The FILTER function will compare every date in the DATEKEY column of the virtual DateTable to the date calculated by *"MAX (DateTable[DATEKEY)"* and therefore will filter all the dates that are before or equal to this date.

Filters Using Measures

In the last of our "surprising results" examples, we must use a measure and not an expression. For example, it could transpire that you want to calculate the number of transactions where the "Total Sales" value for each transaction is greater than $10,000. To remind you, this is the expression that is used in the "Total Sales" measure:

```
Total Sales =
SUMX ( Winesales, Winesales[CASES SOLD] *
                  RELATED ( Wines[PRICE PER CASE] )
```

You may be tempted to use this expression (highlighted) to calculate the number of sales that are greater than $10,000:

```
No. Of Sales GT 10,000 #1=
VAR MySales =
    SUMX ( Winesales, Winesales[CASES SOLD] *
                  RELATED ( Wines[PRICE PER CASE] ) )
VAR SalesTable =
    FILTER ( Winesales, MySales > 10000 )
RETURN
    CALCULATE ( [No. Of Sales], SalesTable )
```

However, as you can appreciate from Figure 15-14, this measure returns the number of sales, not the number greater than $10,000.

WINE	No. of Sales	No. Of Sales GT 10,000 #1
Bordeaux	180	180
Champagne	132	132
Chardonnay	187	187
Chenin Blanc	200	200
Chianti	148	148
Grenache	182	182
Malbec	170	170
Merlot	157	157
Piesporter	115	115
Pinot Grigio	168	168
Rioja	197	197
Sauvignon Blanc	168	168
Shiraz	203	203
Total	**2,207**	**2,207**

Figure 15-14. *The "No. Of Sales GT 10,000" measure does not return the correct result*

In the "No. Of Sales GT $10,000 #1" measure, we are using SUMX to calculate the total sales. However, the SUMX expression would sum the total sales for all transactions in the Winesales table for each wine in the current filter context, giving us the grand total of "Total Sales" for each wine. In Figure 15-15, the SUMX expression has been placed into a calculated column in the Winesales table to understand its return value in memory (only showing total sales for "Bordeaux"), which is the total sales for the wine, not the total sales for each transaction. When FILTER iterates the Winesales table, this value will always be greater than $10,000.

```
1 Total Sales Column = SUMX ( Winesales, Winesales[CASES SOLD] *
2                  RELATED ( Wines[PRICE PER CASE] ) )
```

SALE DATE	WINESALES NO	SALESPERSON ID	CUSTOMER ID	WINE ID	CASES SOLD	Total Sales Column
18/01/2017	18	1	12	1	327	$4,055,250
19/01/2017	22	1	19	1	386	$4,055,250
27/01/2017	29	1	26	1	401	$4,055,250
07/02/2017	43	1	30	1	266	$4,055,250
19/02/2017	58	4	2	1	168	$4,055,250
08/04/2017	94	1	19	1	284	$4,055,250
11/04/2017	98	5	11	1	347	$4,055,250
26/06/2017	156	2	18	1	394	$4,055,250
13/08/2017	187	1	36	1	297	$4,055,250
03/10/2017	233	6	10	1	376	$4,055,250
04/10/2017	235	1	26	1	232	$4,055,250
28/10/2017	253	6	44	1	232	$4,055,250
05/11/2017	260	1	16	1	439	$4,055,250

***Figure 15-15.** The SUMX expression evaluated in a calculated column returns the grand total sales for the wine in the current filter*

When we write the expression to find the number of sales greater than $10,000, we must therefore use the "Total Sales" *measure* (highlighted) inside FILTER as follows:

```
No. Of Sales GT 10,000 #2 =
VAR MyTable =
    FILTER ( Winesales, [Total Sales] > 10000 )
RETURN
    CALCULATE ( [No. Of Sales], MyTable )
```

As you can now see in Figure 15-16, the "No. Of Sales GT 10,000 #2" measure using the nested measure "Total Sales" returns the correct value.

WINE	No. of Sales	No. Of Sales GT 10,000 #2
Bordeaux	180	170
Champagne	132	131
Chardonnay	187	186
Chenin Blanc	200	
Chianti	148	35
Grenache	182	15
Malbec	170	151
Merlot	157	
Piesporter	115	90
Pinot Grigio	168	
Rioja	197	
Sauvignon Blanc	168	122
Shiraz	203	37
Total	**2,207**	**937**

Figure 15-16. *Using a nested measure inside FILTER and not an expression returns the correct results for the number of sales greater than $10,000*

We must again investigate the reason why our second attempt at this calculation using the nested measure works. If we put the "Total Sales" measure into a calculated column, this will reveal what FILTER returns when it iterates the Winesales table in memory. We can see that it is the total sales for each transaction because it's using context transition to filter each row in memory; see Figure 15-17.

```
1 Total Sales Column = [Total Sales]
```

SALE DATE	WINESALES NO	SALESPERSON ID	CUSTOMER ID	WINE ID	CASES SOLD	Total Sales Column
18/01/2017	18	1	12	1	327	$24,525
19/01/2017	22	1	19	1	386	$28,950
27/01/2017	29	1	26	1	401	$30,075
07/02/2017	43	1	30	1	266	$19,950
19/02/2017	58	4	2	1	168	$12,600
08/04/2017	94	1	19	1	284	$21,300
11/04/2017	98	5	11	1	347	$26,025
26/06/2017	156	2	18	1	394	$29,550
13/08/2017	187	1	36	1	297	$22,275
03/10/2017	233	6	10	1	376	$28,200
04/10/2017	235	1	26	1	232	$17,400
28/10/2017	253	6	44	1	232	$17,400
05/11/2017	260	1	16	1	439	$32,925

***Figure 15-17.** The "Total Cases" measure evaluated in a calculated column filters the Winesales table to the single row that's being evaluated*

When FILTER iterates the Winesales table, it can use this value to find values greater than $10,000.

I think we've made our point regarding the "surprising results" to which Marco Russo and Alberto Ferrari alluded, and in doing so, you now understand the concept of context transition. This is where the row context is transitioned into a filter context because of the presence of CALCULATE within an iteration, and these filters propagate through the data model, just as all filters do. No matter how long you've been using DAX, these are challenging calculations to get your head around. In what follows in this chapter, we'll explore why it's so important that you take up the challenge to understand the strange behaviors and surprising results that context transition throws at you because in doing so, you will begin to reap the real benefits of using DAX.

Aggregating Totals Using Context Transition

The power of context transition comes when you use it to calculate averages, maximums, and minimums of totals as opposed to row-level values, and in this section, we will be exploring why this is mandatory knowledge in advanced calculations. This is also where the importance of having clearly defined dimension tables comes to the fore because to achieve this type of calculation, we will be passing context transition across dimension tables both real and virtual. What you will discover in the following section is that context transition can just as equally be passed into virtual tables generated by table expressions as it can be passed into actual dimensions within the data model.

Aggregating in Dimensions

We will begin by exploring how context transition works when used in expressions that reference dimension tables.

For example, take this simple measure:

```
Max Cases =
MAX ( Winesales[CASES SOLD] )
```

The "Max Cases" measure can tell us the maximum number of cases in any single transaction in the Winesales table for each wine. For example, for "Bordeaux", the maximum number of cases sold in any single transaction is **500** cases. This is a row-level calculation, but this is not what we want.

This measure, we know, will sum the cases sold values in the Winesales table:

```
Total Cases =
SUM ( Winesales[CASES SOLD] )
```

Our goal here is to calculate the maximum of this "Total Cases" measure, not the maximum of the individual transactions, as shown in Figure 15-18.

WINE	Max Cases	Total Cases
Bordeaux	500	54,070
Champagne	500	49,158
Chardonnay	250	42,030
Chenin Blanc	150	24,739
Chianti	299	27,323
Grenache	350	35,965
Malbec	326	34,290
Merlot	200	23,084
Piesporter	162	10,253
Pinot Grigio	200	23,449
Rioja	200	33,951
Sauvignon Blanc	350	47,415
Shiraz	150	17,497
Total	**500**	**423,224**

We want to calculate the maximum of the Total Cases

Figure 15-18. *The "Max Cases" measure is a row-level calculation, but we want to calculate across totals*

To do this, we need to use context transition. We know that context transition happens when a measure is iterated over a table. We looked earlier at creating a calculated column in the Wines dimension ("Wine Total Cases 3") that used the measure "Total Cases" to evoke context transition and so found the total cases sold value for each wine in the Wines dimension; see Figure 15-19.

```
1 Wine Total Cases 3 =
2 [Total Cases]
```

WINE ID	WINE	SUPPLIER	TYPE	WINE COUNTRY	PRICE PER CASE	COST PRICE	Wine Total Cases 3
1	Bordeaux	Laithwaites	Red	France	$75.00	$25.00	54,070
2	Champagne	Laithwaites	White	France	$150.00	$100.00	49,158
3	Chardonnay	Alliance	White	France	$100.00	$75.00	42,030
4	Malbec	Laithwaites	Red	Germany	$85.00	$40.00	34,290
5	Grenache	Redsky	Red	France	$30.00	$10.00	35,965
6	Piesporter	Redsky	White	Germany	$135.00	$50.00	10,253
7	Chianti	Redsky	Red	Germany	$40.00	$10.00	27,323
8	Pinot Grigio	Majestic	White	Italy	$30.00	$5.00	23,449
9	Merlot	Majestic	Red	France	$39.00	$15.00	23,084
10	Sauvignon Blanc	Majestic	White	Italy	$40.00	$20.00	47,415
11	Rioja	Majestic	Red	Italy	$45.00	$15.00	33,951
12	Chenin Blanc	Alliance	White	France	$50.00	$10.00	24,739
13	Shiraz	Alliance	Red	France	$78.00	$30.00	17,497
14	Lambrusco	Alliance	White	Italy	$20.00	$15.00	

Figure 15-19. *Creating a calculated column in the Wines dimension using the "Total Cases" measure evokes context*

Rather than putting this measure into a calculated column to witness the context transition, we could nest this measure *in another measure* using MAXX, and this will iterate the Wines dimension, just as the calculated column does. If we do this, context transition will happen in memory, and we can find the maximum of the values that you can see in the calculated column. Let's now author this measure:

```
Max of Totals =
MAXX ( Wines, [Total Cases] )
```

The MAXX function iterates the Wines dimension in memory to calculate the "Total Cases" *measure* for every row in the dimension, just like the calculated column in Figure 15-19. It then finds the maximum of these values.

Note We are using the MAXX function here because it's clearer to understand the evaluation – you can easily see which value is the largest. However, the AVERAGEX function would work better because you must calculate the average of the totals to know what that value is.

However, when this measure is placed in the Table visual in Figure 15-20, why does it return the same result as the "Total Cases" measure in all rows except in the Total row, where the value is correct?

WINE	Max Cases	Total Cases	Max of Totals
Bordeaux	500	54,070	54,070
Champagne	500	49,158	49,158
Chardonnay	250	42,030	42,030
Chenin Blanc	150	24,739	24,739
Chianti	299	27,323	27,323
Grenache	350	35,965	35,965
Malbec	326	34,290	34,290
Merlot	200	23,084	23,084
Piesporter	162	10,253	10,253
Pinot Grigio	200	23,449	23,449
Rioja	200	33,951	33,951
Sauvignon Blanc	350	47,415	47,415
Shiraz	150	17,497	17,497
Total	**500**	**423,224**	**54,070**

Figure 15-20. *"Max of Totals" measure is correct in the Total row, which is 54,070, the cases sold for "Bordeaux"*

Let's now answer this question. In this Table visual, the first evaluation is for "Bordeaux" wine, and so in the current filter, there is only one value for "Total Cases", and that is the value of the total cases for "Bordeaux". The maximum of only one value is that value, and that is why we see the same value for "Total Cases" and for "Max of Totals". It's not until the measure reaches the evaluation of the Total row, where there is no filter on the Wines dimension, that it can then find the maximum of all the wines, which is **54,070** for "Bordeaux".

What is important to emphasize here is that context transition always works within filters placed on the data model. For example, if we add a Salesperson slicer to the canvas and filter "Abel", we can now see the maximum cases for "Abel" (**10,993**). Because the Winesales fact table is now filtered for "Abel's" sales, context transition now calculates these values in the new filter context; see Figure 15-21.

WINE	Total Cases	Max of Totals
Bordeaux	8,531	8,531
Champagne	10,993	10,993
Chardonnay	8,099	8,099
Chenin Blanc	2,769	2,769
Chianti	3,699	3,699
Grenache	6,123	6,123
Malbec	4,738	4,738
Merlot	4,520	4,520
Piesporter	2,064	2,064
Pinot Grigio	4,211	4,211
Rioja	5,669	5,669
Sauvignon Blanc	5,318	5,318
Shiraz	3,137	3,137
Total	**69,871**	**10,993**

SALESPERSON
- ■ Abel
- ☐ Blanchet
- ☐ Charron
- ☐ Denis
- ☐ Leblanc
- ☐ Reyer

Figure 15-21. *The "Max of Totals" measure calculated in a different filter context*

However, because the "Max of Totals" measure returns the same value as "Total Cases" for each of the wines, the "Max of Totals" measure does not really work in a visual that filters the wine names. This measure is more fitting when placed in a visual that filters a different dimension. In Figure 15-22, we have used the "Max of Totals" measure in a Matrix and a Table visual that shows the maximum value for each salesperson. We have focused on the maximum cases value for "Abel", which is **10,993** for "Champagne". We've also placed this measure in a Card visual to show the maximum for all wines.

SALESPERSON	Total Cases	Max of Totals
Abel	**69,871**	**$10,993**
Bordeaux	8,531	$8,531
Champagne	10,993	$10,993
Chardonnay	8,099	$8,099
Chenin Blanc	2,769	$2,769
Chianti	3,699	$3,699
Grenache	6,123	$6,123
Malbec	4,738	$4,738
Merlot	4,520	$4,520
Piesporter	2,064	$2,064
Pinot Grigio	4,211	$4,211
Total	**423,224**	**$54,070**

SALESPERSON	Total Cases	Max of Totals
Abel	69,871	$10,993
Blanchet	65,581	$10,345
Charron	68,137	$11,052
Denis	84,018	$12,224
Leblanc	69,304	$9,293
Reyer	66,313	$8,881
Total	**423,224**	**$54,070**

$54,070
Max of Totals

Figure 15-22. *The "Max of Totals" measure works better if placed in visuals that filter dimensions other than the Wines dimension*

Let's now consider another scenario. Rather than calculating the maximum of the total cases, perhaps you want to programmatically identify which wine has the maximum total ("Bordeaux" in our case).

In this situation, for the evaluation of each wine, we must calculate the maximum of the totals for all the wines. We can then compare the maximum against each wine's total. Therefore, we must remove the filter from the Wines dimension by using ALL or ALLSELECTED as in this example:

```
Max of Totals #2 =
MAXX (
    ALL ( Wines ) , [Total Cases] )
```

In this measure, we are using ALL to generate a virtual table containing all the rows in the Wines dimension, and therefore, MAXX will iterate all the rows in this temporary table. Context transition calculates the total cases for every row, and MAXX finds the largest of these. If you have a slicer filtering the wines, you must use ALLSELECTED which will output to a virtual table containing the wines filtered in the slicer.

We can now write the measure that specifically returns the name of the wine that has the maximum cases sold, using the expression in the "Max of Totals #2" measure as a variable:

```
Wine with Max =
VAR MyMax =
```

```
    MAXX ( ALL ( Wines ), [Total Cases] )
RETURN
    CALCULATE ( VALUES (Wines[WINE] ),
    FILTER ( Wines, [Total Cases] = MyMax ) )
```

Note the use of the VALUES function to return the name of the wine that will be filtered according to the filter expression.

WINE	Total Cases	Max of Totals #2
Bordeaux	54,070	54,070
Champagne	49,158	54,070
Chardonnay	42,030	54,070
Chenin Blanc	24,739	54,070
Chianti	27,323	54,070
Grenache	35,965	54,070
Lambrusco		54,070
Malbec	34,290	54,070
Merlot	23,084	54,070
Piesporter	10,253	54,070
Pinot Grigio	23,449	54,070
Rioja	33,951	54,070
Sauvignon Blanc	47,415	54,070
Shiraz	17,497	54,070
Total	**423,224**	**54,070**

Bordeaux

Wine with Max

Figure 15-23. *The "Max of Totals #2" returns the maximum value for all wines. We can then use this expression to find the wine that has the maximum*

You can see the results of these measures in Figure 15-23. Note how we use the "Wine with Max" measure in a Card visual that displays the scalar value returned by VALUES.

In the preceding examples, we're using ALL to generate a virtual table containing all the rows and all the columns in the Wines dimension. We'll see in the next section that we could equally use ALL to generate a virtual table containing only the WINE column.

Aggregating in Virtual Tables

So far, we've looked at finding the maximum of the total values using context transition with a dimension table. We then used ALL to generate a virtual Wines dimension to find the maximum of the totals of all the wines. Therefore, we know that context transition can be generated in virtual tables too. We are now ready to explore examples of using ALL to build virtual tables that contain only the columns that we require for the expression, not all the columns in the table. Because ALL will return a table containing a column or columns of distinct values, we are essentially using ALL to group our data so that context transition can calculate totals across these ad hoc groups.

Using ALL to Group Columns in the Same Table

For example, we've been asked to calculate the variance between the total sales for each wine and the average of these totals. Consider the following measure that uses context transition to find the average of "Total Sales" for our wines. However, this time we're using the ALL function on the WINE column rather than ALL on the Wines table:

```
Average of Totals =
AVERAGEX ( ALL ( Wines[WINE] ), [Total Sales] )
```

As mentioned before, depending on the filters in your report, you may need to use ALLSELECTED in place of ALL.

Let's look at the three steps in the evaluation of this measure:

1. The ALL function creates a virtual table comprising a single column holding a list of unique values in the WINE column.

2. AVERAGEX then iterates the virtual table and using context transition calculates the "Total Sales" measure for each of the wines in the virtual table generated by ALL.

3. AVERAGEX finds the average of these values.

In Figure 15-24, you can see the virtual table generated by ALL and how the average of the total values is calculated.

WINE	Total Sales
Bordeaux	$4,055,250
Champagne	$7,373,700
Chardonnay	$4,203,000
Malbec	$2,914,650
Grenache	$1,078,950
Piesporter	$1,384,155
Chianti	$1,092,920
Pinot Grigio	$703,470
Merlot	$900,276
Sauvignon Blanc	$1,896,600
Rioja	$1,527,795
Chenin Blanc	$1,236,950
Shiraz	$1,364,766
Lambrusco	

Average = 2,287,114

Figure 15-24. *The ALL function creates a virtual one-column table of the WINE column. The table is iterated by AVERAGEX, and context transition calculates the "Total Sales" for each row. AVERAGEX finds the average of these values*

Now we can edit this measure to calculate how each wine's total sales vary from the average and visualize the data; see Figure 15-25.

```
Variance from Average of Totals =
VAR AvgOfTotals =
  AVERAGEX ( ALL ( Wines[WINE] ), [Total Sales] )
RETURN
[Total Sales] -  AvgOfTotals
```

WINE	Total Sales	Average of Totals	Variance from Average of Totals
Bordeaux	$4,055,250	$2,287,114	$1,768,136
Champagne	$7,373,700	$2,287,114	$5,086,586
Chardonnay	$4,203,000	$2,287,114	$1,915,886
Chenin Blanc	$1,236,950	$2,287,114	($1,050,164)
Chianti	$1,092,920	$2,287,114	($1,194,194)
Grenache	$1,078,950	$2,287,114	($1,208,164)
Malbec	$2,914,650	$2,287,114	$627,536
Merlot	$900,276	$2,287,114	($1,386,838)
Piesporter	$1,384,155	$2,287,114	($902,959)
Pinot Grigio	$703,470	$2,287,114	($1,583,644)
Rioja	$1,527,795	$2,287,114	($759,319)
Sauvignon Blanc	$1,896,600	$2,287,114	($390,514)
Shiraz	$1,364,766	$2,287,114	($922,348)
Total	**$29,732,482**	**$2,287,114**	**$27,445,368**

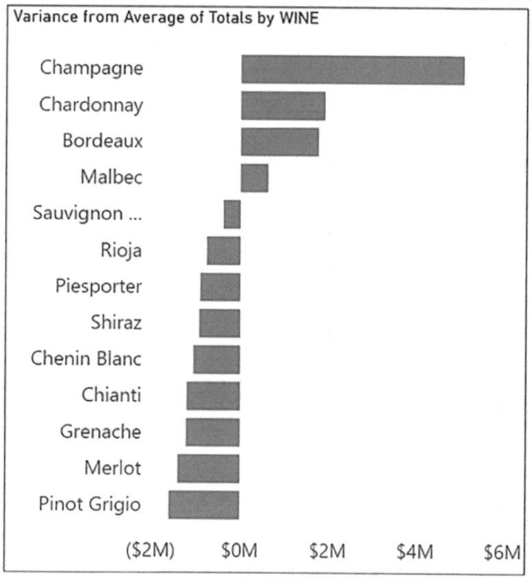

Variance from Average of Totals by WINE
Champagne
Chardonnay
Bordeaux
Malbec
Sauvignon ...
Rioja
Piesporter
Shiraz
Chenin Blanc
Chianti
Grenache
Merlot
Pinot Grigio
($2M) $0M $2M $4M $6M

***Figure 15-25.** Using context transition to calculate the average of the totals and then we can find the variance*

Here, we have been using ALL to group by values in the WINE column. Indeed, we could use ALL to group by salespeople, customers, or regions by generating tables containing just a list of the names of the entities in these dimensions accordingly.

However, if we want to pass context transition into the DateTable, it becomes a little more problematic. For example, the expression

```
Average Daily Sales For Dates =
AVERAGEX (DateTable, [Total Sales])
```

would pass context transition to every row in the DateTable, therefore aggregating the total sales for each day. Therefore, this measure would calculate the average daily sales. However, this may not be the date granularity in which you are interested. Perhaps you would like to calculate the average quarterly sales in each year, as shown in Figure 15-26. Here, we have authored the measure "Average Quarterly for Each Year" and placed this in a Matrix visual. Note that this measure works best if the visual only shows the YEAR column from the DateTable.

YEAR	Total Sales	Average Quarterly for Each Year
2017	**$4,649,285**	**$1,162,321**
Qtr 1	$1,238,163	$1,162,321
Qtr 2	$921,562	$1,162,321
Qtr 3	$1,165,083	$1,162,321
Qtr 4	$1,324,477	$1,162,321
2018	**$4,207,871**	**$1,051,968**
Qtr 1	$939,780	$1,051,968
Qtr 2	$1,097,410	$1,051,968
Qtr 3	$1,072,199	$1,051,968
Qtr 4	$1,098,482	$1,051,968
2019	**$4,710,744**	**$1,177,686**
Qtr 1	$529,926	$1,177,686
Total	**$29,732,482**	**$7,433,121**

YEAR	Total Sales	Average Quarterly for Each Year
2017	$4,649,285	$1,162,321
2018	$4,207,871	$1,051,968
2019	$4,710,744	$1,177,686
2020	$7,900,864	$1,975,216
2021	$8,263,718	$2,065,930
Total	**$29,732,482**	**$7,433,121**

Figure 15-26. *The "Average Quarterly for Each Year" measure works best in a Matrix visual that only shows years*

In the Matrix visuals in Figure 15-26, on the evaluation sales in "2017", there is a filter on the YEAR column in the DateTable. We must now generate a virtual one-column table that retains the filter on the YEAR column but lists all four quarters in that year, and we can use the ALL function to do that, referencing the QTR column. AVERAGEX can then iterate this table and using context transition can calculate the total sales for each of the quarters in "2017", finding the average of these values. This is the code we have used in the measure:

```
Average Quarterly for Each Year =
AVERAGEX ( ALL ( DateTable[QTR] ), [Total Sales] )
```

The virtual table generated by ALL in the evaluation of the "2017" average would look like Figure 15-27.

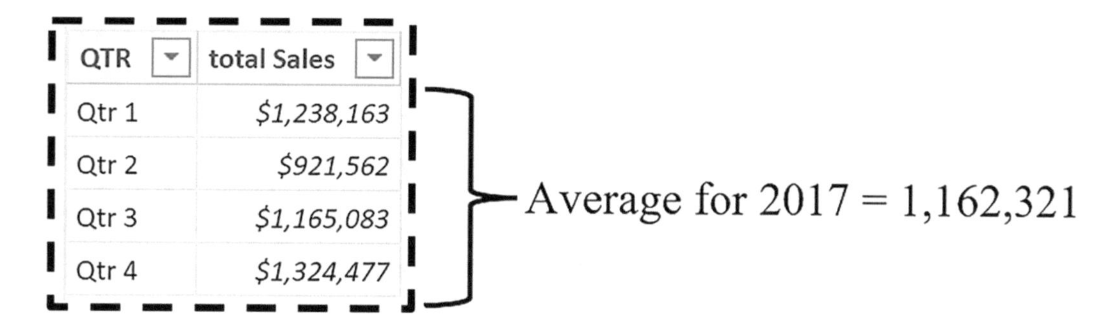

QTR	total Sales
Qtr 1	$1,238,163
Qtr 2	$921,562
Qtr 3	$1,165,083
Qtr 4	$1,324,477

Figure 15-27. *The ALL function builds a table containing all the quarters in the filtered year*

The ALL function can also generate virtual tables comprising unique *combinations* of columns from the same table to enable you to group by these combinations. For example, you could generate a virtual table using ALL containing a distinct list of years and quarters from the DateTable using this table expression:

```
ALL ( DateTable[YEAR], DateTable[QTR] )
```

Such a virtual table generated by ALL is shown in Figure 15-28.

YEAR	QTR
2017	Qtr 1
2018	Qtr 1
2019	Qtr 1
2020	Qtr 1
2021	Qtr 1
2017	Qtr 2
2018	Qtr 2
2019	Qtr 2

Figure 15-28. *Using ALL referencing multiple columns from the same table generates a table containing the distinct combination of those values*

Using context transition and the ALL expression that generates the table in Figure 15-28, you could find the average quarterly total sales across all years as in this measure:

```
Average Quarterly for All Years =
```

```
AVERAGEX (
   ALL ( DateTable[YEAR], DateTable[QTR] ),
   [Total Sales] )
```

Here, the ALL function creates an in-memory table that generates a distinct list combining the YEAR column and the QTR column, and then AVERAGEX, using context transition, finds the average of the "Total Sales" values; see Figure 15-29.

YEAR	Total Sales	Average Quarterly for All Years
2017	**$4,649,285**	**$1,486,624**
Qtr 1	$1,238,163	$1,486,624
Qtr 2	$921,562	$1,486,624
Qtr 3	$1,165,083	$1,486,624
Qtr 4	$1,324,477	$1,486,624
2018	**$4,207,871**	**$1,486,624**
Qtr 1	$939,780	$1,486,624
Qtr 2	$1,097,410	$1,486,624
Qtr 3	$1,072,199	$1,486,624
Qtr 4	$1,098,482	$1,486,624
2019	**$4,710,744**	**$1,486,624**
Qtr 1	$529,926	$1,486,624
Qtr 2	$969,874	$1,486,624
Qtr 3	$1,897,917	$1,486,624
Qtr 4	$1,313,027	$1,486,624
Total	**$29,732,482**	**$1,486,624**

YEAR	QTR	Total Sales
2017	Qtr 4	$1,324,477
2017	Qtr 2	$921,562
2017	Qtr 1	$1,238,163
2017	Qtr 3	$1,165,083
2018	Qtr 3	$1,072,199
2018	Qtr 2	$1,097,410
2018	Qtr 4	$1,098,482
2018	Qtr 1	$939,780
2019	Qtr 2	$969,874
2019	Qtr 1	$529,926
2019	Qtr 4	$1,313,027
2019	Qtr 3	$1,897,917
2020	Qtr 3	$2,220,593
2020	Qtr 1	$1,261,506
2020	Qtr 2	$2,298,464
2020	Qtr 4	$2,120,301
2021	Qtr 3	$2,169,735
2021	Qtr 2	$2,013,616
2021	Qtr 4	$2,011,618
2021	Qtr 1	$2,068,749

Average = 1,486,624

***Figure 15-29.** The ALL function generates a virtual table containing the distinct combination of values from multiple columns from the same table. Context transition can be passed to this table to calculate averages*

We can appreciate that viewing this measure in a Matrix comprising the YEAR and QTR columns, where the values returned are repeated, will not do much for people viewing your report. Just as in the "Max of Totals" measure before, the "Average Quarterly for All Years" measure works best when you have no filter on the DateTable as in Figure 15-30.

SALESPERSON	Total Sales	Average Quarterly for All Years
Denis	$5,431,390	$271,570
Abel	$5,265,266	$263,263
Charron	$5,147,366	$257,368
Blanchet	$4,860,044	$243,002
Leblanc	$4,792,407	$239,620
Reyer	$4,236,009	$211,800
Total	**$29,732,482**	**$1,486,624**

Figure 15-30. *The "Average Quarterly for All Years" measure works well when analyzing entities other than those from the DateTable*

Here, we are analyzing our salespeople's sales performance by comparing their average quarterly total sales value.

Using SUMMARIZE to Group Columns from Related Tables

We can normally use ALL or ALLSELECTED to group columns into virtual tables so we can perform calculations across ad hoc groups using context transition. It's only occasionally that you will require another function called SUMMARIZE to do this job, and that's when you need to group columns from *different* tables.

The SUMMARIZE function allows you to retrieve combinations of columns from the same table or from *one or more related tables.* As we've seen before, we can usually use ALL to group columns from the same table, so SUMMARIZE normally need only be used to group columns from different related tables.

The SUMMARIZE function has the following syntax:

= **SUMMARIZE (table, group by column1, group by column2 etc., name, expression)**

where:

table is the table or table expression containing the columns you want to group by.

group by columns are the columns by which you want to group your data. These can be columns from the same table or from related tables.

name (optional) is the name of the expression you want to generate in the expression argument later. This is a nonmandatory argument.

expression (optional) is an expression that will be calculated for every row in the virtual table. This is a nonmandatory argument.

Mostly you will use SUMMARIZE only with the first two arguments, specifying a table and the group by columns as follows:

=SUMMARIZE (Winesales, Wines[WINE], DateTable[YEAR])

This table expression builds a virtual table grouping by the WINE column and then by the YEAR column and can do this because the Wines table and the DateTable are related to Winesales.

However, there is one big difference between using ALL to generate ad hoc groups of columns and using SUMMARIZE, and that is that SUMMARIZE builds a virtual table comprising the values in the current filter context. Therefore, often, the ALL function is required, nested inside SUMMARIZE, to remove these filters.

With two of the arguments inside SUMMARIZE (i.e., "name" and "expression"), you can optionally create calculations in the virtual table, and we will look at an example of this in the next chapter. However, usually, to create calculations for these groups using context transition, you can nest the table generated by SUMMARIZE inside functions such as MAXX and AVERAGEX that will then perform the calculations.

Using SUMMARIZE, you can group columns from related tables. For instance, if the table you reference inside SUMMARIZE is a fact table, then you can group by any columns from the dimensions related to the fact table.

If you use the **New Table** button on the Modeling tab in Power BI, you can create *calculated tables* using table functions. This is a convenient way to see the output of table expressions such as those involving SUMMARIZE. You can view calculated tables in Data view just as you would any tables in your data model. However, when the SUMMARIZE expression is nested inside a measure, its output will be filtered in memory according to the filter context, and this is something that you can't see in the calculated table in Data view.

We can use SUMMARIZE to group the WINE column from the Wines dimension and the YEAR column from the DateTable using this table expression:

```
Wine and Year Table =
SUMMARIZE ( Winesales,
   Wines[WINE], DateTable[YEAR] )
```

In Figure 15-31, you can see the result of this table expression when used in a calculated table using the **New table** button.

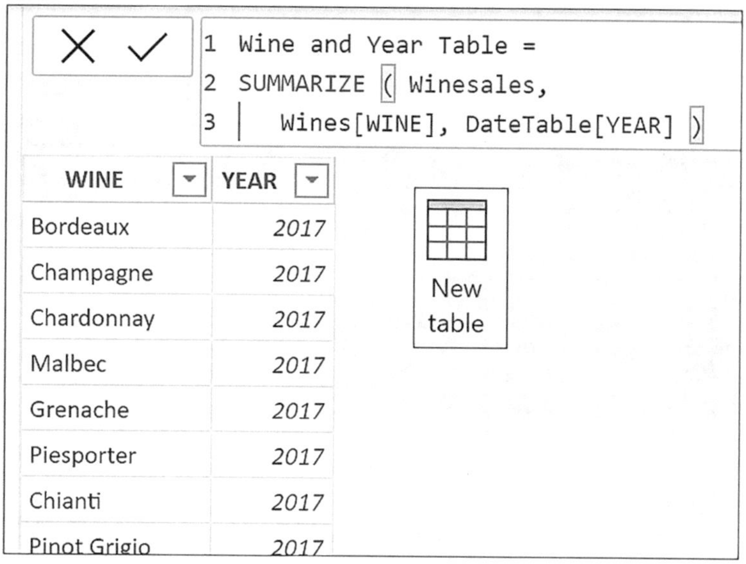

```
1 Wine and Year Table =
2 SUMMARIZE ( Winesales,
3     Wines[WINE], DateTable[YEAR] )
```

WINE	YEAR
Bordeaux	2017
Champagne	2017
Chardonnay	2017
Malbec	2017
Grenache	2017
Piesporter	2017
Chianti	2017
Pinot Grigio	2017

New table

Figure 15-31. *Using SUMMARIZE to generate a table containing the WINE and YEAR columns*

Let's look at a scenario where we may need to generate this virtual table using SUMMARIZE, remembering that such a table will be built in the current filter context. We want to calculate the yearly average total sales for all our wines and display this in a Card visual. This is the measure we will author using SUMMARIZE to group by both WINE and YEAR (note the use of a variable to store the virtual table).

```
Yearly Average =
VAR SummaryTable = SUMMARIZE ( ALL ( Winesales ), Wines[WINE],
DateTable[YEAR] )
RETURN
AVERAGEX ( SummaryTable, [Total Sales] )
```

You can see in Figure 15-32 that on average, the yearly sales for our wines is **$457,423**. To understand this average, you could create a Clustered Column chart visual plotting wines sales in each year. If you then display an average analytical line,[4] this will show the same value that you have calculated in the measure.

[4] For information on working with the analytical lines, visit https://docs.microsoft.com/en-us/power-bi/transform-model/desktop-analytics-pane

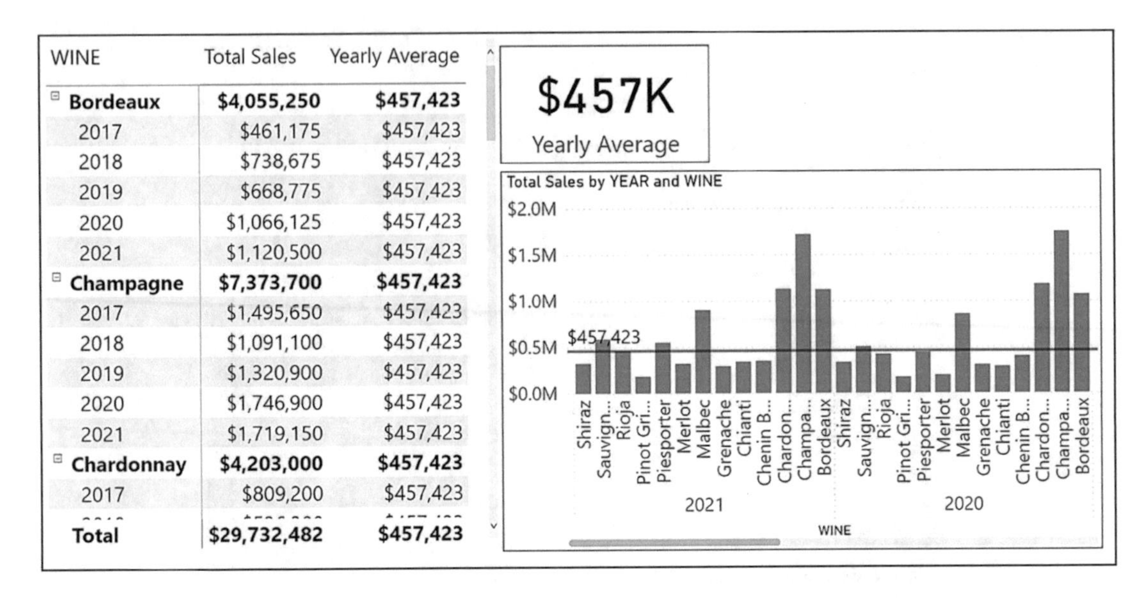

Figure 15-32. *Using SUMMARIZE and context transition to calculate the yearly average for all wines. This would be the average calculated by the "Analytics" average line*

In the "Yearly Average" measure, the "Total Sales" nested measure uses context transition to calculate sales for each combination of WINE and YEAR in the virtual table generated by SUMMARIZE. The AVERAGEX function will find the average of the values returned by the context transition. Note the use of the ALL function on the Winesales table to remove the filter coming from the WINE column and the YEAR column in the Matrix visual. Remember that unless you use ALL, the SUMMARIZE function creates a summary table of values *within the current filter context.*

Alternatively, if we put this measure into a table that didn't use the WINE or YEAR column, we would not require the ALL function, as you can see in Figure 15-33 where we are using the REGION column instead. Here, we are analyzing the average of the total yearly sales of wines in each region.

```
Yearly Average =
VAR SummaryTable = SUMMARIZE (  Winesales , Wines[WINE], DateTable[YEAR] )
RETURN
AVERAGEX ( SummaryTable, [Total Sales] )
```

REGION	Total Sales	Yearly Average
Argentina	$1,327,724	$29,505
Australia	$1,069,007	$27,410
Canada	$453,086	$18,123
China	$1,841,757	$35,418
Czech Republic	$2,465,286	$44,823
England	$1,610,288	$35,784
France	$795,084	$18,931
Germany	$1,216,223	$28,284
India	$2,258,602	$39,625
Ireland	$115,120	$28,780
Italy	$2,354,198	$41,302
Total	**$29,732,482**	**$457,423**

Figure 15-33. *The "Yearly Average" measure calculated for regions*

Having calculated the average yearly sales for all wines in all the years, you may want to calculate the average yearly sales for each wine. Perhaps again, this is to calculate the variance from the average. If this is the case, this is the code you would require:

```
Yearly Average Each Wine =
VAR Summarytable =
    SUMMARIZE ( ALLEXCEPT ( Winesales, Wines[WINE] ), DateTable[YEAR] )
RETURN
    AVERAGEX ( Summarytable, [Total Sales] )
```

The reason that we can use the ALLEXEPT function in this context will be explained in Chapter 18 when we explore the concept of table expansion. All we need to note here is that by using ALLEXCEPT, the filter has been removed from the YEAR column leaving the filter on the WINE column. Therefore, this enables us to pass the average across sales in every year for each wine. This measure would then calculate the variance, but only at the YEAR grain:

```
Variance from Average Each Yr =
IF (
    HASONEVALUE ( DateTable[YEAR] ),
    [Total Sales] - [Yearly Average Each Wine] )
```

You can see the outcomes of the "Yearly Average Each Wine" and "Variance from Average Each Yr" measures in Figure 15-34.

WINE	Total Sales	Yearly Average Each Wine	Variance from Average Each Yr
Bordeaux	**$4,055,250**	**$811,050**	
2017	$461,175	$811,050	($349,875)
2018	$738,675	$811,050	($72,375)
2019	$668,775	$811,050	($142,275)
2020	$1,066,125	$811,050	$255,075
2021	$1,120,500	$811,050	$309,450
Champagne	**$7,373,700**	**$1,474,740**	
2017	$1,495,650	$1,474,740	$20,910
2018	$1,091,100	$1,474,740	($383,640)
2019	$1,320,900	$1,474,740	($153,840)
2020	$1,746,900	$1,474,740	$272,160
2021	$1,719,150	$1,474,740	$244,410
Chardonnay	**$4,203,000**	**$840,600**	
2017	$809,200	$840,600	($31,400)
2018	$536,300	$840,600	($304,300)
2019	$550,300	$840,600	($290,300)
Total	**$29,732,482**	**$5,946,496**	

Figure 15-34. *The "Yearly Average Each Wine" and "Variance from Average Each Yr" measures*

In this chapter, you have learned to use context transition to produce aggregations on measures as opposed to aggregations on row-level values. You now also appreciate the importance of dimension tables in these calculations, that they are used to group and aggregate the data at the dimension granularity. You have also learned that virtual tables play a significant part in these calculations, enabling you to generate ad hoc summary groups over which to harness the power of context transition.

CHAPTER 16

Leveraging Context Transition

In the last chapter, you learned how context transition enables you to programmatically aggregate data into dimensions and virtual tables. You could then author expressions that grouped and aggregated data at this higher granularity. Once you have learned the skill of using DAX in this way, the world of DAX opens up to you considerably. You will now be able to author more complex calculations that enable you to gain deeper data insights. In this chapter, you will be applying your knowledge of context transition to solving the following data analysis questions:

How do I

- Rank entities?
- Bin measures into numeric ranges?
- Calculate top or bottom N percent using dynamic parameters?
- Find like for like sales across my customer base?
- Calculate running totals in a table using a calculated column?
- Calculate differences in values in the previous row in a calculated column?

In generating these insights, you will learn transferrable skills and techniques that you can take on board, extend the ideas, and apply them to your own data.

© Alison Box 2022

A. Box, *Up and Running with DAX for Power BI*, https://doi.org/10.1007/978-1-4842-8188-8_16

Ranking Data: Looking at RANKX

You have learned that by combining iterating functions with measures (which implicitly call CALCULATE), you can reap the benefits of context transition. Let's take this opportunity to look at another iterating function, RANKX.

The RANKX function has the following syntax:

= RANKX (table, expression, value, order, ties)

where:

table is the table that you want to iterate to rank items. This table is often generated by the ALL function, so ranking is performed on all the rows of the table, not just those in the current filter.

expression is the measure or expression to be used to rank the items.

value is optional and is used to compare items to be ranked (rarely used).

order is optional - ASC (1 is the lowest rank) or DESC (1 is the highest rank). The default is DESC.

ties is optional and is either Skip or Dense as follows:

> **Skip** where the next rank value after a tie is the rank value of the tie plus the count of tied values. For example, if 5 values are tied with a rank of 11, then the next value will receive a rank of 16 (11 + 5). This is the default value when the ties parameter is omitted.
>
> **Dense** where the next rank value after a tie is the next rank value. For example, if 5 values are tied with a rank of 11, then the next value will receive a rank of 12.

Here is an example of RANKX syntax:

= RANKX (ALL (Wines), [Total Sales] , , ASC)

As its name suggests, we can use this function to rank our entities by a specific measure, for example, to rank our wines by the "Total Sales" measure; see Figure 16-1.

WINE	Total Sales	Rank Wine ▲
Champagne	$7,373,700	1
Chardonnay	$4,203,000	2
Bordeaux	$4,055,250	3
Malbec	$2,914,650	4
Sauvignon Blanc	$1,896,600	5
Rioja	$1,527,795	6
Piesporter	$1,384,155	7
Shiraz	$1,364,766	8
Chenin Blanc	$1,236,950	9
Chianti	$1,092,920	10
Grenache	$1,078,950	11
Merlot	$900,276	12
Pinot Grigio	$703,470	13

Figure 16-1. *Ranking wines by "Total Sales"*

This is DAX code for the "Rank Wine" measure:

```
Rank Wine =
IF ( [Total Sales],
RANKX ( ALL ( Wines ), [Total Sales] ) )
```

This measure first checks that there is a value for "Total Sales"; otherwise, items with blank values will be considered in the evaluation, such as "Lambrusco" wine that has no sales. If a sales value is present, the measure builds a virtual Wines table containing all the rows from the table using ALL. It then uses context transition to calculate the "Total Sales" value, iterating every row. Finally, it ranks the sales value in the current filter against all the values in the table returned by ALL, returning their rank value.

Let's take a look at another example of using RANKX. You may, for instance, want to rank your financial quarters by sales in each year as shown in Figure 16-2.

YEAR	Total Sales	Rank by Qtr	Rank by Qtr #2
2017	**$4,649,285**	**1**	
Qtr 1	$1,238,163	2	2
Qtr 2	$921,562	4	4
Qtr 3	$1,165,083	3	3
Qtr 4	$1,324,477	1	1
2018	**$4,207,871**	**1**	
Qtr 1	$939,780	4	4
Qtr 2	$1,097,410	1	1
Qtr 3	$1,072,199	2	2
Qtr 4	$1,098,482	3	3
2019	**$4,710,744**	**1**	
Qtr 1	$529,926	4	4
Qtr 2	$969,874	3	3
Qtr 3	$1,897,917	1	1
Total	**$29,732,482**	**1**	

***Figure 16-2.** Using RANKX to rank financial quarters*

In the Matrix visual in Figure 16-2, the first ranking evaluation is for "Qtr 1" in "2017", filters being applied to the DateTable accordingly. Here, you must use the ALL function to generate a virtual table containing a column of all four values in the QTR column of the DateTable (i.e., "Qtr 1", "Qtr 2", "Qtr 3", "Qtr 4") for "2017". This is so that the sales values for all four quarters in that year can be ranked using context transition. This is the measure you can create here:

```
Rank by Qtr =
RANKX ( ALL ( DateTable[QTR] ), [Total Sales] )
```

However, if you put this measure into a Matrix visual, you will notice that the YEAR column is ranked as "1" as there is only one subtotal value to rank. To avoid this irrelevant value, you can use the HASONEVALUE function to return only a value for the QTR column:

```
Rank by Qtr #2 =
 IF ( HASONEVALUE(DateTable[QTR] ),
      RANKX ( ALL ( DateTable[QTR] ), [Total Cases] ))
```

Note We are using the ALL function inside RANKX in our examples shown before, but remember that you may require the ALLSELECTED function instead if you have slicers on your canvas.

We will be meeting the RANKX function again later in this chapter when we use it to rank our customers. However, the most important takeaway from this section is how RANKX, as an iterating function, is used with a measure and, therefore, evokes context transition.

Binning Measures into Numeric Ranges

A common requirement when analyzing data in Power BI is binning the results of a measure into numeric ranges. Consider the visual on the left in Figure 16-3. It is telling us that we have nine customers whose "Total Sales" values are greater than 800,000. In the Table visual on the right, we can see who these customers are.

Range	No. of Customers with these Total Sales
1 - 50,000	10
50,001 - 200,000	32
200,001 - 400,000	10
400,001 - 600,000	4
600,001 - 800,000	18
800,001 and >	9
Total	**83**

CUSTOMER NAME	Total Sales
Erlangen & Co	$1,104,291
Martinsville Bros	$923,865
Burningsuit Ltd	$893,230
Eilenburg Ltd	$891,382
Plattsburgh Ltd	$821,556
Castle Rock Ltd	$815,730
El Cajon & Sons	$813,175
Melbourne Ltd	$812,608
Rhodes Ltd	$809,243
Spokane Ltd	$794,764
Lavender Bay Ltd	$791,084
Port Hammond Bros	$787,172
Clifton Ltd	$764,300
Port Orchard & Sons	$757,784
Littleton & Sons	$729,349
Warrnambool Ltd	$726,492
Chandler & Sons	$725,413
Total	**$29,732,482**

Figure 16-3. *Binning "Total Sales" into numeric ranges*

The starting point for this analysis is to generate a parameter table that defines the ranges you require. You learned how to create parameter tables in Chapter 12 when we explored the SELECTEDVALUE function. To generate the parameter table, use the **Enter Data** button on the **Home** tab. We've called this table "Bins for Sales", and you can see it in Figure 16-4. As with all parameter tables, it's not related to any other tables in the data model.

Range	MinValue	MaxValue	Sort
1 - 50,000	1	50000	1
50,001 - 200,000	50001	200000	2
200,001 - 400,000	200001	400000	3
400,001 - 600,000	400001	600000	4
600,001 - 800,000	600001	800000	5
800,001 and >	800001	999999999	6

Figure 16-4. *The "Bins for Sales" parameter table*

Next, as a *calculated column* in Data view, in the "Bins for Sales" table, we could author this expression that will count the number of customers whose sales fall between the range values:

```
No. of Customers Column =
COUNTROWS (
        FILTER (
            Customers,
            [Total Sales] >= 'Bins for Sales'[MinValue]
                && [Total Sales] <= 'Bins for Sales'[MaxValue] ) )
```

You can see the results of this expression in Figure 16-5.

```
1 No. of Customers Column = COUNTROWS (
2         FILTER (
3             Customers,
4             [Total Sales] >= 'Bins for Sales'[MinValue]
5                 && [Total Sales] <= 'Bins for Sales'[MaxValue] ) )
```

Range	MinValue	MaxValue	Sort	No. of Customers Column
1 - 50,000	1	50000	1	10
50,001 - 200,000	50001	200000	2	32
200,001 - 400,000	200001	400000	3	10
400,001 - 600,000	400001	600000	4	4
600,001 - 800,000	600001	800000	5	18
800,001 and >	800001	999999999	6	9

Figure 16-5. *Start by binning the customers into a calculated column*

Let's take a closer look at the evaluation of "No. of Customers Column". Because we are using a calculated column, the "Bins for Sales" table is iterated, and the values for "MinValue" and "MaxValue" in the current row will be used in the calculation. The "Total Sales" measure used by the FILTER function evokes context transition in the Customers table whereby it returns each customer's total sales value in memory, and it is this value that is used to compare to the range value sitting in the current row of the "Bins for Sales" table. The COUNTROWS function then counts the number of rows in the virtual Customers table generated by FILTER. In Figure 16-6, we step through the evaluation of this expression.

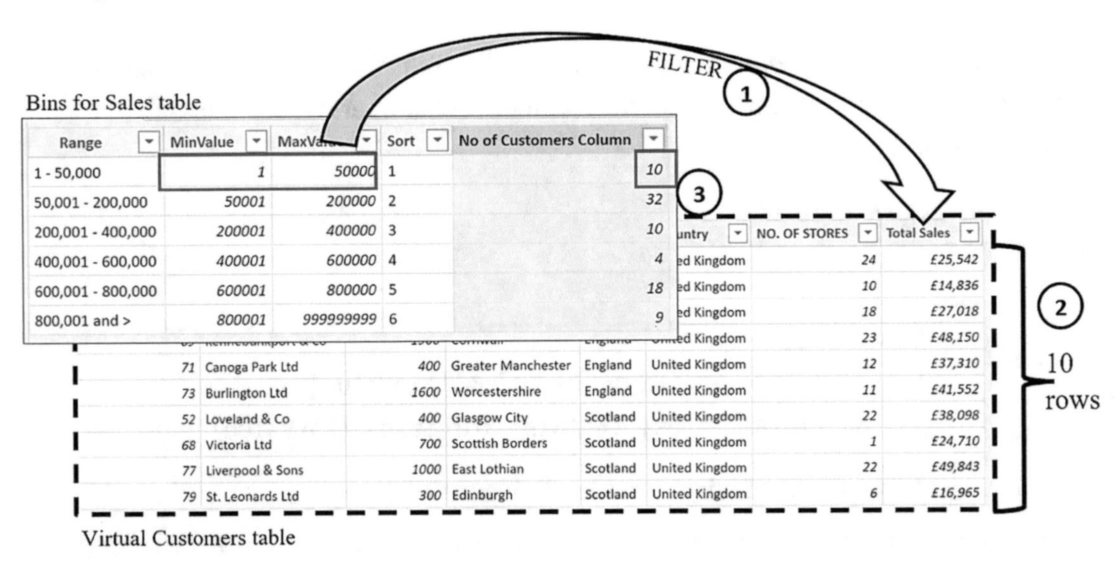

***Figure 16-6.** The evaluation of the "No. of Customers Column" calculated column*

1. The expression iterates the "Bins for Sales" table. FILTER generates a virtual Customers table that is filtered by using the range values in the current row of the "Bins for Sales" table.
2. COUNTROWS counts the rows in the virtual Customers table.
3. The value returned by COUNTROWS is calculated in the current row of the "Bins for Sales" table.

However, we don't want these values sitting in a calculated column; we want them in a measure that we can put into a Table visual so we can slice and dice the data. We learned in Chapter 5 how we can often convert an expression evaluated in a column into an expression evaluated as a measure. There, we took this expression,

*"Winesales[CASESSOLD] * RELATED (Wines[PRICEPERCASE]))"*, and wrapped it inside SUMX. We can do the same with our calculated column, remembering that it is the "Bins for Sales" table that must be iterated by SUMX:

```
No. of Customers with these Total Sales =
SUMX (
    'Bins for Sales',
    COUNTROWS (
        FILTER (
            Customers,
            [Total Sales] >= 'Bins for Sales'[MinValue]
                && [Total Sales] <= 'Bins for Sales'[MaxValue] ) ) )
```

You can then place a Table visual on your canvas and populate it with the "Range" column from the "Bins for Sales" table. Next, place the "No. of Customers with these Total Sales" measure into this table as in Figure 16-3.

Calculating TopN Percent

In this example, you will put into practice all the knowledge of DAX you've learned so far and author a complex measure.

The challenge is to find a way to dynamically browse your best and worst performing customers. The requirement is to do this by finding the topN and bottomN percent of customers by sales, where the "top" and "bottom" and "N" are dynamically selected via slicers. You would also like to browse customers' sales by any entities from dimension tables such as by salespeople or by regions.

You can see in Figure 16-7 that we've solved this scenario. In the Table visual, you can see that we are looking at the bottom 10% of customers by sales for salesperson "Abel".

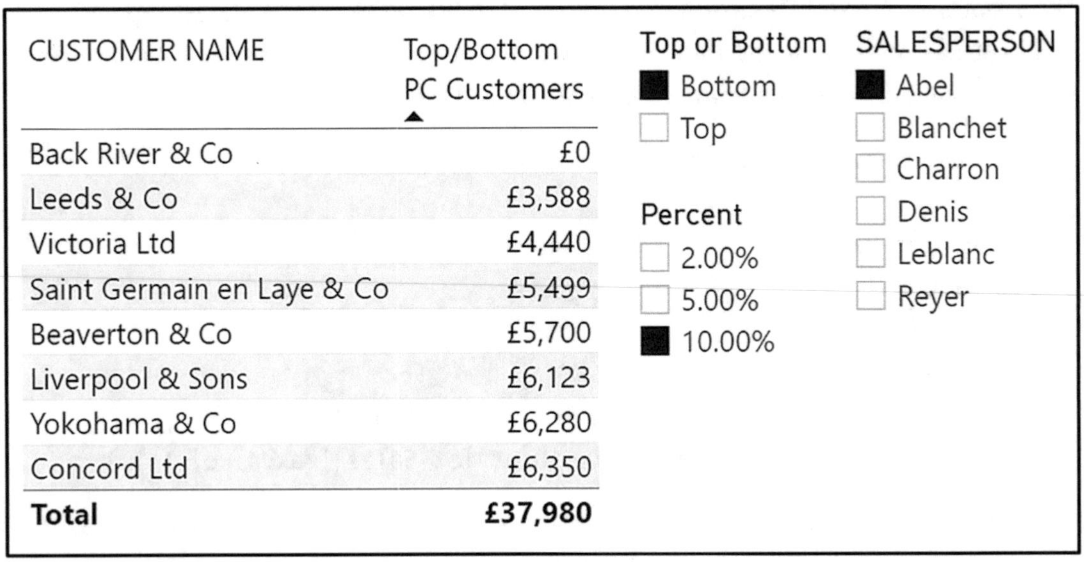

CUSTOMER NAME	Top/Bottom PC Customers ▲
Back River & Co	£0
Leeds & Co	£3,588
Victoria Ltd	£4,440
Saint Germain en Laye & Co	£5,499
Beaverton & Co	£5,700
Liverpool & Sons	£6,123
Yokohama & Co	£6,280
Concord Ltd	£6,350
Total	**£37,980**

Top or Bottom
■ Bottom
□ Top

Percent
□ 2.00%
□ 5.00%
■ 10.00%

SALESPERSON
■ Abel
□ Blanchet
□ Charron
□ Denis
□ Leblanc
□ Reyer

Figure 16-7. *Top or bottom percent of customers by sales*

The measure "Top/Bottom PC Customers" is a compelling example of using context transition within a DAX expression to gain insights into your data, and you can now discover how to re-create this example for yourself.

There are two steps to setting up this analysis:

1. Create the slicers to select which percentage and whether top or bottom.

2. Create the measure to find the top or bottom percent selected in the slicers that will also respond to the Salesperson slicer.

Create the Slicers

The "Top or Bottom" and "Percent" slicers use parameter tables. We've called these tables "Select Percent" and "Select Top or Bottom", and both tables contain just a single column, "Top or Bottom" and "Percent", as shown in Figure 16-8.

Select Top or Bottom Table

Top or Bottom
Top
Bottom

Select Percent Table

Percent
2.00%
5.00%
10.00%

Figure 16-8. *The parameter tables used for top/bottom and percent*

Use the columns from these two tables to populate two slicers.

Create the Measure to Find the Top or Bottom Percent Selected in Slicers

The measure used in the "Top/Bottom PC Customers" uses many skills you have learned so far in this book. Let's think through what will be required of you to arrive at the correct DAX code for this measure.

- You will use variables throughout the expression to separate each part of the evaluation.
- You will use the SELECTEDVALUE function to harvest the values selected in the slicers, either "Top" or "Bottom", and the percentage to be calculated.
- The percentage selected is used to find the base rank. For example, if 10% is chosen in the slicer and there are 84 customers who have sales, you must find customers whose rank is less than 8.4. You will rank customers descending for top ranked customers (top = 1) and ascending for bottom ranked customers (bottom = 1). Therefore, you will be finding a rank less than 8.4 in both cases.
- Using context transition and the RANKX function, you will rank the customers, top or bottom, according to their "Total Sales" value.
- Because there are customers with no sales that will be ranked by default when finding bottom percent, you must filter the Customers table so only customers who have sales are ranked.

- Using the FILTER function, you will filter top or bottom customers whose rank is, for example, less than 8.4, if finding 10%.
- Because the measure must return a scalar value, you must now calculate the "Total Sales" measure for the filtered customers.
- Lastly, you must write a calculation that returns "Total Sales" for either the top or the bottom ranked customers depending on the slicer selection.

This is the measure that you can now author (we have added a comment under each part of the expression to explain the purpose of the code):

```
Top/Bottom PC Customers =

VAR PercentToFind =
    COUNTROWS ( ALL ( Customers ) ) * SELECTEDVALUE ( 'Select
    Percent'[Percent] )
-- Harvest the percent using the slicer selection

VAR TopOrBottom =
    SELECTEDVALUE ( 'Select Top or Bottom'[Top or Bottom] )
-- Harvest whether top or bottom using the slicer selection

VAR RankCustsTop =
    RANKX ( ALL ( Customers ), [Total Sales] )
-- Rank the customers descending by Total Sales value (Top = 1)

VAR RankCustsBottom =
    RANKX (FILTER( ALL ( Customers ),NOT(ISBLANK([Total Sales]))), [Total
    Sales],, ASC )
-- Rank the customers ascending by Total Sales value (Bottom = 1) but only
if they have sales

VAR FindCustsTop =
    FILTER ( Customers, RankCustsTop <= PercentToFind )
-- Filter top customers whose rank is less than or equal to the
PerCentToFind

VAR FindCustsBottom =
    FILTER ( Customers, RankCustsBottom <= PercentToFind )
```

```
-- Filter bottom customers whose rank is less than or equal to the
PerCentToFind

VAR CalcSalesTop =
    CALCULATE ( [Total Sales], FindCustsTop )
-- Calculate "Total Sales" for top ranked customers

VAR CalcSalesBottom =
    CALCULATE ( [Total Sales], FindCustsBottom )
-- Calculate "Total Sales" for bottom ranked customers

RETURN

IF ( HASONEVALUE ( Customers[CUSTOMER NAME] ),
-- This tests that the evaluation is not for the Total Row.
IF ( TopOrBottom = "top",  CalcSalesTop, CalcSalesBottom  ),
--The calculation for rows not in the Total row

        CALCULATE ( [Total Sales],
        ALLSELECTED (Customers[CUSTOMER NAME] ) ) )
--The calculation for the Total Row
```

Note the "RETURN" expression that executes different code if the calculation is for the Total row. This is to resolve the problem of users selecting "Bottom" percent and no value showing in the Total row. This is because the Total row is evaluated in the same way as the evaluation for each customer. Therefore, the Total row value, which is always greater than the individual sales values, is given a bottom ranking of 85 (if there are 84 customers with sales), because the bottom ranking is ascending (higher values get a larger ranked number). The Total row, therefore, fails the ranking bottom test performed by FILTER, and so there is no data to show in the Total row.

You must, therefore, author a different expression for the Total row to ensure that the Total row sums the total sales for the customers shown in the visual. To test that the evaluation is *not* for the Total row, you can use the HASONEVALUE function. You can then use the ALLSELECTED function to calculate the "Total Sales" value for just the customers shown in the visual.

However, we have not yet resolved the problem, because you will note that at this stage, the Total row shows the total sales for *all* customers for "Abel"; see Figure 16-9.

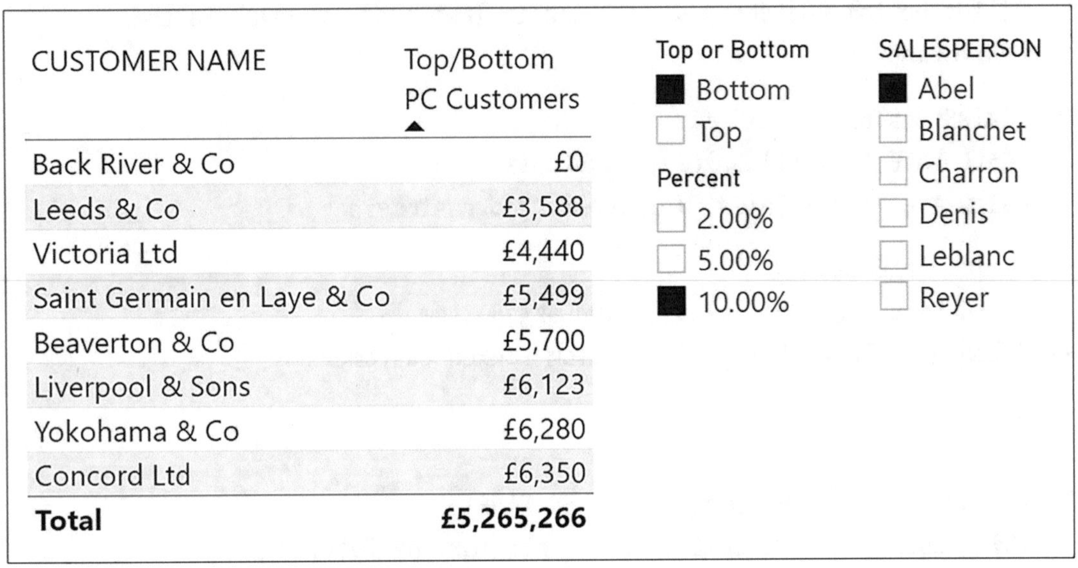

CUSTOMER NAME	Top/Bottom PC Customers
Back River & Co	£0
Leeds & Co	£3,588
Victoria Ltd	£4,440
Saint Germain en Laye & Co	£5,499
Beaverton & Co	£5,700
Liverpool & Sons	£6,123
Yokohama & Co	£6,280
Concord Ltd	£6,350
Total	**£5,265,266**

Top or Bottom
Bottom
Top
Percent
2.00%
5.00%
10.00%

SALESPERSON
Abel
Blanchet
Charron
Denis
Leblanc
Reyer

Figure 16-9. *The Total row is not correct*

This is because the ALLSELECTED expression calculates the "Total Sales" measure independently of the ranking calculation, and so there is no filter on the Customers table for ALLSELECTED to remove. Therefore, to place a filter on the Customers table, you can use a visual-level filter, populate it with the "Total Sales" measure, and set the filter to "Show items when the value is not blank"; see Figure 16-10.

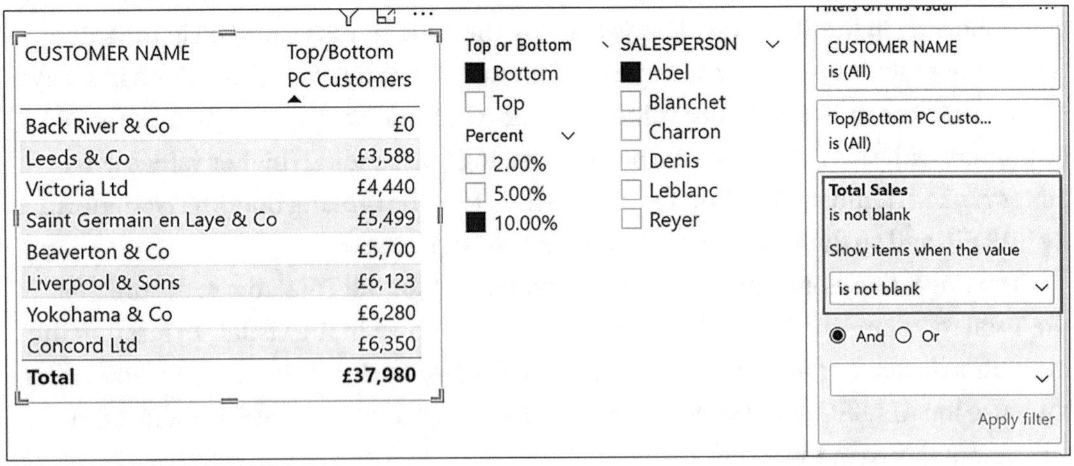

CUSTOMER NAME	Top/Bottom PC Customers
Back River & Co	£0
Leeds & Co	£3,588
Victoria Ltd	£4,440
Saint Germain en Laye & Co	£5,499
Beaverton & Co	£5,700
Liverpool & Sons	£6,123
Yokohama & Co	£6,280
Concord Ltd	£6,350
Total	**£37,980**

Top or Bottom
Bottom
Top
Percent
2.00%
5.00%
10.00%

SALESPERSON
Abel
Blanchet
Charron
Denis
Leblanc
Reyer

Filters on this visual
CUSTOMER NAME
is (All)
Top/Bottom PC Custo...
is (All)
Total Sales
is not blank
Show items when the value
is not blank
And Or
Apply filter

Figure 16-10. *The Total row is correct if you provide a visual-level filter for ALLSELECTED to remove*

Finally, you can change the slicer selections, and the measure recalculates accordingly, finding your best and worst customers using our great DAX friend, context transition.

You may feel that the dynamic ranking of customers that we have achieved here has been quite a daunting experience. It would appear that once you have "cracked" the obvious calculation of ranking the customers, there were then unexpected problems that arose, such as how the Total row must be evaluated. Let me tell you now, this is par for the course. This is true DAX in action, and you are beginning to appreciate that what you must do above all else is *think it through*. Why is my expression returning correct results most of the time but then odd results only sometimes? Always think through exactly how your measure is being evaluated and, particularly, the evaluation context in which it has been placed.

Calculating "Like for Like" Yearly Sales Using SUMMARIZE

We have been analyzing our customer sales values in a variety of ways throughout this book. One of the more insightful metrics, however, we have yet to explore is calculating like for like sales to make more accurate comparisons between our customers.

Let's start by setting up the scenario. We want to analyze our customers' sales of "Chianti" wine in the years 2019, 2020, and 2021. The problem with multiselecting years in a slicer is that our "Total Sales" measure will filter customers with sales of "Chianti" in any of the selected years and not sales in all of them; see Figure 16-11.

WINE

- [ ] Bordeaux
- [ ] Champagne
- [ ] Chardonnay
- [ ] Chenin Blanc
- [x] Chianti
- [ ] Grenache
- [ ] Lambrusco
- [ ] Malbec
- [ ] Merlot
- [ ] Piesporter
- [ ] Pinot Grigio
- [ ] Rioja

YEAR

- [ ] 2017
- [ ] 2018
- [x] 2019
- [x] 2020
- [x] 2021

CUSTOMER NAME	2019	2020	2021	**Total**
Ballard & Sons		$5,640		**$5,640**
Barstow Ltd			$11,600	**$11,600**
Burningsuit Ltd	$12,800	$21,880	$7,840	**$42,520**
Canoga Park Ltd			$9,080	**$9,080**
Cape Canaveral Ltd		$5,240	$10,920	**$16,160**
Castle Rock Ltd		$5,560	$4,920	**$10,480**
Chandler & Sons			$5,880	**$5,880**
Charleston Ltd		$4,840	$8,960	**$13,800**
Charlottesville & Co			$3,800	**$3,800**
Chatou & Co	$16,040		$6,120	**$22,160**
Cheney & Co		$2,880		**$2,880**
Clifton Ltd	$9,400	$10,840	$17,840	**$38,080**
Columbus & Sons		$5,000		**$5,000**
East Orange & Co		$8,840		**$8,840**
Eilenburg Ltd		$4,720	$16,160	**$20,880**
El Cajon & Sons			$8,840	**$8,840**
Total	**$189,440**	**$285,800**	**$341,400**	**$816,640**

Figure 16-11. *Multiselecting years returns customers with sales in any of the selected years, not all the selected years*

However, we'd like to select a range of years in a slicer and find out which customers bought "Chianti" in *all* the selected years so we can compare like for like on the total. For instance, in Figure 16-11, we can see that in the years 2019, 2020, and 2021, "Burningsuit Ltd" had sales in all three years for "Chianti" but "Ballard & Sons" only had sales in 2020 and "Barstow Ltd" in 2021. Therefore, the total sales for those three years would not be like for like when considering these three customers' sales of "Chianti".

The visual that provides the analysis we require is shown in Figure 16-12. Here, we have selected "Chianti" wine and years 2019, 2020, and 2021 in the slicers, and the table visual shows sales for only customers who have sales of "Chianti" in all those years.

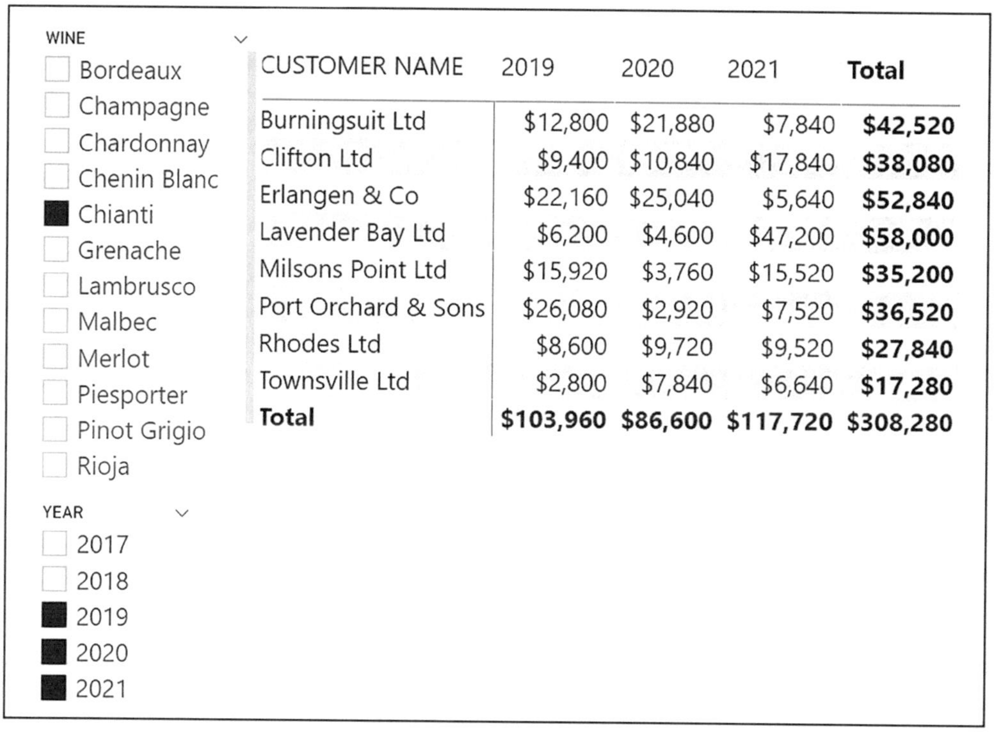

WINE
- Bordeaux
- Champagne
- Chardonnay
- Chenin Blanc
- Chianti (selected)
- Grenache
- Lambrusco
- Malbec
- Merlot
- Piesporter
- Pinot Grigio
- Rioja

YEAR
- 2017
- 2018
- 2019 (selected)
- 2020 (selected)
- 2021 (selected)

CUSTOMER NAME	2019	2020	2021	Total
Burningsuit Ltd	$12,800	$21,880	$7,840	**$42,520**
Clifton Ltd	$9,400	$10,840	$17,840	**$38,080**
Erlangen & Co	$22,160	$25,040	$5,640	**$52,840**
Lavender Bay Ltd	$6,200	$4,600	$47,200	**$58,000**
Milsons Point Ltd	$15,920	$3,760	$15,520	**$35,200**
Port Orchard & Sons	$26,080	$2,920	$7,520	**$36,520**
Rhodes Ltd	$8,600	$9,720	$9,520	**$27,840**
Townsville Ltd	$2,800	$7,840	$6,640	**$17,280**
Total	**$103,960**	**$86,600**	**$117,720**	**$308,280**

Figure 16-12. *Calculating like for like sales in 2019 to 2021 for "Chianti" wine*

To understand the code we must author that calculates such sales, we will pick the calculation apart into its constituent steps:

1. Identify customers who have sales in the selected years of the selected wine.

2. Calculate in how many of those years selected in the slicer the customer has sales.

3. Filter customers who have sales in the same number of years as the number of years selected in the slicer.

Let's take step #1 and explore how we identify those customers that have sales in the selected years. For this, we must digress a little and revisit the SUMMARIZE function to learn more. In the previous chapter, you learned how you can use SUMMARIZE to generate a virtual table grouping columns from different tables. However, as one of the arguments inside SUMMARIZE, you can optionally include an expression to be evaluated for the rows returned in the virtual table. Therefore, to identify in which of the selected years our customers have sales, we could write the following measure where we have highlighted the two arguments used for calculating the total sales for each customer in each year:

```
No. of Years that Customers have Sales =
    COUNTROWS (
        SUMMARIZE (
            Winesales,
            Customers[CUSTOMER NAME],
            DateTable[Year],
            "Sales", [Total Sales]
        )
    )
```

We will now work through the details of the "No. of Years that Customers have Sales" measure. We are using SUMMARIZE to create the virtual table shown in Figure 16-13. We don't see all the years for every customer because this table is evaluated in the current filter of the Matrix visual that it occupies; for instance, "Ballard & Sons" only has sales in 2020; see Figure 16-14.

CUSTOMER NAME	YEAR	Sales
Ballard & Sons	2020	5640
Barstow Ltd	2021	11600
Burningsuit Ltd	2021	7840
Burningsuit Ltd	2020	21880
Burningsuit Ltd	2019	12800
Canoga Park Ltd	2021	9080
Cape Canaveral Ltd	2021	10920
Cape Canaveral Ltd	2020	5240
Castle Rock Ltd	2021	4920
Castle Rock Ltd	2020	5560

***Figure 16-13.** The virtual table generated by SUMMARIZE in the "No. of Years that Customers have Sales" measure*

You can see that the SUMMARIZE function includes an expression called "Sales" which will return the "Total Sales" measure. The name that you give to this column inside SUMMARIZE (e.g., "Sales") is purely arbitrary.

We can now put this measure into a Matrix visual with CUSTOMER NAME in rows and YEAR in columns (Figure 16-14). We are also slicing by "Chianti" wine and years 2019, 2020, and 2021. You can see that it returns "1" for every customer that has sales of "Chianti" in the selected years.

WINE
- [ ] Bordeaux
- [ ] Champagne
- [ ] Chardonnay
- [ ] Chenin Blanc
- [x] Chianti
- [ ] Grenache
- [ ] Lambrusco
- [ ] Malbec
- [ ] Merlot
- [ ] Piesporter
- [ ] Pinot Grigio
- [ ] Rioja

YEAR
- [ ] 2017
- [ ] 2018
- [x] 2019
- [x] 2020
- [x] 2021

CUSTOMER NAME	2019	2020	2021	Total
Ballard & Sons		1		**1**
Barstow Ltd			1	**1**
Burningsuit Ltd	1	1	1	**3**
Canoga Park Ltd			1	**1**
Cape Canaveral Ltd		1	1	**2**
Castle Rock Ltd		1	1	**2**
Chandler & Sons			1	**1**
Charleston Ltd		1	1	**2**
Charlottesville & Co			1	**1**
Chatou & Co	1		1	**2**
Cheney & Co		1		**1**
Clifton Ltd	1	1	1	**3**
Columbus & Sons		1		**1**
East Orange & Co		1		**1**
Eilenburg Ltd		1	1	**2**
El Cajon & Sons			1	**1**
Total	**15**	**31**	**30**	**76**

Figure 16-14. *The "No. of Years that Customers have Sales" measure in a Matrix visual*

Now for step #2 where we must calculate in how many of those years selected in the slicer a customer has sales. Remembering that the columns WINE, CUSTOMER, and YEAR are providing the filter context, we must remove the filter from YEAR so we can look at our customers' sales of "Chianti" for all the years selected in the slicer. We can use CALCULATE with ALLSELECTED on the DateTable to do this job and simply nest our SUMMARIZE expression inside CALCULATE:

```
No. of Years that Customers have Sales #2=
CALCULATE (
    COUNTROWS (
        SUMMARIZE (
            Winesales,
            Customers[CUSTOMER NAME],
            DateTable[Year],
            "Sales", [Total Sales]
        )
    ),
    ALLSELECTED ( DateTable[Year] )
)
```

We can see the values this measure returns in Figure 16-15.

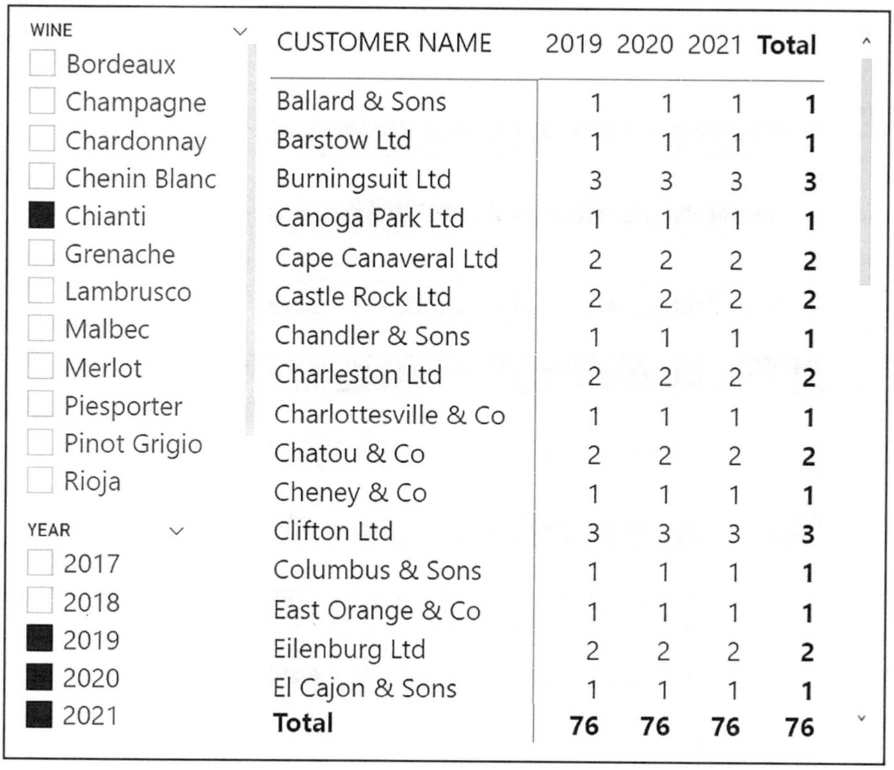

WINE

- [ ] Bordeaux
- [ ] Champagne
- [ ] Chardonnay
- [ ] Chenin Blanc
- [x] Chianti
- [ ] Grenache
- [ ] Lambrusco
- [ ] Malbec
- [ ] Merlot
- [ ] Piesporter
- [ ] Pinot Grigio
- [ ] Rioja

YEAR

- [ ] 2017
- [ ] 2018
- [x] 2019
- [x] 2020
- [x] 2021

CUSTOMER NAME	2019	2020	2021	**Total**
Ballard & Sons	1	1	1	**1**
Barstow Ltd	1	1	1	**1**
Burningsuit Ltd	3	3	3	**3**
Canoga Park Ltd	1	1	1	**1**
Cape Canaveral Ltd	2	2	2	**2**
Castle Rock Ltd	2	2	2	**2**
Chandler & Sons	1	1	1	**1**
Charleston Ltd	2	2	2	**2**
Charlottesville & Co	1	1	1	**1**
Chatou & Co	2	2	2	**2**
Cheney & Co	1	1	1	**1**
Clifton Ltd	3	3	3	**3**
Columbus & Sons	1	1	1	**1**
East Orange & Co	1	1	1	**1**
Eilenburg Ltd	2	2	2	**2**
El Cajon & Sons	1	1	1	**1**
Total	**76**	**76**	**76**	**76**

Figure 16-15. *The "No. of Years that Customers have Sales #2" measure evaluated in the Matrix visual*

We already know from Figure 16-13 that "Ballard & Sons" has only bought "Chianti" in 2020 so they only have sales in one of the years selected in the slicer.

To complete the calculation in step #3, we can filter the Customers table to contain only those customers whose number of years returned by the "No. of Years that Customers have Sales #2" measure equals the number of years filtered in the slicer and return the "Total Sales" value for these customers. This is the "Like for Like Sales" measure that we've used in the visual in Figure 16-14 that returns the result we need:

```
Like for Like Sales =
CALCULATE (
    [Total Sales],
    FILTER (
        Customers,
```

```
        [No. of Years that Customers have Sales #2] =
        COUNTROWS ( ALLSELECTED ( DateTable[Year] ) )
    )
)
```

Figure 16-16 shows this measure evaluated in a Matrix visual.

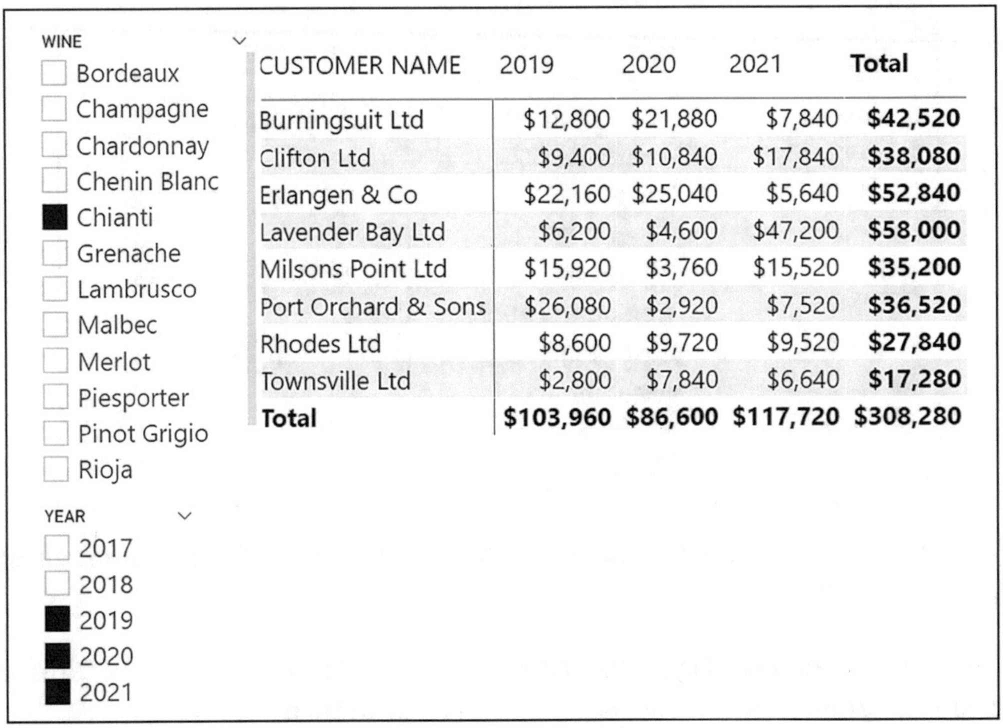

WINE
- ☐ Bordeaux
- ☐ Champagne
- ☐ Chardonnay
- ☐ Chenin Blanc
- ■ Chianti
- ☐ Grenache
- ☐ Lambrusco
- ☐ Malbec
- ☐ Merlot
- ☐ Piesporter
- ☐ Pinot Grigio
- ☐ Rioja

YEAR
- ☐ 2017
- ☐ 2018
- ■ 2019
- ■ 2020
- ■ 2021

CUSTOMER NAME	2019	2020	2021	Total
Burningsuit Ltd	$12,800	$21,880	$7,840	**$42,520**
Clifton Ltd	$9,400	$10,840	$17,840	**$38,080**
Erlangen & Co	$22,160	$25,040	$5,640	**$52,840**
Lavender Bay Ltd	$6,200	$4,600	$47,200	**$58,000**
Milsons Point Ltd	$15,920	$3,760	$15,520	**$35,200**
Port Orchard & Sons	$26,080	$2,920	$7,520	**$36,520**
Rhodes Ltd	$8,600	$9,720	$9,520	**$27,840**
Townsville Ltd	$2,800	$7,840	$6,640	**$17,280**
Total	**$103,960**	**$86,600**	**$117,720**	**$308,280**

Figure 16-16. *The "Like for Like Sales" measure evaluated for "Chianti" wine*

In the preceding scenario, where we have calculated like for like sales, you may have noticed the absence of any reference to context transition when working through the evaluation of the measures we built using SUMMARIZE. In fact, these measures do not use context transition. SUMMARIZE is not an iterating function, and in the absence of an iteration, context transition cannot occur. The method that SUMMARIZE uses to calculate its "expression" argument is complex, and its explanation is beyond the scope of this book. However, the behavior of the "Total Sales" measure in the expressions using SUMMARIZE is indistinguishable from context transition to most DAX users. That is, we have generated a summary table, and the "Total Sales" measure is calculated at that granularity. This is why I have included this example in this chapter.

Using Context Transition in Calculated Columns

Understanding context transition allows you to write more challenging calculated columns too. What you will learn in this section is that by using CALCULATE in calculated columns, you are released from the constraints of the row context where you can only calculate values for the current row. We can now harness the power of context transition to *programmatically create filters on tables* and so pass calculations across these filtered rows in calculated columns.

Calculating Running Totals

You have already learned how to calculate cumulative totals using measures in Chapter 9 (see Figure 9-10) and Chapter 15 (see Figure 15-10). However, we now have a different cumulative total we would like to find, and that is a running total of the quantity in the CASES SOLD column; see Figure 16-17. Using variables and context transition makes this calculation straightforward. This is the DAX calculated column you can create:

```
CUMULATIVE TOTAL =
VAR MyDate = Winesales[SALE DATE]
VAR MyFilter =
    FILTER ( Winesales, Winesales[SALE DATE] <= MyDate )
RETURN
    CALCULATE ( SUM ( Winesales[CASES SOLD] ), MyFilter )
```

The variable "MyDate" finds the value in the SALE DATE column sitting in the current row. The variable "MyFilter" uses the FILTER function to create a virtual table filtering the rows where the SALE DATE is on or before this date. Using context transition, CALCULATE can use this new filter generated by the virtual table to sum the CASES SOLD for these filtered rows.

SALE DATE	WINESALES NO	SALESPERSON ID	CUSTOMER ID	WINE ID	CASES SOLD	CUMULATIVE TOTAL
01/01/2017	2	6	16	10	213	539
01/01/2017	1	3	16	4	326	539
02/01/2017	3	4	20	5	70	609
03/01/2017	4	1	12	10	264	873
07/01/2017	5	2	17	3	147	1020
08/01/2017	6	3	45	11	155	1175
09/01/2017	7	6	11	7	173	1348
10/01/2017	8	2	75	13	106	1454
12/01/2017	10	4	16	13	136	1738
12/01/2017	9	4	14	13	148	1738
13/01/2017	11	1	22	3	228	1966
14/01/2017	12	5	13	9	120	2086

Figure 16-17. *The "CUMULATIVE TOTAL" in a calculated column*

Notice the use of the variable "MyDate" to find the date in the current row. Before variables were introduced into DAX in 2015, we had to use a function called EARLIER to do this job, as follows:

```
CUMULATIVE TOTAL =
CALCULATE (
    SUM ( Winesales[CASES SOLD] ),
    FILTER ( Winesales, Winesales[SALE DATE] <=
     EARLIER ( Winesales[SALE DATE] ) )
)
```

I think you'll agree that the calculated column using the variable is a lot easier to create and understand.

Calculating the Difference from the Value in the Previous Row

You have learned that calculated columns use the row context in their evaluation where the values used by the expression are the values sitting in the current row. However, a common question that is often asked is how to find values in another row. For example, you may be asked to calculate the number of days between sales transactions as in the "DAYS DIFFERENCE" calculated column in Figure 16-18.

SALE DATE	WINESALES NO	SALESPERSON ID	CUSTOMER ID	WINE ID	CASES SOLD	DAYS DIFFERENCE
01/01/2017	2	6	16	10	213	
01/01/2017	1	3	16	4	326	
02/01/2017	3	4	20	5	70	1
03/01/2017	4	1	12	10	264	1
07/01/2017	5	2	17	3	147	4
08/01/2017	6	3	45	11	155	1
09/01/2017	7	6	11	7	173	1
10/01/2017	8	2	75	13	106	1
12/01/2017	10	4	16	13	136	2
12/01/2017	9	4	14	13	148	2
13/01/2017	11	1	22	3	228	1
14/01/2017	12	5	13	9	120	1
15/01/2017	14	5	32	3	246	1

***Figure 16-18.** The "DAYS DIFFERENCE" calculated column*

To do this calculation, we need to find the SALE DATE that is in the previous row. This is the expression for the calculated column:

```
DAYS DIFFERENCE =
VAR MyDate = Winesales[SALE DATE]
VAR PreviousDate =
    CALCULATE (
        MAX ( Winesales[SALE DATE] ),
        FILTER ( WineSales, Winesales[SALE DATE] < MyDate ) )
RETURN
    IF ( PreviousDate, MyDate - PreviousDate )
```

The variable "MyDate" finds the value of SALE DATE sitting in the current row, for example, **7 January 2017**. The variable "PreviousDate" uses CALCULATE and so invokes context transition that will apply a filter to the rows. Using the FILTER function, a virtual table is created filtering the rows where the SALE DATE is before "MyDate" (i.e., all the rows with dates up to and including **6 January 2017**). CALCULATE then calculates the latest date (using the MAX function) in the virtual table (**6 January 2017**). Therefore, this date is the date immediately before the date in the current row. The RETURN statement checks for the presence of a previous date and then subtracts the date in the current row from the date generated by "PreviousDate". The value returned is a date, so the last step is to change the data type to a whole number.

By working through the examples contained in this and the previous chapter, you have learned how to use context transition to author more complex and challenging expressions. However, you are still sitting on the tip of the iceberg of calculations that can be achieved using context transition. You'll find your own reasons to benefit from using this aspect of DAX, and you will no longer find the behavior of context transition in any way "strange" or "surprising," and that's because you now understand it.

CHAPTER 17

Virtual Relationships: The LOOKUPVALUE and TREATAS Functions

Our data model comprises well-defined physical relationships between the tables, generating a star schema. However, there is another type of relationship we can create, and that's a "virtual" relationship. A virtual relationship is a DAX expression that simulates the behavior of a physical relationship defined in the data model. In this chapter, you will learn to create virtual relationships that can resolve problems created by anomalies in the data model. Such anomalies can exist for the following reasons:

- When a relationship does not exist, for example, when using a lookup table.
- The relationship between tables is not part of a star or snowflake schema.
- When a relationship cannot be created because there are duplicate values in both of the columns you want to relate.

Specifically, we will delve into the outcomes of using two functions that create virtual relationships: LOOKUPVALUE and TREATAS. In fact, these two functions are very different. LOOKUPVALUE returns a value, usually from a different table, that is looked up based on search criteria that are provided by the function. TREATAS, on the other hand, is a table function that returns a virtual table that can be used to filter another table. However, they're both used in situations where it's not possible to use a physical relationship, and that's why we've consolidated them into this chapter.

© Alison Box 2022
A. Box, *Up and Running with DAX for Power BI*, https://doi.org/10.1007/978-1-4842-8188-8_17

LOOKUPVALUE Function

We've already learned that we can use the RELATED function to pull values through from the one side of the relationship into the many, just in the same way that the VLOOKUP function works in Excel. However, RELATED only works if you have a many-to-one relationship in place. Let's look at a situation where it would not be possible to use RELATED.[1]

The situation is this; currently, our wines have a single price per case, but we now want our wines to have different prices according to different price bands. We've added another table to our model that records the price bands of the wines in a table called "Prices", shown in Figure 17-1.

WINE ID	PRICE BAND	PRICE PER CASE
1	A	$63.00
1	B	$90.00
1	C	$189.00
1	D	$156.00
1	E	$155.00
1	F	$179.00
2	A	$105.00
2	B	$104.00
2	C	$106.00
2	D	$171.00

***Figure 17-1.** The Prices table records the price band and price per case for each wine*

Now, when we make a sale of any wine, the price band is also recorded in the transaction in the Winesales fact table; see Figure 17-2.

[1] To follow along with the examples, use the Power BI Desktop file "4 DAX LOOKUPVALUE.pbix".

SALE DATE	WINESALES NO	SALESPERSON ID	CUSTOMER ID	WINE ID	CASES SOLD	PRICE BAND
01/01/2017	2	6	16	10	213	C
01/01/2017	1	3	16	4	326	C
02/01/2017	3	4	20	5	70	D
03/01/2017	4	1	12	10	264	C
07/01/2017	5	2	17	3	147	E
08/01/2017	6	3	45	11	155	D
09/01/2017	7	6	11	7	173	A
10/01/2017	8	2	75	13	106	C
12/01/2017	9	4	14	13	148	B
12/01/2017	10	4	16	13	136	B
13/01/2017	11	1	22	3	228	B
14/01/2017	12	5	13	9	120	A
15/01/2017	16	5	32	6	70	B

***Figure 17-2.** Each transaction records the price band*

In Figure 17-3, you can see that relating wines to their prices in a many-to-one relationship using the WINE ID column is straightforward.

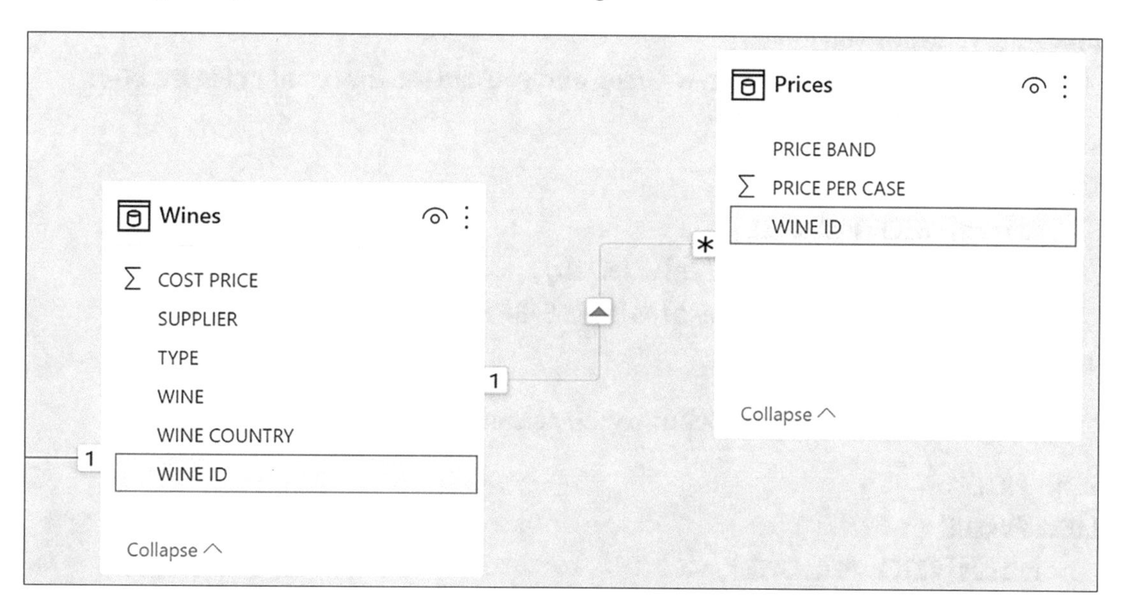

***Figure 17-3.** The Prices table can be related to the Wines table in a many-to-one relationship*

However, how would you find the price of each transaction in the Winesales table? You can't use RELATED because this function can only populate values from the "one" side of the relationship into the "many" side and the Prices table sits on the "many" side. But more importantly, the price depends on *two* criteria: the wine and the price band. In this scenario, the relationship between the tables isn't going to help you. In fact, you don't need the relationship between Wines and Prices at all. What you can do here is create a "virtual" relationship using LOOKUPVALUE in a calculated column.

The LOOKUPVALUE function has the following syntax:

= LOOKUPVALUE(result column name , search column name1, search value1, search column name2, search value2 etc.)

result column name is the column whose value you want to be returned.

search column name is the column where you want to match the first "search value." Usually, this is a column from a different table, but it can be in the same table.

search value is the value to search for in "search column name." This can be a value in a column or any single value.

The "search column name" and "search value" can be repeated for as many pairs of matching values as you need.

This is the calculated column we need and you can see the result in Figure 17-4:

```
WINE PRICE =
LOOKUPVALUE (
    Prices[PRICE PER CASE],
    Prices[WINE ID], Winesales[WINE ID],
    Prices[PRICE BAND], Winesales[PRICE BAND]
)
```

This is the same calculated column with comments:

```
WINE PRICE =
LOOKUPVALUE (
    Prices[PRICE PER CASE],
--the price to return into the Winesales table from the prices table
    Prices[WINE ID], Winesales[WINE ID],
--look in the WINE ID column of the Prices table to match the WINE ID in
the current row of the Winesales table
    Prices[PRICE BAND], Winesales[PRICE BAND]
```

```
-- AND look in the PRICE BAND column of the Prices table to match the PRICE
BAND in the current row of the Winesales table
)
```

OMER ID	WINE ID	CASES SOLD	PRICE BAND	WINE PRICE
16	10	213	C	$193.00
16	4	326	C	$73.00
20	5	70	D	$130.00
12	10	264	C	$193.00
17	3	147	E	$119.00
45	11	155	D	$120.00
11	7	173	A	$85.00
75	13	106	C	$146.00
14	13	148	B	$116.00
16	13	136	B	$116.00

Figure 17-4. *The WINE PRICE calculated column using LOOKUPVALUE*

Notice in the calculated column, we need to match both the WINE ID and the PRICE BAND, and this is where LOOKUPVALUE becomes particularly useful. The LOOKUPVALUE function allows you to find values in unrelated tables by matching values in any number of columns.

At this juncture, we must let you know that the code you have just written using the LOOKUPVALUE function is now a little outdated. Prior to the introduction of variables, it was the simplest way to achieve this outcome. However, the following code using variables and CALCULATE is an alternative approach:

```
WINE PRICE #2 =
VAR currentwine = Winesales[WINE ID]
VAR priceband = Winesales[PRICE BAND]
RETURN
CALCULATE ( VALUES ( Prices[PRICE PER CASE] ),
        Prices[PRICE BAND] = priceband,
        Prices[WINE ID] = currentwine )
```

There is no discernable difference in the performance of "WINE PRICE #2", so it is a personal choice as to which expression you prefer to use.

Finally, let's give the last word to Alberto Ferrari in his blog on the LOOKUPVALUE function here: `www.sqlbi.com/articles/introducing-lookupvalue/`

"If your search list is made up of only one-column, then LOOKUPVALUE is pretty much never your best option. Indeed, when searching for a single column, a relationship is always better: it is faster and provides a clearer structure to the model. When on the other hand you search for multiple columns, then LOOKUPVALUE comes in handy.

Another scenario where LOOKUPVALUE is preferable over a relationship in the model is when the condition you set is not a single column, but instead a more complex condition based on multiple columns. In that case, LOOKUPVALUE provides greater flexibility than a relationship."

The TREATAS Function

To understand the requirement for the TREATAS function, we must consider the following problem that has now arisen in our data model.[2] We have added a Targets table to our model that records each salesperson's yearly targets; see Figure 17-5.

[2] To follow along with the examples, use the Power BI Desktop file "5 DAX TREATAS.pbix".

YEAR	TARGET	SALESPERSON ID
2021	$1,458,759	4
2021	$1,216,703	1
2021	$1,404,567	2
2021	$1,181,700	6
2021	$1,611,780	5
2021	$1,284,953	3
2017	$846,828	2
2017	$1,353,526	3
2017	$1,424,657	5
2017	$1,350,491	1
2017	$1,154,295	4
2017	$1,377,016	6

Figure 17-5. *The Targets table*

We would like to compare our salespeople's yearly sales with their targets, as in Figure 17-6 where we are looking at sales in 2021.

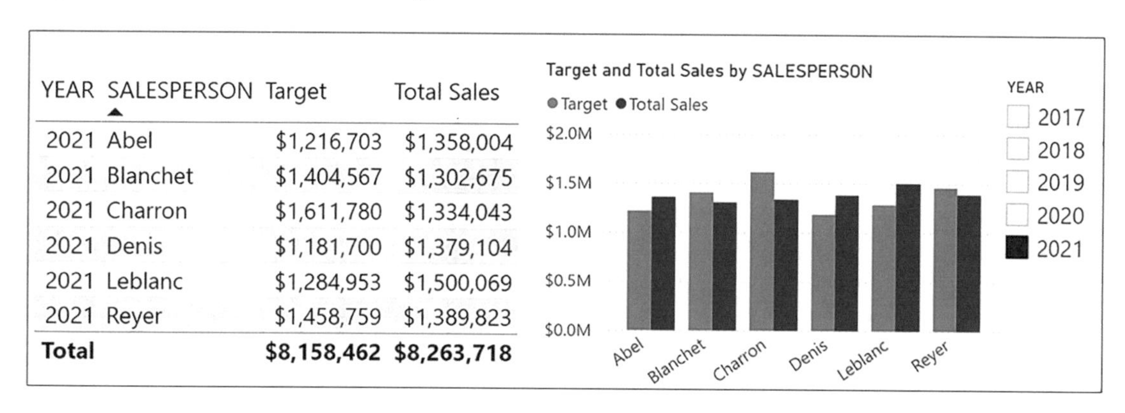

YEAR	SALESPERSON	Target	Total Sales
2021	Abel	$1,216,703	$1,358,004
2021	Blanchet	$1,404,567	$1,302,675
2021	Charron	$1,611,780	$1,334,043
2021	Denis	$1,181,700	$1,379,104
2021	Leblanc	$1,284,953	$1,500,069
2021	Reyer	$1,458,759	$1,389,823
Total		**$8,158,462**	**$8,263,718**

Figure 17-6. *Reporting on salespeople's yearly targets*

The Targets table is related to the SalesPeople table (using the SALESPERSON ID column from both tables) in a many-to-one relationship as shown in Figure 17-7. Because we will be using the Winesales table and the DateTable, we've also shown how these are related in the model.

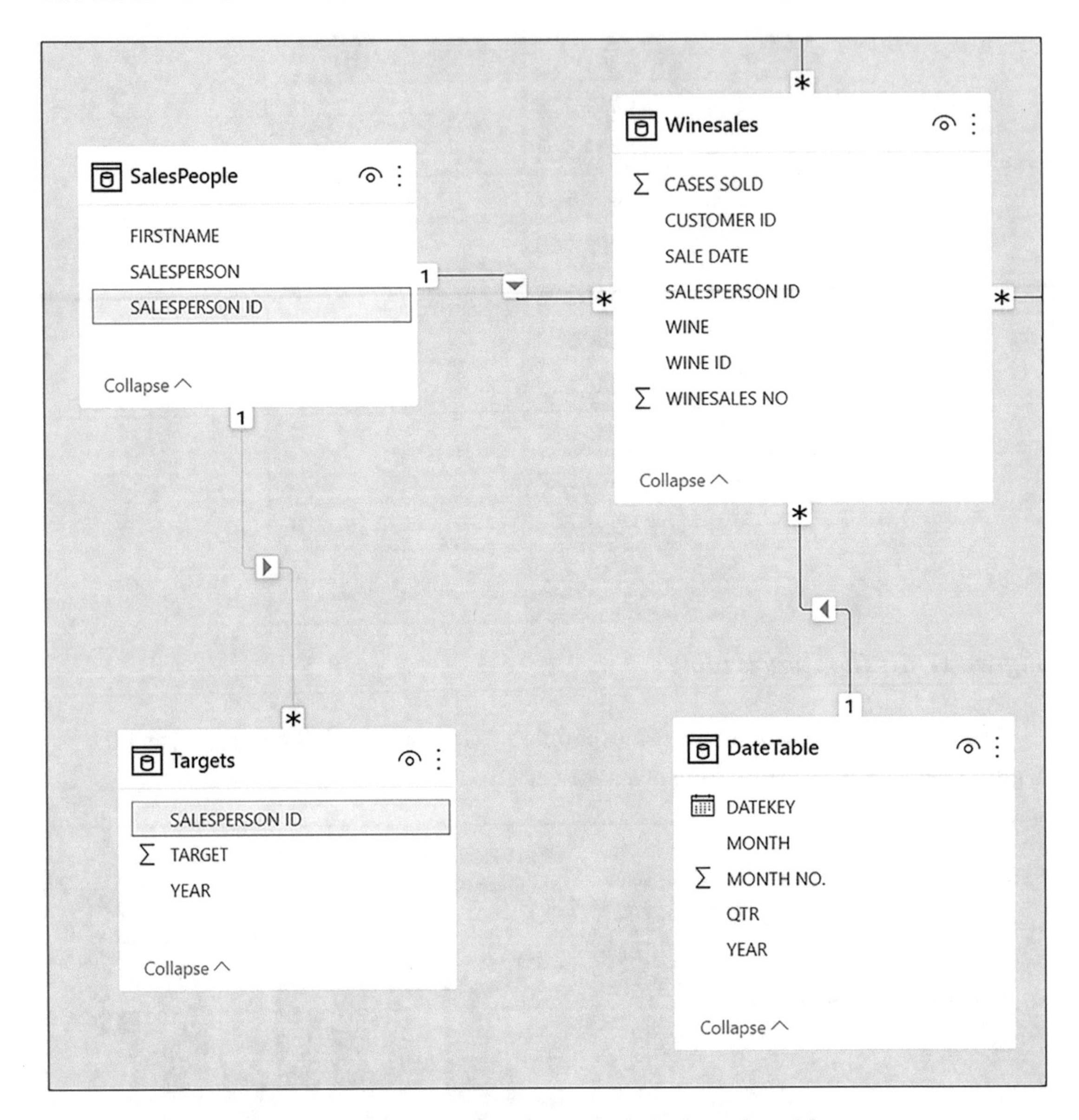

Figure 17-7. *The Targets table is related to the SalesPeople table*

We could create a measure to calculate the target values:

```
Target =
SUM ( Targets[TARGET] )
```

and then show the "Target" and "Total Sales" measure in a visual that includes the SALESPERSON column from the SalesPeople table and the YEAR column. However, from which table will we take the YEAR column, from the DateTable or from the Targets table? It's here that we meet the problem of how to get both the "Target" value and the "Total Sales" value in the same visual against each year. We get different calculations depending on which table the YEAR comes from, as shown in Figure 17-8.

YEAR from the DateTable

YEAR	SALESPERSON	Target	Total Sales
2017	Abel	$6,006,511	$1,052,606
2018	Abel	$6,006,511	$497,512
2019	Abel	$6,006,511	$852,516
2020	Abel	$6,006,511	$1,504,628
2021	Abel	$6,006,511	$1,358,004
2017	Blanchet	$5,311,862	$562,864
2018	Blanchet	$5,311,862	$606,390
2019	Blanchet	$5,311,862	$1,185,109
2020	Blanchet	$5,311,862	$1,203,006
2021	Blanchet	$5,311,862	$1,302,675
2017	Charron	$6,418,704	$872,902
2018	Charron	$6,418,704	$995,058
2019	Charron	$6,418,704	$792,385
Total		**$36,381,148**	**$29,732,482**

YEAR from the Targets Table

YEAR	SALESPERSON	Target	Total Sales
2017	Abel	$1,350,491	$5,265,266
2018	Abel	$1,534,256	$5,265,266
2019	Abel	$871,904	$5,265,266
2020	Abel	$1,033,157	$5,265,266
2021	Abel	$1,216,703	$5,265,266
2017	Blanchet	$846,828	$4,860,044
2018	Blanchet	$911,762	$4,860,044
2019	Blanchet	$1,168,168	$4,860,044
2020	Blanchet	$980,537	$4,860,044
2021	Blanchet	$1,404,567	$4,860,044
2017	Charron	$1,424,657	$5,147,366
2018	Charron	$1,242,696	$5,147,366
2019	Charron	$1,029,944	$5,147,366
Total		**$36,381,148**	**$29,732,482**

Figure 17-8. *Taking the YEAR column from either the DateTable or the Targets table won't work*

If the YEAR column comes from the DateTable, the "Total Sales" measure is correct but not the "Target" measure. If the YEAR column comes from the Target table, the targets are correct but not the total sales. If we now consider our data model in Figure 17-9, we can identify the problem.

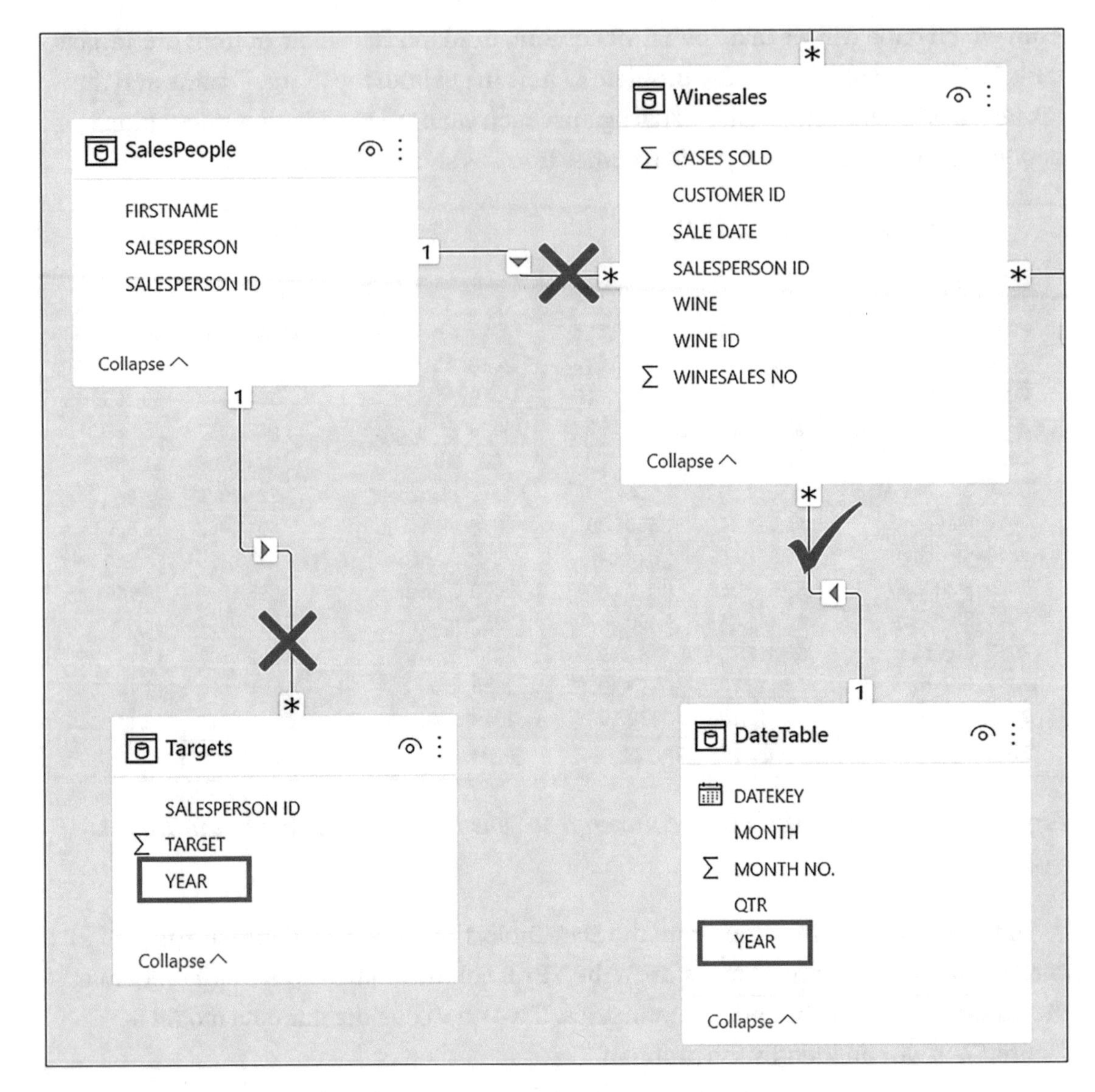

Figure 17-9. *Filtering YEAR in the DateTable filters the fact table, but filtering YEAR in the Targets table does not filter any other tables*

If we take YEAR from the DateTable, the YEAR filter is propagated to the Winesales fact table filtering "Total Sales" for each year (shown by the tick), but this filter is not propagated onward to the Targets table via the SalesPeople table (shown by the crosses) to filter the targets in each year. If we take YEAR from the Targets table, this filters the YEAR in the Targets table but won't propagate to the Winesales table to filter sales (shown by the crosses).

One solution would be to create a relationship between the YEAR field in the Targets table and the YEAR field in the DateTable. If we do this, filtering the YEAR in the Targets table would filter the YEAR in the DateTable, and this would propagate to the fact table.

The issue, however, is that in both the DateTable and the Targets table, values in the YEAR column are duplicated, so if we attempt to make this relationship, we will generate a many-to-many relationship prompting this warning message, as shown in Figure 17-10.

> ! This relationship has cardinality Many-Many. This should only be used if it is expected that neither column (Year and Year) contains unique values, and that the significantly different behavior of Many-many relationships is understood. Learn more

Figure 17-10. *You will get a warning if you attempt to create a many-to-many relationship*

We are told that such a relationship will have a "significantly different behavior" and it should not be used unless you understand the consequences of your actions. Be that as it may, this would resolve the problem because it would set a bidirectional filter. However, now is the time to take on board the conclusions at which we arrived in Chapter 13 regarding bidirectional filtering. Any changes to your data model that push it further away from the star schema structure are never to be recommended. Besides, there is another, much simpler approach, and that is to resolve the problem using DAX and the TREATAS function. This function will take the result of a table expression and use it to filter a column (or columns) from an unrelated table and this filter expression can be used in the filter argument of CALCULATE.

TREATAS has the following syntax:

= TREATAS (table expression , column1, column2 etc.)

table expression is any expression that returns a table.

column1, column2 etc. is one or more existing columns that must match the columns in the table expression that will receive the filter from the table expression.

We can now create this measure:

```
Target #2 =
CALCULATE (
    SUM ( Targets[TARGET] ),
    TREATAS ( VALUES ( DateTable[YEAR] ), Targets[YEAR] )
)
```

Notice the VALUES function used as a table expression to create a one-column table (often with only one row) containing the YEAR value from the DateTable in the current filter context, which is "2017" in the first evaluation. This one-row, one-column table is used to filter the YEAR column in the Targets table to equal "2017" and this is the filter used by CALCULATE. In Figure 17-11, you can see how this plays out in memory. The virtual one-column, one-row table (or multirow table in the evaluation of the Total row) containing the YEAR from the DateTable in the current filter context is used to filter the YEAR column in the Targets table. It's important therefore that we use the YEAR column from the DateTable in the visual.

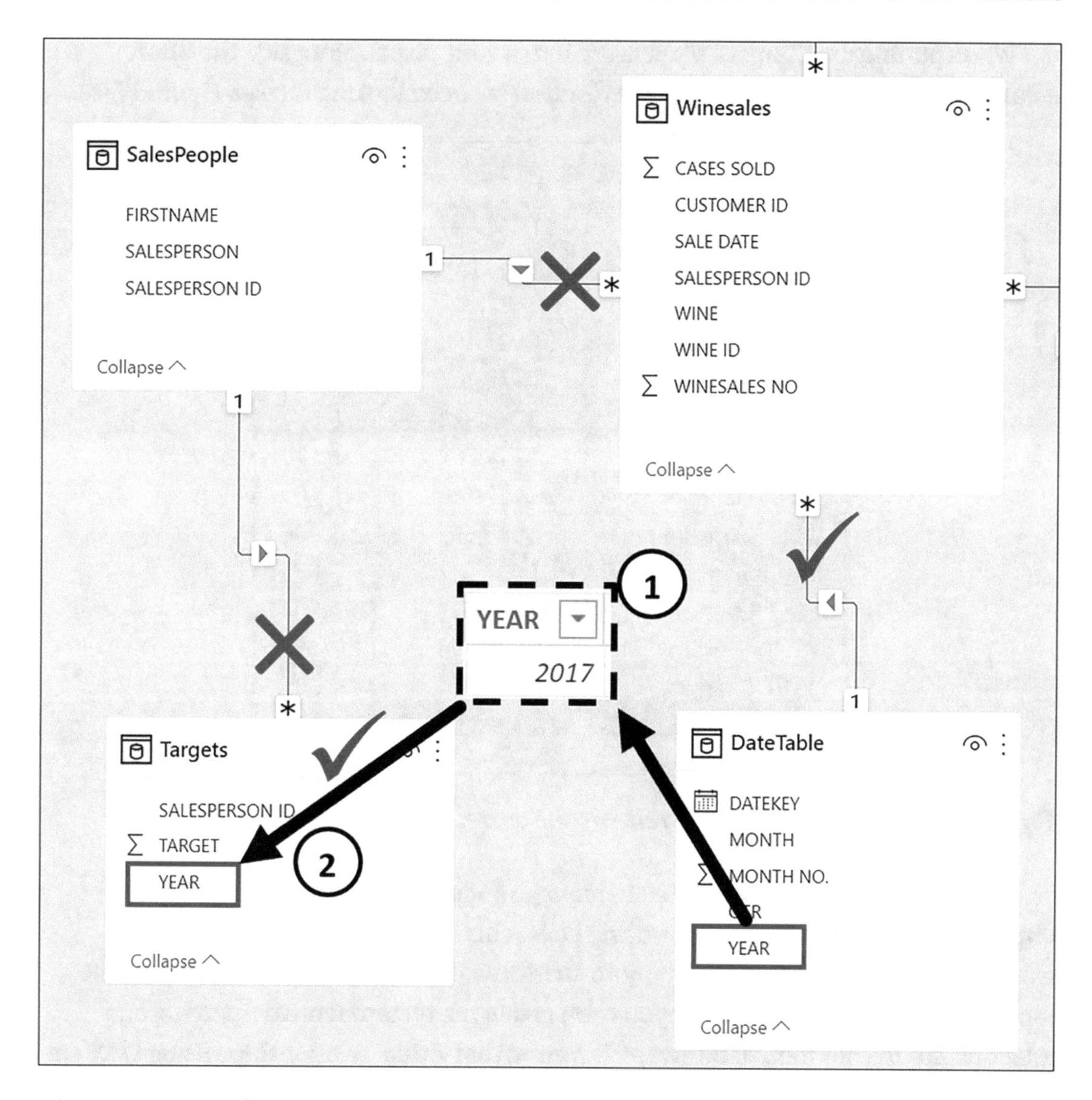

Figure 17-11. *The evaluation of the TREATAS function*

1. The first argument in TREATAS uses the VALUES function to create a virtual table containing the YEAR column from the DateTable in the current filter context, for example, "2017".

2. The second argument in TREATAS defines the YEAR column in the Targets table as the column to receive the filter from the virtual table generated by VALUES.

When putting the "Target #2" measure into a table visual, alongside the YEAR column from the DateTable, we get the result we've been looking for; see Figure 17-12.

YEAR	SALESPERSON	Target #2	Total Sales
2017	Abel	$1,350,491	$1,052,606
2018	Abel	$1,534,256	$497,512
2019	Abel	$871,904	$852,516
2020	Abel	$1,033,157	$1,504,628
2021	Abel	$1,216,703	$1,358,004
2017	Blanchet	$846,828	$562,864
2018	Blanchet	$911,762	$606,390
2019	Blanchet	$1,168,168	$1,185,109
2020	Blanchet	$980,537	$1,203,006
2021	Blanchet	$1,404,567	$1,302,675
2017	Charron	$1,424,657	$872,902
2018	Charron	$1,242,696	$995,058
2019	Charron	$1,029,944	$792,385
2020	Charron	$1,109,627	$1,152,978
Total		**$36,381,148**	**$29,732,482**

***Figure 17-12.** Using TREATAS returns the correct result for the target value*

In this chapter, you have learned to manage anomalies in the data model by implementing virtual relationships using DAX. This is always a better strategy than using bidirectional filtering and many-to-many relationships. Therefore, you need no longer be daunted by the fact that you can't create the recommended many-to-one relationships in your model. Be aware, however, that virtual relationships using DAX are never better than "real" many-to-one relationships and should only be used where no other option is possible.

CHAPTER 18

Table Expansion

In this chapter, you will learn how to reference expanded tables in your DAX code and explore how this knowledge can help you manage the limitations imposed on you by the structure of your tables within the star schema. The concept of table expansion is the final piece in the jigsaw of understanding how DAX works.[1] This implies there is some precedence in the importance of DAX concepts. However, just as in a jigsaw, it's only when all the pieces have been fitted do you see the whole picture, and we can at last reveal to you the truth about how DAX works, and any misconceptions you currently hold can now be dispelled.

The starting point in understanding table expansion is to remind you of the DAX verity; filters only propagate from the one side of a relationship to the many, unless you use the CROSSFILTER function to programmatically change the filter direction. Within this verity, you have also probably assumed, although it has never been stated unequivocally, that relationships between tables use a "primary" and a "foreign" key to perform a "lookup" from the dimension table to the fact table to enable filtering. For example, a filter on the Wines dimension will use the WINE ID column in the Wines table to "lookup" the same value in the WINE ID column of the Winesales table. This is probably how you think filter propagation works. It's not that this theory is wrong; it's just that it's not complete, and it's this misunderstanding that we will resolve in this chapter.

Before we move forward, however, we must take a closer look at the data model in the companion file for this chapter, "6 DAX Expanded Tables.pbix". You will notice there is an additional table related to the Regions table called Region Group, and this table will become important in the following sections; see Figure 18-1.

[1] To follow along with the examples, use the Power BI Desktop file "6 DAX Expanded Tables.pbix".

© Alison Box 2022
A. Box, *Up and Running with DAX for Power BI*, https://doi.org/10.1007/978-1-4842-8188-8_18

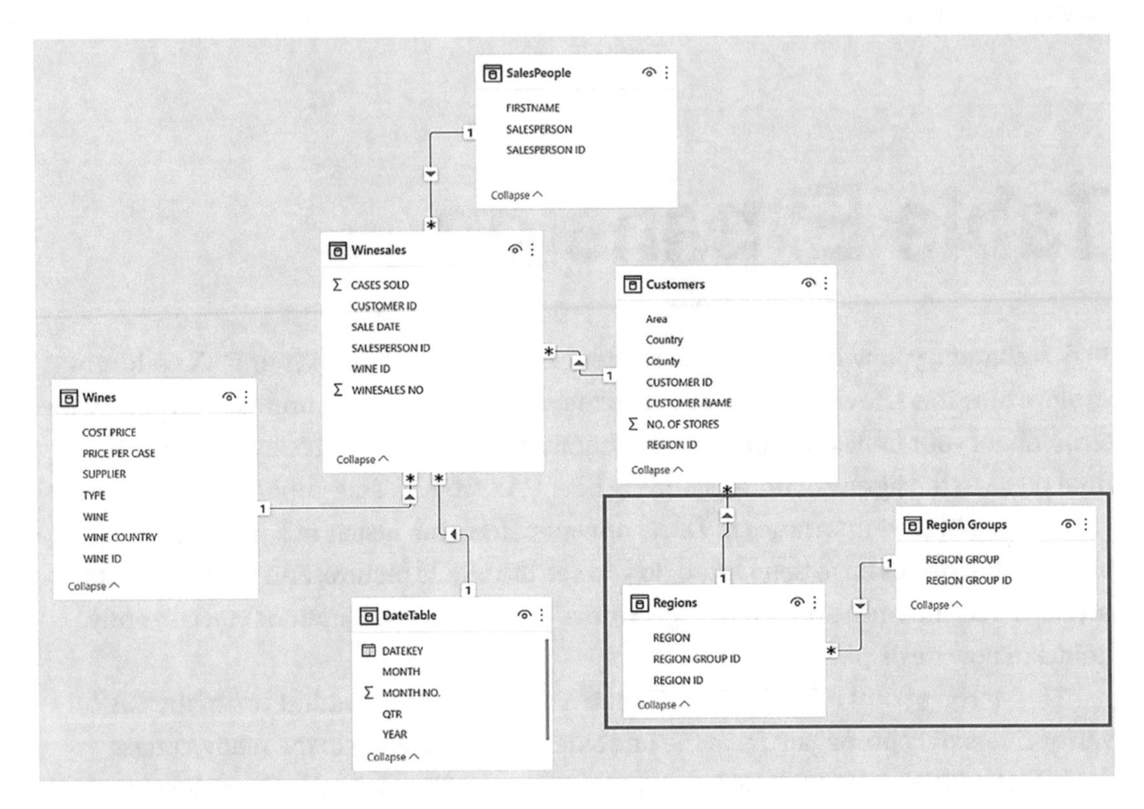

***Figure 18-1.** Please note there is an additional table, Region Group, in the data model that is related to Regions*

The importance of understanding table expansion lies in the fact that we can, at last, explain to you how filters in a data model *really* work and not an approximation of how they work. Armed with this knowledge, you will learn how to leverage table expansion to resolve the inherent problem in the data model of how to "reach" dimension and snowflake tables to perform aggregations at the larger grain. We will also be explaining why using functions such as RELATED and CROSSFILTER that can do a similar job is not always fit for purpose.

However, prior to tackling the challenging ideas behind table expansion, we must first revisit the knowledge already gained regarding the context in which filters are evaluated. If we do this, you will discover that there are some details behind filter propagation that may currently be eluding you.

Revisiting Filters

Despite rigorous explanations in this book, there remain some aspects of filters generated by DAX that remain nonsensical. Consider these two questions:

1. How is it possible that you can filter the fact table by using values in dimensions that don't exist in the fact table?
2. How can the ALL function inside CALCULATE when it's applied to the fact table remove filters that aren't placed on the fact table?

Let's start by considering the first of these incongruities; we place filters on columns in dimensions that don't exist in the fact table. For this, we need to revisit what we already know regarding column filters.

Column Filters Revisited

Throughout this book, you have authored measures using CALCULATE similar to this:

```
Abel's Cases =
CALCULATE ( [Total Cases],  Salespeople[SALESPERSON] = "abel"
 )
```

Did you ever stop to ask: How can this measure cross-filter the Winesales table using the SALESPERSON column in the SalesPeople dimension, when Winesales only contains the SALESPERSON ID? To answer this question, we must delve deeper into the nature of column filters.

In Chapter 7, you learned that column filters are more efficient than table filters and should always be used in preference where possible. However, at that stage in your knowledge of DAX, we weren't able to tell you the complete story of column filters and therefore gave you only an approximation of how column filters work.

Now in this chapter, we do not hide anything from you and state this fact: in DAX, *all filters are table filters*. This statement may come as a surprise to you considering that we took such pains to distinguish between column filters and table filters in that earlier chapter. Now we are saying that column filters are table filters too!

The complete explanation as to why column filters are more efficient than table filters is not that you are placing a filter directly on a column, but that the virtual table generated by a column filter is more efficient than the virtual table generated by an explicit filter expression. This is quite a challenging concept, and so we must again dig more deeply.

Let's start by considering this measure that generates a filter on the SALESPERSON column of the SalesPeople table:

```
Abel's Cases =
CALCULATE ( [Total Cases], SalesPeople[SALESPERSON] = "abel" )
```

In the evaluation of this measure, the DAX engine in memory converts this column filter to this expression:

```
Abel's Cases Real =
CALCULATE (
    [Total Cases],
    FILTER ( ALL ( SalesPeople[SALESPERSON] ),
                                   SalesPeople[SALESPERSON] = "abel" )
)
```

If we look at this code, we can see that DAX, using the ALL function, generates a one-column table comprising a distinct list of salespeople's names. This table is then iterated by FILTER to find the value that equates to "Abel", and this filtered table is then used to filter the Winesales table accordingly; see Figure 18-2.

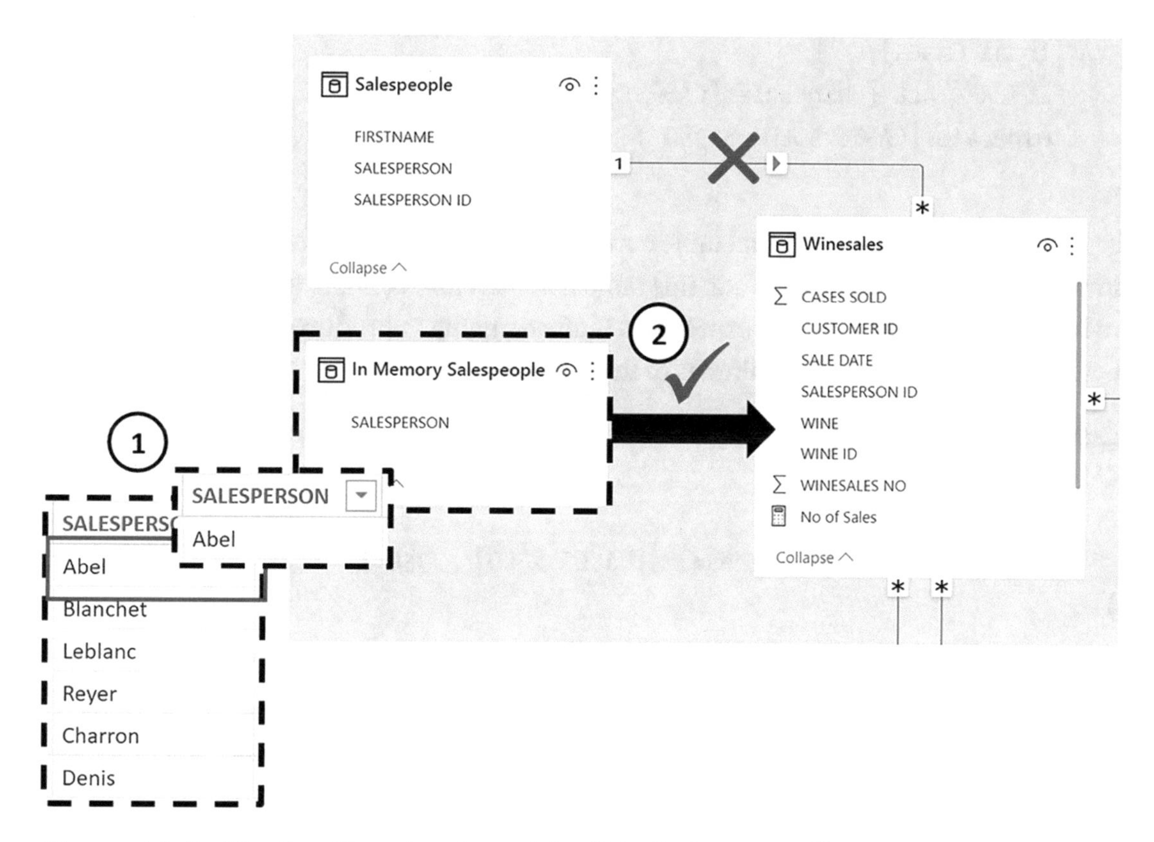

Figure 18-2. *The "real" evaluation of the "Abel's Cases Real" measure*

1. The DAX engine uses FILTER to generate a one-column table containing the distinct values in the SALESPERSON column. FILTER iterates this table to filter "Abel".

2. The filtered virtual table generated by FILTER is used to filter the Winesales table.

Let's look at another example of a column filter by exploring the evaluation of this measure:

```
Cases GT 350 =
CALCULATE ( [Total Cases],  Winesales[CASES SOLD] > 350 )
```

DAX converts this filter to the following:

```
Cases GT 350 Real =
CALCULATE (
```

```
    [Total Cases],
    FILTER ( ALL ( Winesales[CASES SOLD] ),
     Winesales[CASES SOLD] > 350 )
)
```

This code generates a virtual table containing a distinct list of the cases sold values in the Winesales table. In our data, this table will therefore contain **409** rows for FILTER to iterate. We can see how this expression is always going to produce a more efficient evaluation than using a table filter as in this measure:

```
Cases GT 350  =
CALCULATE (
    [Total Cases],
    FILTER ( Winesales, Winesales[CASES SOLD] > 350 )
)
```

Here, FILTER must iterate all the rows in the fact table, which will be **2,207** iterations of our Winesales fact table (the fact table often contains millions of rows).

At this juncture, we can also revisit the "Sales for Red or French #1" measure that we authored in Chapters 6 and 7:

```
Sales for Red or French #1=
CALCULATE (
    [Total Sales],
    Wines[TYPE] = "red"
        || Wines[WINE COUNTRY] = "France"
)
```

We noticed that the problem with this measure was that if there were filters on either the TYPE or the WINECOUNTRY column, the filter didn't work (refer to Figure 6-10). We can, at last, explain why. It's because DAX converts the measure internally to this:

```
Sales for Red or French #1=
CALCULATE (
    [Total Sales],
ALL ( Wines[TYPE], Wines[WINE COUNTRY]),
FILTER( Wines,
    Wines[TYPE] = "red"
```

```
        || Wines[WINE COUNTRY] = "France"
)
)
```

Therefore, filters are always removed from the TYPE or WINECOUNTRY column because of the presence of ALL.

Now that you understand that column filters are converted to table filters and that all filters are table filters, we seem no further on in answering the question we posed before. In the "Abel's Cases Real" measure, we are filtering the SALESPERSON column in the SalesPeople table, but the Winesales table only contains the SALESPERSON ID column, so how can the filter propagate from the SalesPeople table to the Winesales table? We'll leave you hanging onto this thought while we explore the second example of nonsensical filters. How can the ALL function applied to the fact table remove filters that aren't placed on the fact table?

The ALL Function Revisited

In Figure 18-3, on the evaluation of the "Total Cases" measure, we know filters have been placed on the WINE and SALESPERSON columns, propagating filters from the Wines and the SalesPeople dimensions to the Winesales fact table, respectively. We've then used the "All Winesales" measure to remove these filters:

```
All Winesales =
CALCULATE ( [Total Cases], ALL ( Winesales ) )
```

WINE	Total Cases	ALL Winesales
Bordeaux	8,531	423,224
Champagne	10,993	423,224
Chardonnay	8,099	423,224
Chenin Blanc	2,769	423,224
Chianti	3,699	423,224
Grenache	6,123	423,224
Lambrusco		423,224
Malbec	4,738	423,224
Merlot	4,520	423,224
Piesporter	2,064	423,224
Pinot Grigio	4,211	423,224
Rioja	5,669	423,224
Sauvignon Blanc	5,318	423,224
Shiraz	3,137	423,224
Total	**69,871**	**423,224**

SALESPERSON

- ■ Abel
- ☐ Blanchet
- ☐ Charron
- ☐ Denis
- ☐ Leblanc
- ☐ Reyer

Figure 18-3. *Filters have been placed on the WINE and SALESPERSON columns, not on the fact table. The ALL function removes filters from the fact table*

We learned in Chapter 8 that the ALL function, when nested inside CALCULATE, *removes* filters. But there are no filters on the Winesales fact table to remove, only cross-filters. The filters have been placed on columns in the dimensions, so how can ALL remove filters from the Winesales table when there are no filters to remove?

Expanded Tables Explained

To answer these probing questions and to truly grasp the behaviors of DAX filters, you must understand table expansion**.** When a measure is evaluated, many-to-one relationships allow table expansion to take place. Table expansion results in the creation of virtual tables by the DAX engine that include the columns of the base table and then expand into all the columns from related tables on the one side of the relationship. The DAX engine then uses the expanded table to group by values in the expanded table's columns and apply filters accordingly. Therefore, every table has a matching expanded version of itself that is generated in memory that contains all its *own columns plus any* *columns from tables that are related to it,* which are on the one side of the relationship either directly or indirectly. Relationships only exist to generate expanded tables.

Therefore, we can now talk about both *base* tables and *expanded* tables in our data model. Base tables are just our tables. Expanded tables are our base tables that also contain all the columns from tables that are related to them. In our model, for example, we have three tables that will expand: Winesales, Customers, and Regions. The Winesales expanded table will contain all the columns from all the tables in the model. The Customers expanded table will include all the columns from the Regions dimension and the Region Groups dimension. The Regions expanded table will include all the columns from the Region Groups dimension. In Figure 18-4, we have redesigned our data model to show what it might look like in memory on the evaluation of a measure. Notice there are no relationships between the tables because relationships only exist to generate expanded tables.

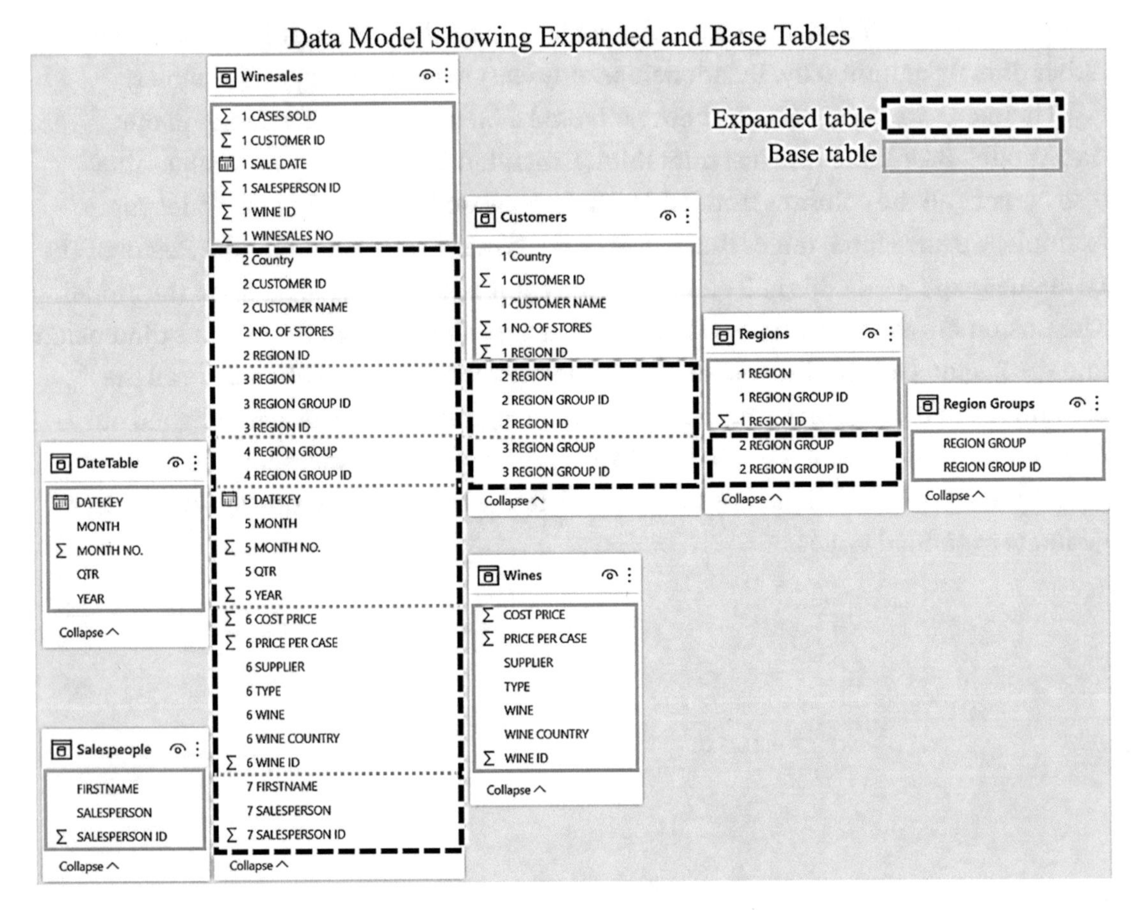

***Figure 18-4.** The Winesales, Customers, and Regions tables all expand on the evaluation of measures*

Once a filter is applied to a column, all the expanded tables containing that column are also filtered. Consider Figure 18-5, which shows the virtual expanded tables and base tables in Model view. We're looking at what happens when we filter the SALESPERSON column from the SalesPeople base table or the REGION GROUP column from the Region Groups base table.

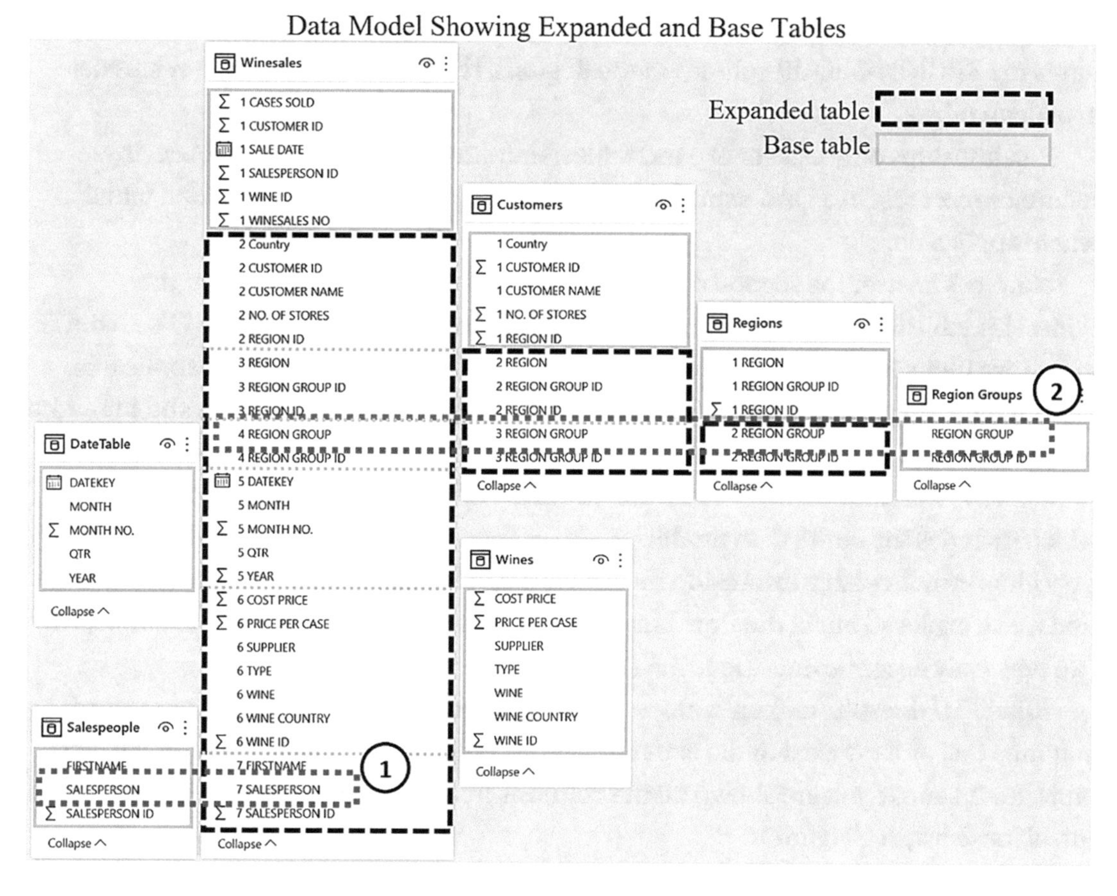

***Figure 18-5.** How tables are expanded in the data model*

1. Filtering SALESPERSON from the SalesPeople base table filters the Winesales expanded table.

2. Filtering REGION GROUP from the Region Groups base table filters the Regions expanded table, the Customers expanded table, and the Winesales expanded table.

So now we can answer the first of the questions we posed. How can a value in the SALESPERSON column in the SalesPeople dimension filter the Winesales fact table when that value doesn't exist in the Winesales table? Now you understand that it does exist in the Winesales table. It exists in the Winesales *expanded* table. When we place a filter on the SALESPERSON column, both the SalesPeople base table and the Winesales expanded table are filtered accordingly. Another example would be a filter on the REGION GROUP column in the Region Groups base table. Notice that this filters the REGION GROUP column in the Regions, the Customers, and the Winesales expanded tables.

Relationships only exist to expand tables; they are not used to filter tables. Any reference to a table in a DAX expression is always a reference to the expanded table, where applicable.

Now let's answer the second question. How can filters be removed from the Winesales table when it has no direct filters on it? When we use ALL inside CALCULATE to remove filters from a table, it removes filters from the *expanded table,* if applicable. This includes any columns from dimensions related to the expanded table and therefore includes columns where the filter was originally generated. So the expression *"ALL (Winesales)"* will remove any filters from any of the base tables related to Winesales, which includes the entire data model.

Understanding table expansion means we can now clarify certain behaviors in DAX that we've explored but at the time have not been able to fully explain. For example, we can now truly describe how the RELATED function works.

RELATED doesn't "lookup" values in related tables but instead allows you to find columns that already exist in the expanded table. When you use RELATED on the fact table, for instance, you are shown all the columns from the expanded fact table in the IntelliSense list; see Figure 18-6.

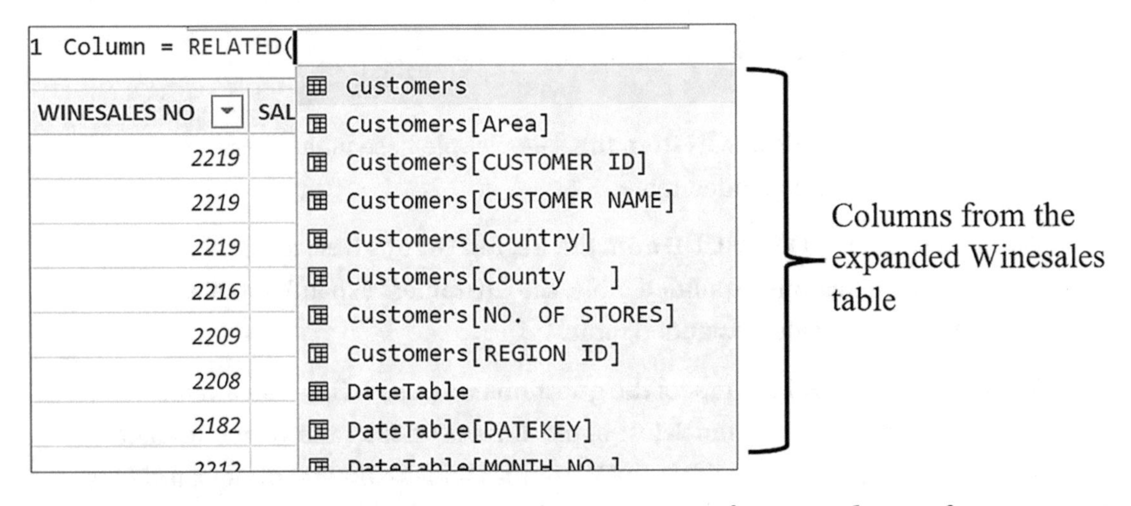

Figure 18-6. *The RELATED function allows you to reference columns from expanded tables*

Like RELATED, the ALLEXCEPT and SUMMARIZE functions also allow you to use the columns in expanded tables. When constructing an expression using these functions, if you reference a fact table or a snowflake dimension, you are again presented with all the columns from the expanded table in the IntelliSense list; see Figure 18-7.

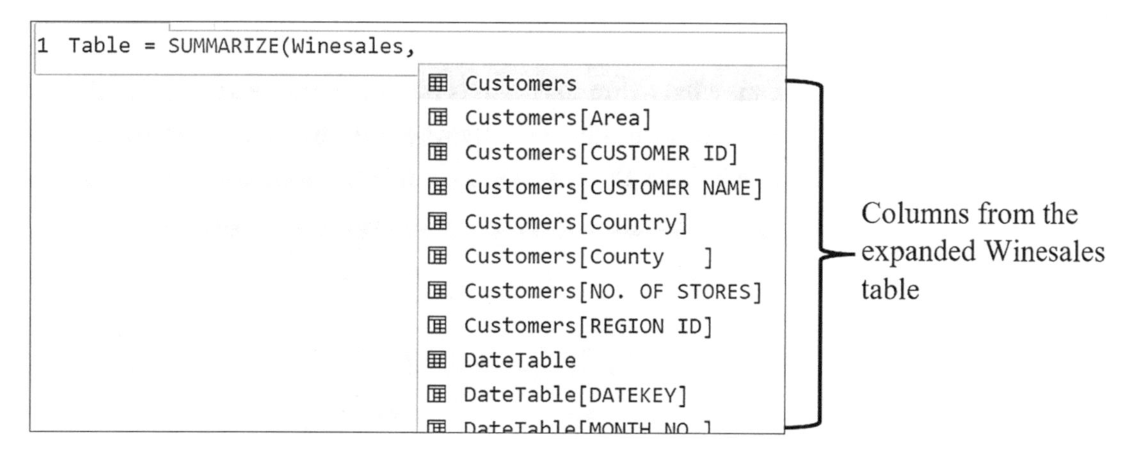

Figure 18-7. *SUMMARIZE will also reference expanded tables*

You may be thinking that knowledge of table expansion is purely theoretical. It explains certain behaviors regarding filter propagation but doesn't lead you forward in constructing more complex DAX expressions. Now is the time to change that perception of table expansion and to learn how to put your knowledge of expanded tables to beneficial use.

Leveraging Expanded Tables

For the most part, the reason you will use table expansion in your expressions is to "reach" dimensions to perform aggregations on columns within them. You may think that we've already covered this scenario when we looked at the CROSSFILTER function that enabled you to reverse the direction of filter propagation. The RELATED function also allows you to pull values from dimensions and snowflake tables into the fact table to enable such aggregations. However, both these approaches are not best practices for reasons we will elucidate as you read on.

"Reaching" Dimensions

Let's see how table expansion can allow you to break free from the limitations imposed on you by star and snowflake schemas. For this, as we've often done before, we'll work through a scenario.

You have been asked to calculate in how many different regions you've sold each wine. The Regions table is a snowflake dimension. It is related to the Customers table that's in turn related to Winesales. Currently, the only way you can deduce in which region a transaction was made is through the Customers table; see Figure 18-8.

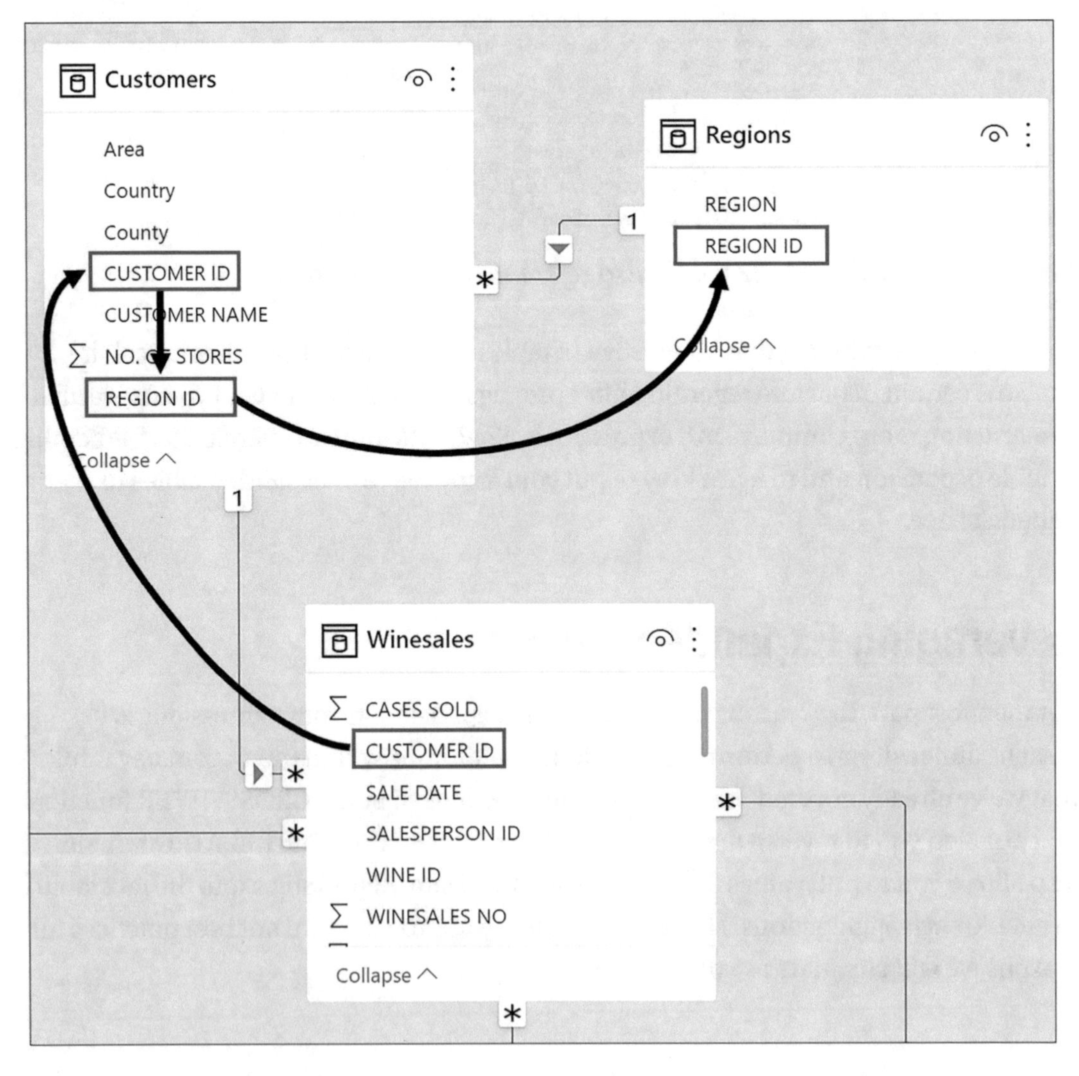

***Figure 18-8.** The region in which a transaction was made can only be found through the Customers table*

You now know, however, that all the columns in the Regions table are in the expanded Winesales table. Therefore, one approach would be to create a calculated column in the Winesales base table using RELATED to find the REGION column from the expanded Winesales table as shown in Figure 18-9.

```
1 REGION = RELATED(Regions[REGION])
```

WINESALES NO	SALESPERSON ID	CUSTOMER ID	WINE ID	CASES SOLD	REGION
2219	6	25	5	168	India
2219	6	25	5	168	India
2219	6	25	5	168	India
2216	6	36	7	123	Nigeria
2209	6	80	13	115	Argentina
2208	6	24	10	331	United States
2182	6	7	11	150	United States
2212	6	3	11	175	India
2153	6	5	10	313	Nigeria
2150	6	29	4	236	China

Figure 18-9. *You can use RELATED in the fact table to show the region name*

You could then write the measure "Distinct Regions" using DISTINCTCOUNT on this calculated column, as shown in the following:

```
Distinct Regions =
DISTINCTCOUNT ( Winesales[REGION] )
```

However, all that's happening here is that you are accessing the REGION column in the expanded Winesales table. You also know that calculated columns should be avoided if possible. There is a better way to calculate the distinct number of regions, and that is to use CALCULATE with a table filter that will filter the Regions table. To do this, we first must remind ourselves how we construct the filter arguments in CALCULATE.

You've learned that the filter arguments inside CALCULATE can contain a table expression. But the filter argument doesn't have to be a table expression; it can just be a reference to a table. If you reference a table in the filter argument of CALCULATE, this will always be the *expanded table,* where applicable.

Returning to calculating the number of different regions in which you've sold your wines, you can use the expanded Winesales table that contains the REGION column as the filter for CALCULATE. If you do this, you can then use a measure to count the rows of the Regions table that have been filtered via the Winesales expanded table. This would be the measure:

```
Distinct Regions =
CALCULATE ( COUNTROWS ( Regions ), Winesales )
```

You must note the simplicity of this expression but the complexity of the concept that lies behind it and also remember something we stated earlier; with DAX, the devil is in the detail.

We can see the evaluation of this measure in Figure 18-10.

WINE	Distinct Regions
Bordeaux	18
Champagne	19
Chardonnay	19
Chenin Blanc	19
Chianti	18
Grenache	17
Malbec	20
Merlot	18
Piesporter	18
Pinot Grigio	17
Rioja	17
Sauvignon Blanc	20
Shiraz	18
Total	**20**

Figure 18-10. *The evaluation of "Distinct Regions" using the expanded Winesales table*

The "Distinct Regions" measure uses the expanded Winesales table in the filter argument of CALCULATE to filter the Regions base table. In the evaluation of this measure, we know that the filter on the WINE column in the Wines table will filter the WINE column in the expanded Winesales table. We also know that the expanded Winesales table contains all the columns in the Regions table. Therefore, the regions where we've sold each wine in the current filter context will also be filtered. The expanded Winesales table, filtered for each wine in the current filter context, is used to filter the Regions table accordingly. The Regions table now contains only regions where the wine in the current filter context was sold and the rows of the filtered Regions dimension are counted; see Figure 18-11.

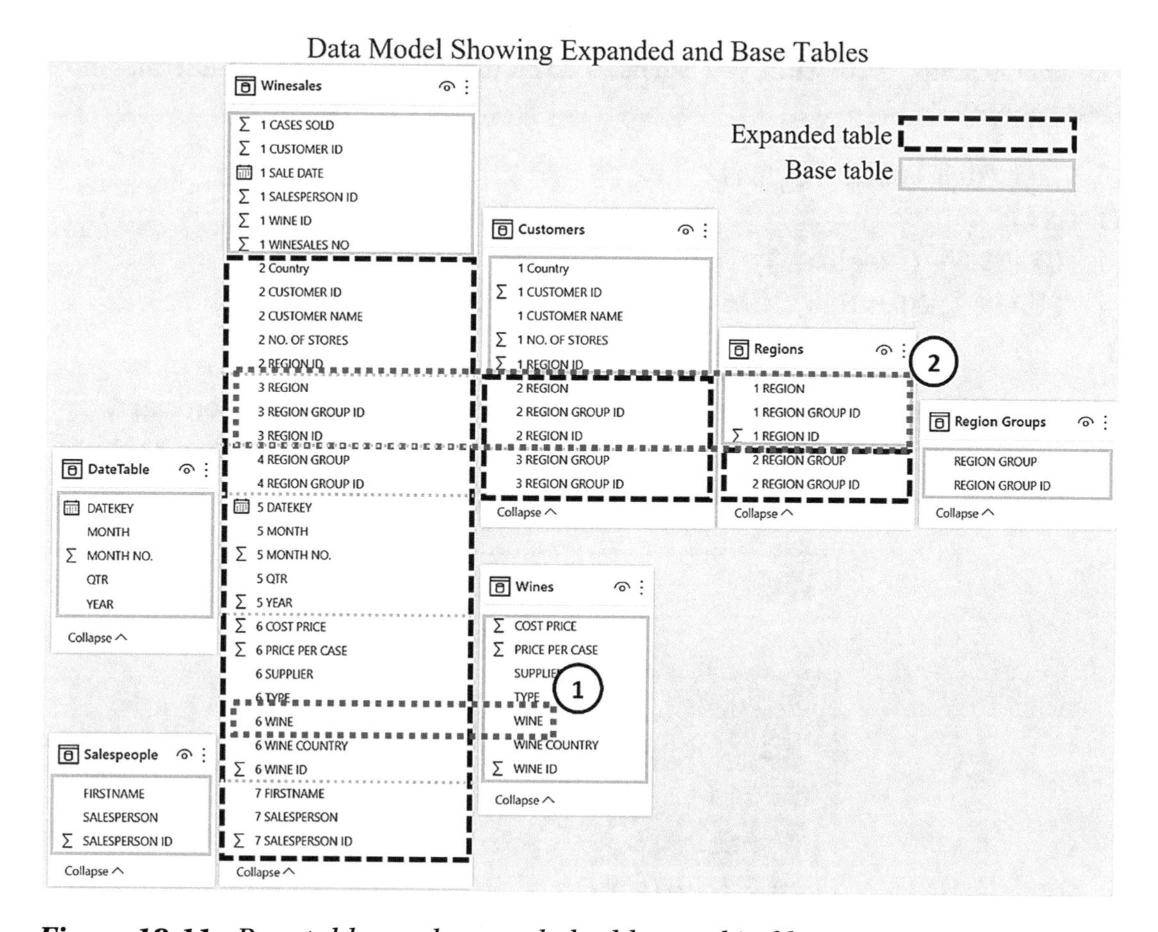

Figure 18-11. *Base tables and expanded tables used in filter propagation*

1. The WINE column in the Wines table filters the WINE column in the expanded Winesales table.
2. The expanded Winesales table contains all the columns in the Regions table. The filter in the expanded Winesales fact table is used to filter the Regions table whose rows are then counted.

What we can conclude from this measure is that with CALCULATE, you can use an expanded table to filter a base table.

Let's look at another example of using expanded tables in our code but this time to author a more challenging calculation. We are going to repeat the scenario before, in that you've been asked to find the number of different regions where you've sold wines, but this time, you must consider only high-volume regions. You've identified that high-volume regions are any regions where transactions of CASES SOLD are greater than 325. To do this calculation, rather than using the entire expanded Winesales table as in the "Distinct Regions" expression, you can use FILTER to filter the expanded Winesales table (highlighted):

```
Distinct High Volume Regions=
CALCULATE (
    COUNTROWS ( Regions ),
    FILTER ( Winesales, Winesales[CASES SOLD] >325 )
)
```

When you put this measure into a Table visual, you will find that for "Bordeaux", there are **18** regions where there are transactions of CASES SOLD greater than 325 but when selling "Grenache", there are only **7** regions; see Figure 18-12.

WINE	Distinct High Volume Regions
Bordeaux	18
Champagne	18
Grenache	7
Malbec	1
Sauvignon Blanc	15
Total	**20**

Figure 18-12. *The "Distinct High Volume Regions" measure evaluated in a Table visual*

If we examine the evaluation of the "Distinct High Volume Regions" measure, in Figure 18-13, you can see that it varies from the "Distinct Regions" measure only in the additional step where the Winesales base table is filtered. The measure filters the WINE column in the Wines dimension and also filters the expanded Winesales table. The FILTER function further filters the Winesales base table to rows where CASES SOLD is greater than 325. The columns from the Regions table are in the expanded Winesales table and so are also filtered. Counting the number of rows in the Regions table reflects only the regions filtered in the expanded Winesales table.

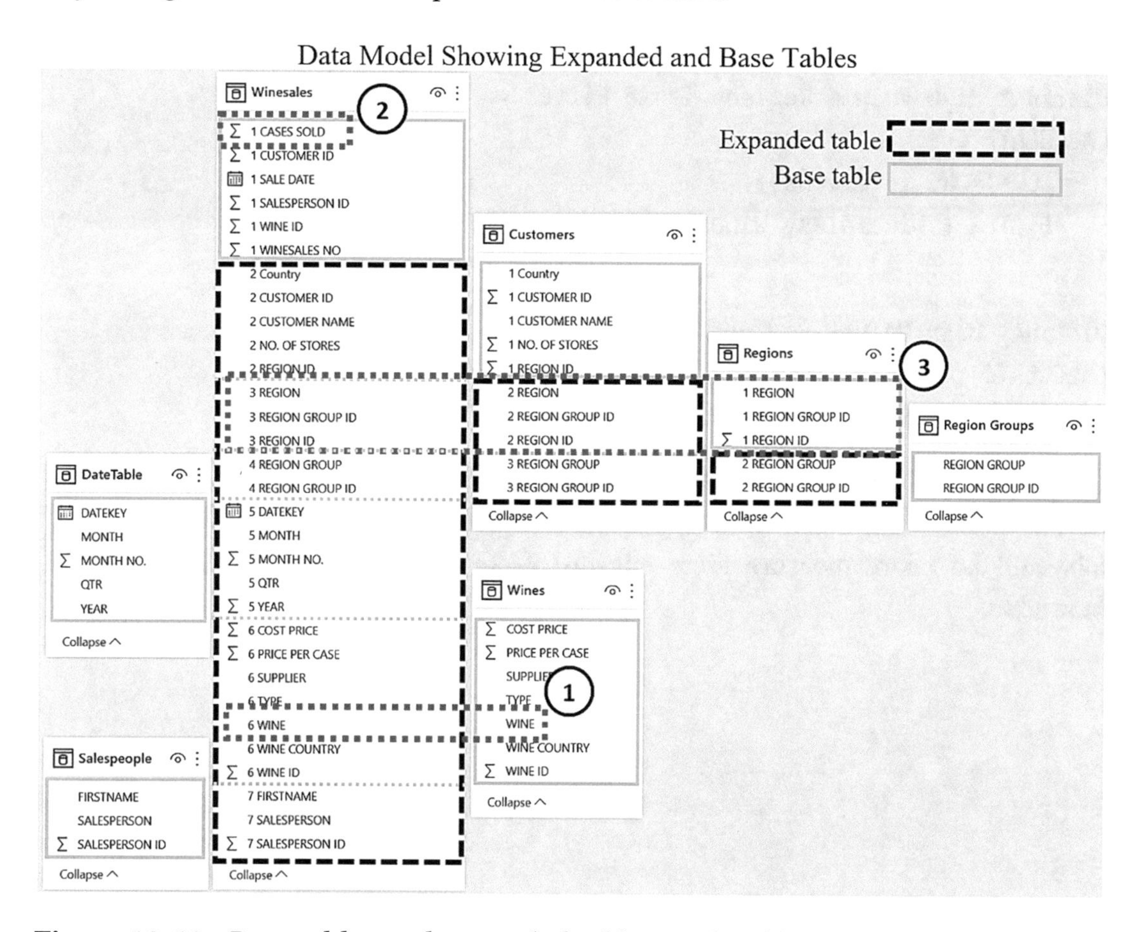

Figure 18-13. *Base tables and expanded tables used in filter propagation*

1. The WINE column filters the WINE column in the expanded Winesales table.
2. The CASES SOLD column in the Winesales base table is filtered for greater than 325.
3. The expanded Winesales table contains all the columns in the Regions table. The filter in the expanded Winesales fact table is used to filter the Regions table.

Knowledge of table expansion also helps to clarify a premise that we have explored a number of times throughout this book, and that is the difference between table filters and column filters. Now that we know that table filters will often involve expanded tables, let's take the measure we have just authored and compare it with another measure that looks almost identical. However, one uses a table filter, using an expanded table, and the other uses a column filter, as shown in the following:

```
Distinct High Volume Regions Table Filter =
CALCULATE (
    COUNTROWS ( Regions ),
    FILTER ( Winesales, Winesales[CASES SOLD] > 325 )
)
```

```
Distinct High Volume Regions Column Filter =
CALCULATE (
    COUNTROWS ( Regions ), Winesales[CASES SOLD] > 325 )
```

You can see in Figure 18-14 that we get different values being returned by similar measures. The reason for this is that the first measure filters the Winesales expanded table and the second measure filters only the CASES SOLD column in the Winesales *base* table.

WINE	Distinct High Volume Regions Table Filter ▼	Distinct High Volume Regions Column Filter
Bordeaux	18	21
Champagne	18	21
Sauvignon Blanc	15	21
Grenache	7	21
Malbec	1	21
Chardonnay		21
Chenin Blanc		21
Chianti		21
Lambrusco		21
Merlot		21
Piesporter		21
Total	**20**	**21**

Figure 18-14. *Similar measures can return different results*

The correct calculation, "Distinct High Volume Regions Table Filter", uses the table filter generated by the FILTER function that filters the expanded Winesales table, filtering the CASES SOLD column. This also filters the regions in the expanded table, and this is used to filter the Regions dimension. This measure then counts the rows in the Regions base table that have been filtered by the Winesales expanded table; see Figure 18-13.

The measure "Distinct High Volume Regions Column Filter" generates a filter only on the CASES SOLD column in the Winesales *base* table, and no filters are propagated in the model. It, therefore, counts *all* the rows in the Regions table irrespective of any filters in the Winesales table; see Figure 18-15.

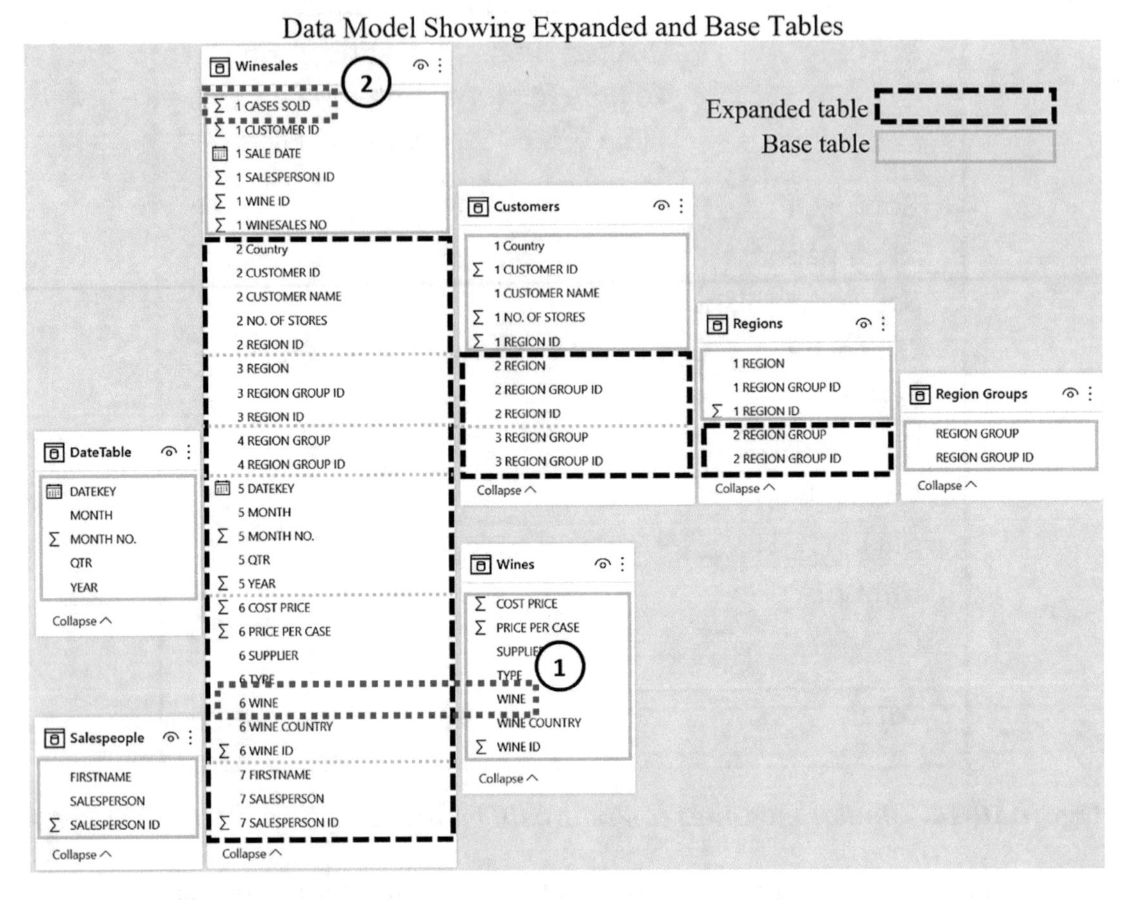

***Figure 18-15.** Base tables and expanded tables used in filter propagation*

1. The WINE column is filtered in both the expanded and base Wines table.
2. Filtering a column in the Winesales base table does not propagate filters to dimension tables.

The takeaway from these examples is that using an expanded table in the filter argument of CALCULATE enables you to pass filters into dimension and snowflake tables, in effect reversing the direction of filter propagation. This is because the expanded table contains the columns from these dimensions that can then be grouped and filtered. However, a question that must now be answered is the following: What is the difference between using expanded tables and using CROSSFILTER. Isn't the end result of using these different methods the same? For instance, we can author this expression using an expanded table:

```
Distinct Regions #1 =
CALCULATE ( COUNTROWS ( Regions ), Winesales )
```

Or we can author this measure using CROSSFILTER that we might assume would return the same result:

```
Distinct Regions #2 =
CALCULATE(COUNTROWS(Regions),
    CROSSFILTER(Winesales[CUSTOMER ID],
       Customers[CUSTOMER ID],both),
    CROSSFILTER(Customers[REGION ID],
       Regions[REGION ID],both))
```

Both these measures will "reach" the Regions table. Clearly, the second measure is a great deal clumsier than the first, but is there a difference in the evaluation? The answer is yes, there is, and we will now explain why.

Table Expansion vs. CROSSFILTER

In Chapter 13, when we explored the CROSSFILTER function, we authored a measure to sum the NO. OF STORES column in the Customers table to calculate the number of stores in which we'd sold our wines. Just to remind you, the problem was that filters don't flow from the Wines dimension through to the Customers dimension so we used the CROSSFILTER function to programmatically change the direction of the filter propagation to a bidirectional filter:

```
Total Stores =
CALCULATE (
    SUM ( Customers[NO. OF STORES] ),
    CROSSFILTER ( Winesales[CUSTOMER ID],
    Customers[CUSTOMER ID], BOTH )
)
```

However, we didn't tell you at the time, and neither would you have noticed, but this measure returns an incorrect value on the Total row; see Figure 18-16.

WINE	Total Stores
Bordeaux	728
Champagne	709
Chardonnay	805
Chenin Blanc	757
Chianti	626
Grenache	685
Malbec	736
Merlot	749
Piesporter	563
Pinot Grigio	696
Rioja	832
Sauvignon Blanc	777
Shiraz	727
Total	**1,181**

Figure 18-16. *The "Total Stores" measure is not correct in the Total row*

Many of the same customers will have bought each wine, so we know that the total of **1,181** will not be the sum of the total values for each wine. However, you might think this value looks about right and so believe it. The value in the Total row should be the total number of stores in which we've sold all our wines. This value is not correct because in the Customers table, we have five customers to whom we've sold no wines. If we "show items with no data" in a Table visual where we calculate the "Total Sales" measure, we can see who they are; see Figure 18-17.

CUSTOMER NAME	Total Sales
Acme & Sons	
Bloxon Bros.	
Jones Ltd	
Sainsbury's	
Smith & Co	
Back River & Co	$0
Palo Alto Ltd	$14,836
St. Leonards Ltd	$16,965
Victoria Ltd	$24,710
Brown & Co	$25,542
Brooklyn Ltd	$27,018
Canoga Park Ltd	$37,310
Loveland & Co	$38,098
Burlington Ltd	$41,552
Total	**$29,732,482**

***Figure 18-17.** There are five customers that have no sales*

The value of **1,181** shown in the Total row includes the stores for these customers. We can see these values in the Customers table in the NO. OF STORES column; see Figure 18-18.

CUSTOMER NAME	REGION ID	County	Area	Country	NO. OF STORES
Bloxon Bros.	2000				11
Acme & Sons	2000				4
Jones Ltd	100				5
Sainsbury's	100				24
Smith & Co	100				25

***Figure 18-18.** Customers with no sales have values in the NO. OF STORES column*

We haven't sold any wine to these customers, so clearly their stores shouldn't be included in the total number of stores in which we've sold our wines. Our total is out by **69**.

What's happening here is that the "Total Stores" measure uses a bidirectional filter. When it arrives at the evaluation of the Total row, the filters are removed from the WINE column of the Wines dimension, and therefore, there is no filter to propagate to the Customers dimension. With no filters propagated, it sums all the values in the NO. OF STORES column. In other words, bidirectional filters are only active if filters are active.

So how do you calculate the correct value of **1,112** in the Total row?

What you must do here is use the expanded Winesales fact table as the filter for the Customers table. This is because, unlike bidirectional filtering, filters from expanded tables are always active. When the Total row is evaluated, the expanded Winesales fact table contains only those customers who have bought wines, and so this will filter the Customers dimension accordingly.

This is the measure that will give you the correct total:

```
Total Stores #2 =
CALCULATE (
    SUM ( Customers[NO. OF STORES] ),
    Winesales
)
```

You can now see in Figure 18-19 that the Total row now shows **1,112**.

WINE	Total Stores #2
Bordeaux	728
Champagne	709
Chardonnay	805
Chenin Blanc	757
Chianti	626
Grenache	685
Malbec	736
Merlot	749
Piesporter	563
Pinot Grigio	696
Rioja	832
Sauvignon Blanc	777
Shiraz	727
Total	**1,112**

Figure 18-19. *Using table expansion returns the correct value in the Total row*

When working with DAX, not only must you have to have an eye for detail and a suspicious mind, but you must also understand table expansion.

Using Snowflake Schemas

Understanding table expansion also explains how we can have problems with "snowflake"-type schemas. This is where there may be a chain of several tables all related in one-to-many relationships through to the fact table. In our data model, we've extended our Regions snowflake by adding another table, Region Groups, which is related to Regions via the REGION GROUP ID. We can see in Figure 18-20 how the Region Groups table is related to the Regions table through the REGION GROUP ID, the Regions table is related to the Customers table through the REGION ID, and the Customers table is related to Winesales through the CUSTOMER ID.

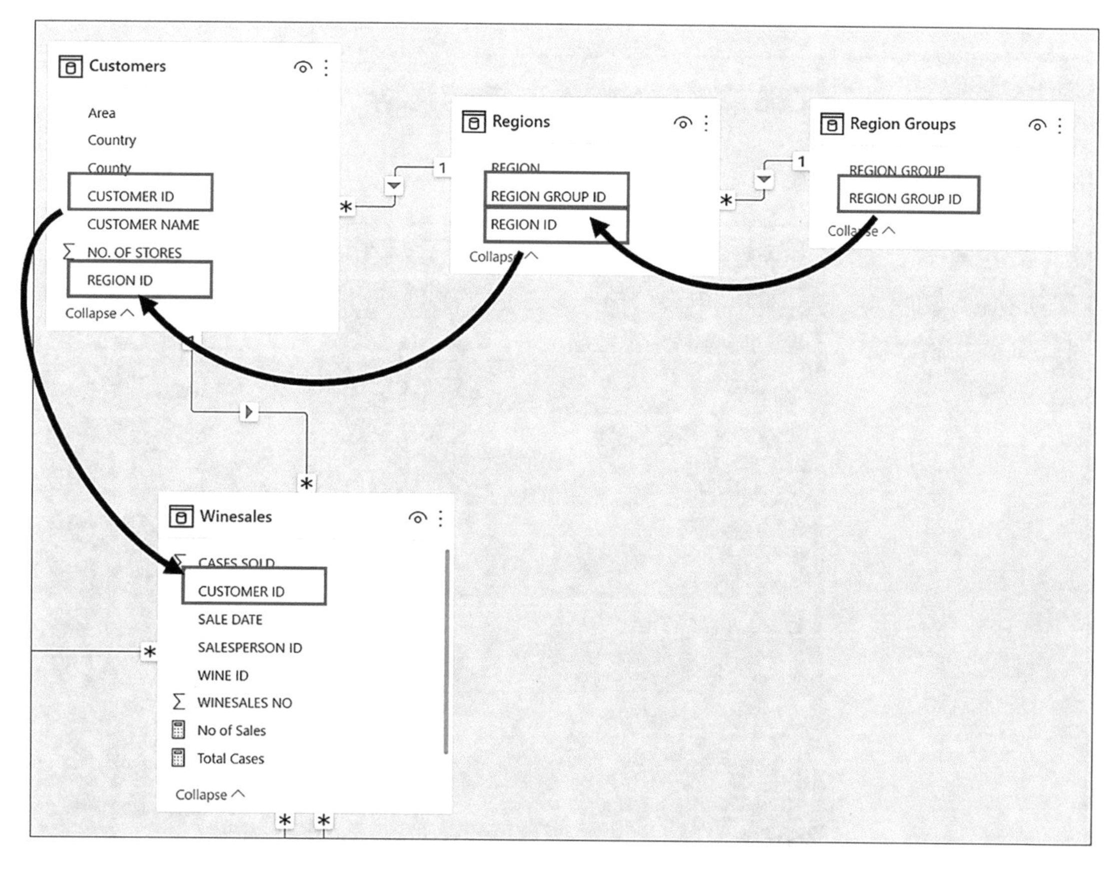

Figure 18-20. *A snowflake schema comprising Region Groups, Regions, and Customers*

In Figure 18-21, we've filtered "South West" Region Group in a slicer and are showing customers in that Region Group in the Table visual. We've attempted to calculate the total sales for these customers (**3,512,539**) so that we can use this value as a denominator to calculate the percentage each customer's sales are of the total for the "South West" region group. This is the measure we have authored using ALL on the Customers table:

```
Total Sales for All Customers in Region Group wrong =
CALCULATE ( [Total Sales], ALL ( Customers ) )
```

As you can see in Figure 18-21, it does not return the correct result, which should be **$3,512,539**. You will also notice that because we are removing all the filters from the Customers table, the Table visual now shows all our customers, not just those in the "South West" region group.

REGION GROUP

- [ ] East
- [ ] North
- [ ] North East
- [ ] South
- [ ] South East
- [x] South West
- [ ] West

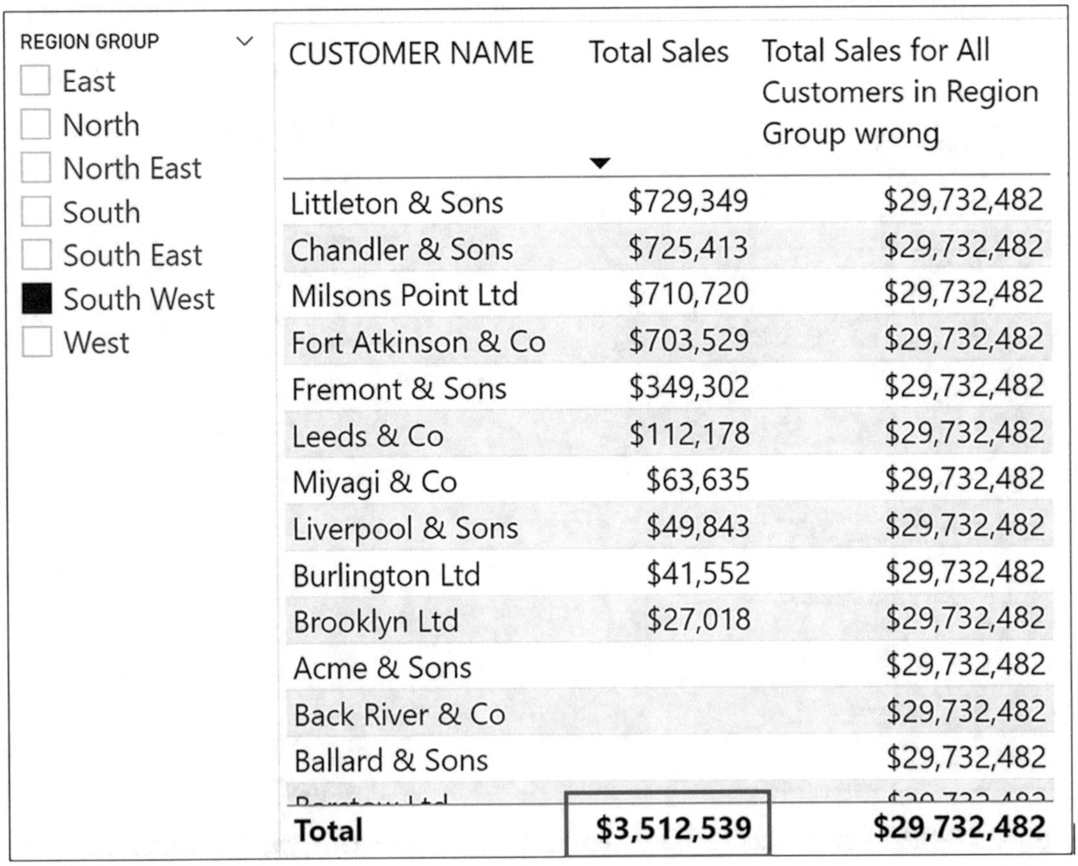

CUSTOMER NAME	Total Sales	Total Sales for All Customers in Region Group wrong
Littleton & Sons	$729,349	$29,732,482
Chandler & Sons	$725,413	$29,732,482
Milsons Point Ltd	$710,720	$29,732,482
Fort Atkinson & Co	$703,529	$29,732,482
Fremont & Sons	$349,302	$29,732,482
Leeds & Co	$112,178	$29,732,482
Miyagi & Co	$63,635	$29,732,482
Liverpool & Sons	$49,843	$29,732,482
Burlington Ltd	$41,552	$29,732,482
Brooklyn Ltd	$27,018	$29,732,482
Acme & Sons		$29,732,482
Back River & Co		$29,732,482
Ballard & Sons		$29,732,482
Total	**$3,512,539**	**$29,732,482**

Figure 18-21. *Calculation of the the total sales for all customers in the region group is not correct*

Let's now explain why we get the wrong calculation. When we use ALL inside CALCULATE to remove filters from a table, it removes filters from the expanded table, if applicable. This measure, therefore, removes filters from the expanded Customers table and so also removes filters from both the Regions table and the Region Groups table. It, therefore, calculates a total for *all* region groups. The Region Groups table is at the end of the snowflake of tables, so this is the same value as the grand total sales.

Figure 18-22 shows how removing filters from the expanded Customers table will also remove filters from Regions and Region Groups.

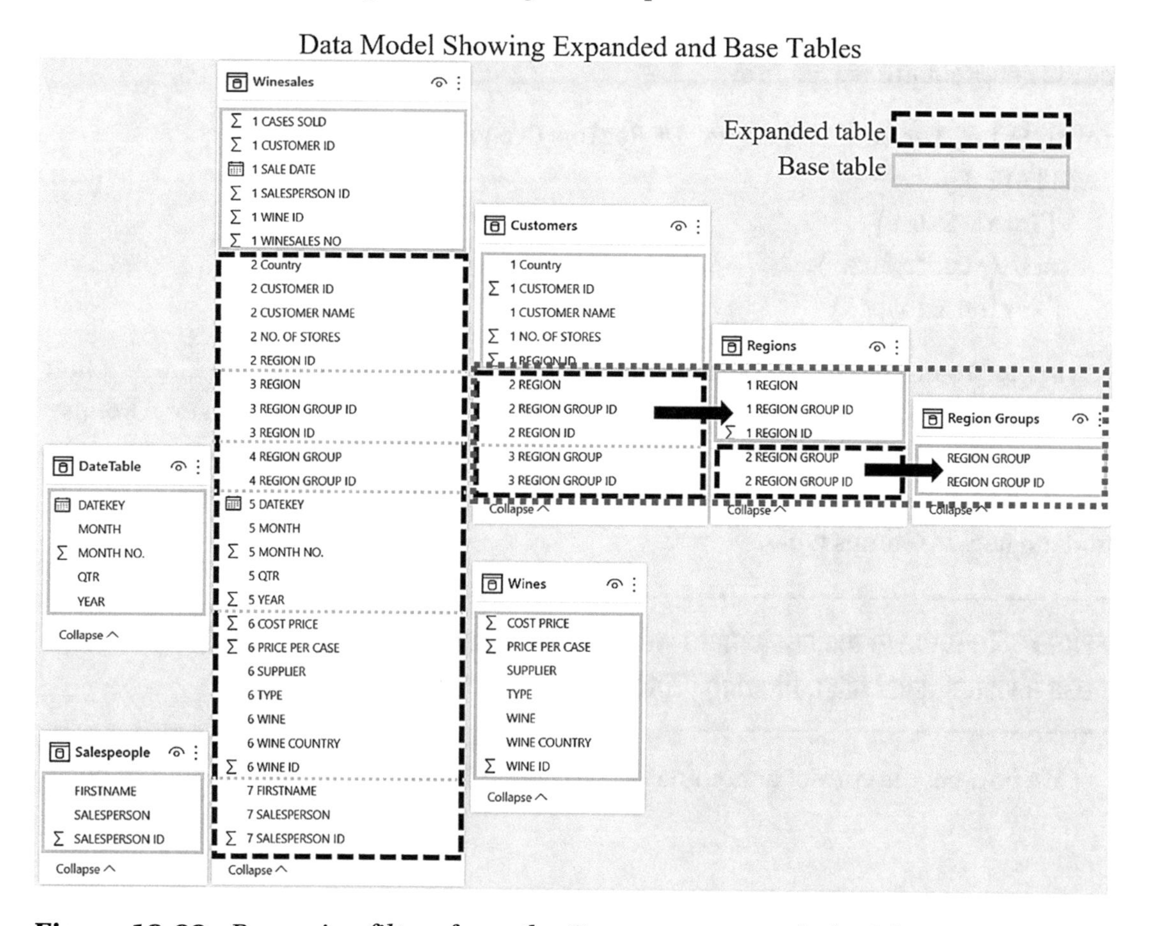

***Figure 18-22.** Removing filters from the Customers expanded table removes filters from Regions and Region Groups*

To calculate the correct denominator, there are several ways to modify the original measure to reapply the filter "lost" on the Customers table. We could, for example, use ALLEXCEPT to remove the filter on the expanded Customers table except for the filter on the REGION GROUP column (because REGION GROUP is contained in the Customers expanded table):

```
Total Sales for All Customers in Region Group #1=
CALCULATE (
    [Total Sales],
    ALLEXCEPT ( Customers, 'Region Groups'[REGION GROUP]))
```

Another approach is to use the filter currently on the Region Groups table that has been generated by the slicer, which currently is “South West”. This measure will also give us the denominator we require:

```
Total Sales for All Customers in Region Group #2 =
CALCULATE (
    [Total Sales],
    ALL ( Customers ),
    'Region Groups')
```

In the “Total Sales for All Customers in Region Group #2” measure, the ALL function removes all the filters from the expanded Customers table, but by using Region Groups as a table filter in the second filter argument in CALCULATE, this reapplies the “South West” filter on the expanded Customers table and therefore also filters the Regions table and the Region Groups table.

Note To remove the customers with no “Total Sales” value from the Table visual, use a visual-level filter, filtering “Total Sales” is not blank.

We now get the correct denominator; see Figure 18-23.

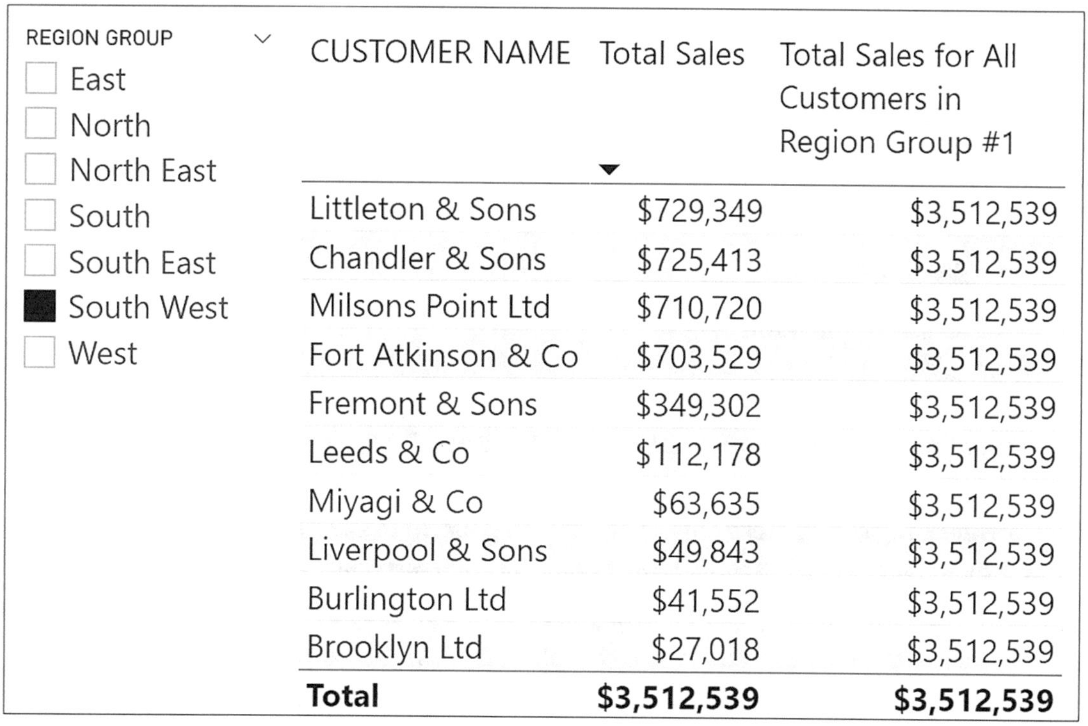

REGION GROUP

- [ ] East
- [ ] North
- [ ] North East
- [ ] South
- [ ] South East
- [x] South West
- [ ] West

CUSTOMER NAME	Total Sales	Total Sales for All Customers in Region Group #1
Littleton & Sons	$729,349	$3,512,539
Chandler & Sons	$725,413	$3,512,539
Milsons Point Ltd	$710,720	$3,512,539
Fort Atkinson & Co	$703,529	$3,512,539
Fremont & Sons	$349,302	$3,512,539
Leeds & Co	$112,178	$3,512,539
Miyagi & Co	$63,635	$3,512,539
Liverpool & Sons	$49,843	$3,512,539
Burlington Ltd	$41,552	$3,512,539
Brooklyn Ltd	$27,018	$3,512,539
Total	**$3,512,539**	**$3,512,539**

Figure 18-23. *You need to reapply "lost" filters when removing filters from expanded tables*

It would also be possible to use this simpler measure using ALLSELECTED.

```
Total Sales for All Customers in Region Group #3 =
CALCULATE ( [Total Sales], ALLSELECTED ( Customers ) )
```

What you are seeing in these examples is the perennial problem with "snowflake"-type schemas. Where you have a chain of tables in many-to-one relationships outward from the fact table, when you remove filters from tables nearer the fact table by using ALL inside CALCULATE, you will also remove all the filters up the chain.

In this chapter, we have delved into the final major concept that underpins DAX, that of table expansion. You have learned that relationships in the data model only serve to generate expanded tables and that filter propagation works by filtering columns inside expanded tables, not by performing lookups from dimensions into the fact table. Knowing about table expansion enables you to author expressions that can use the filter currently placed on the expanded table and therefore pass filters back to dimension tables, in effect reversing the direction of filter propagation.

You are now about to move on to the last chapter in this book. Congratulations on getting this far! It hasn't always been an easy journey, and some DAX expressions we have investigated together would be demanding to any DAX user. However, you now understand the four major concepts that underpin DAX:

- Evaluation context
- Iterators
- Context transition
- Table expansion

According to Alberto Ferrari in his blog "7 reasons DAX is not easy," you are now a DAX guru![2]

However, regarding these concepts, Alberto goes on to say *"The thing is: you need to master them, not only have some basic knowledge of what they are. Moreover, these are foundational concepts: they have nothing to do with specific functions."*

Let this be the best advice. On the completion of this book, you will not be at the end of your journey through learning DAX, but only at the end of the beginning. You must now assimilate your knowledge, work with it, and have the confidence to tackle challenging calculations that will furnish you with the insights into your data that truly inform.

However, you still have one chapter to go. In the next chapter, we will be taking your expert knowledge of DAX to the next level. You will be learning the purpose of the function CALCULATETABLE.

[2] SQLBI.com. 7 reasons DAX is not easy, June 2020. [Online]. Available from `www.sqlbi.com/blog/alberto/2020/06/20/7-reasons-dax-is-not-easy/`

CHAPTER 19

The CALCULATETABLE Function

Now that you are officially a DAX expert, you are ready to confront DAX expressions that will truly test your knowledge and understanding of DAX. One of the DAX functions that can only be understood with a clear grasp of how DAX works is CALCULATETABLE, and this rather obscure function is the last function we will investigate in this book.

CALCULATETABLE operates in all the same ways as CALCULATE except that it returns a table rather than a scalar value. In other words, it returns a table or table expression where the filter on the table has been modified in some way. On the face of it, therefore, CALCULATETABLE should be straightforward to understand. However, because it returns a table, the question that is often asked is the following: How would it be used inside measures? The reason we've left this function till last is because inside measures, it becomes particularly useful when used in conjunction with expanded tables.[1]

The syntax for CALCULATETABLE is

= CALCULATETABLE (table or table expression, filter1, filter2 etc.)

where:

table or table expression is the table you want to be returned by CALCULATETABLE.

filter1, 2 etc. provides the filter for the table returned by **table**.

You may think that this function seems remarkably similar to the FILTER function, and indeed, you can often use CALCULATETABLE in place of FILTER.

[1] To follow along with the examples, use the Power BI Desktop file "6 DAX Expanded Tables.pbix".

© Alison Box 2022
A. Box, *Up and Running with DAX for Power BI*, https://doi.org/10.1007/978-1-4842-8188-8_19

CALCULATETABLE vs. FILTER

However, CALCULATETABLE, unlike FILTER, *modifies* the filter context, and this is the first behavior of this function that we will explore. Let's compare these two measures:

```
Sales of Red Wines Filter =
 CALCULATE ( [Total Sales],
   FILTER ( Wines, Wines[TYPE] = "red" )
)

Sales of Red Wines CalculateTable =
CALCULATE ( [Total Sales],
    CALCULATETABLE ( Wines, Wines[TYPE] = "red" )
)
```

Both these expressions are building a table filter for CALCULATE. The first uses FILTER to build the table containing red wines, and the second uses CALCULATE to build a similar table. These two measures return the same values. However, CALCULATETABLE will modify the filter context. Therefore, if the TYPE column from the Wines dimension is providing the filter context, it will *replace* the filter on TYPE. Therefore, if "White" is the filter, it will be replaced with "Red". FILTER can only filter what's already in the filter context and so returns no value if "White" is the current filter; see Figure 19-1.

SALESPERSON	Sales of Red Wines Filter	Sales of Red Wines CalculateTable
Abel	$2,050,276	$2,050,276
Blanchet	$1,734,279	$1,734,279
Charron	$2,029,616	$2,029,616
Denis	$2,711,085	$2,711,085
Leblanc	$2,232,097	$2,232,097
Reyer	$2,177,254	$2,177,254
Total	**$12,934,607**	**$12,934,607**

TYPE	Sales of Red Wines Filter	Sales of Red Wines CalculateTable
Red	$12,934,607	$12,934,607
White		$12,934,607
Total	**$12,934,607**	**$12,934,607**

Figure 19-1. *CALCULATETABLE will modify the filter context, but FILTER can only filter within the current filter context*

The CALCULATETABLE function, therefore, becomes useful when you must generate an in-memory table where the filter context must be modified. In reality, FILTER and CALCULATETABLE are very different functions even if their output is sometimes the same. The former creates a virtual table by iterating another table within the current filter context. The latter also generates a virtual table but uses a *new filter context* to build the virtual table.

To illustrate this, let's build a measure named "Current No. of Sales" that will calculate the number of sales generated in each region up to the end of the prior month, the year and month being selected in a slicer; see Figure 19-2.

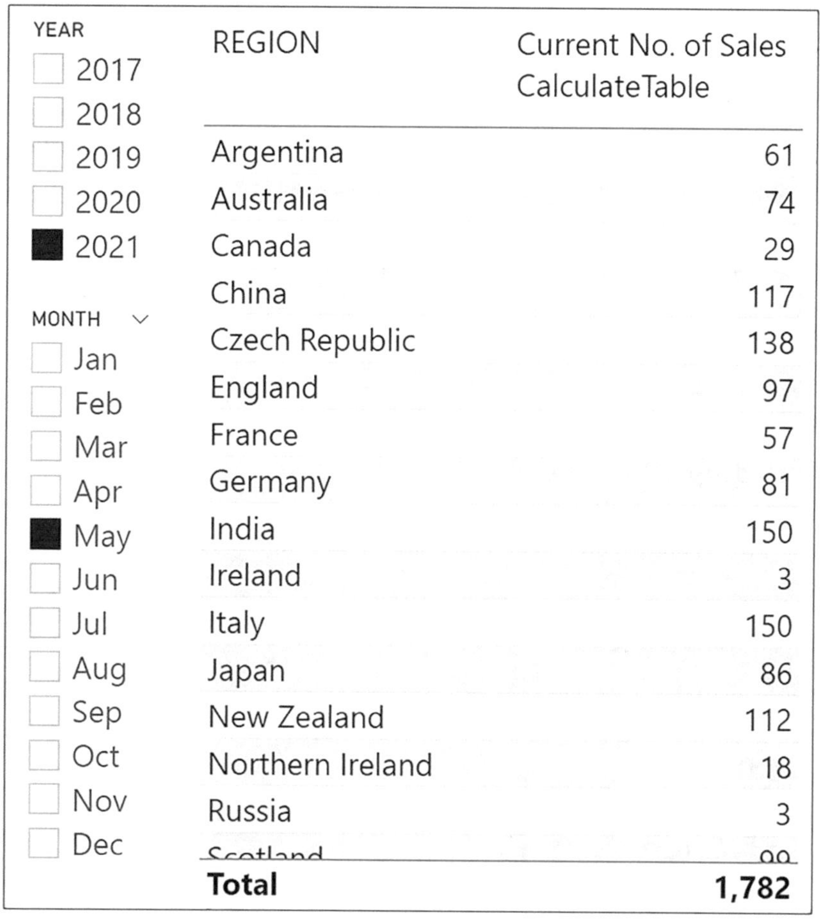

YEAR
- [ ] 2017
- [ ] 2018
- [ ] 2019
- [ ] 2020
- [x] 2021

MONTH
- [ ] Jan
- [ ] Feb
- [ ] Mar
- [ ] Apr
- [x] May
- [ ] Jun
- [ ] Jul
- [ ] Aug
- [ ] Sep
- [ ] Oct
- [ ] Nov
- [ ] Dec

REGION	Current No. of Sales CalculateTable
Argentina	61
Australia	74
Canada	29
China	117
Czech Republic	138
England	97
France	57
Germany	81
India	150
Ireland	3
Italy	150
Japan	86
New Zealand	112
Northern Ireland	18
Russia	3
Scotland	99
Total	**1,782**

Figure 19-2. *Calculating the number of sales up to the end of the prior month*

There are three steps to this calculation:

1. First, we must filter the DateTable for all the dates up to the end of the prior month selected in the slicer e.g., up to but not including the 1st May 2021.

2. The filtered DateTable can then be used to filter the Winesales table to contain only the sales up to the end of the prior month.

3. We can then use COUNTROWS to count how many sales there are in the filtered Winesales table.

The question will be the following: Which filter function are we going to use for step 2 that will generate the DateTable that will filter the Winesales table for the dates we need? Are we going to use FILTER or CALCULATETABLE? We've constructed two versions of the measure, the first using CALCULATETABLE and the second using FILTER (highlighted) where we will then have a second inner FILTER function:

```
Current No. of Sales CalculateTable =
COUNTROWS (
    CALCULATETABLE (
        Winesales,
        FILTER ( ALL ( DateTable ), DateTable[DATEKEY] < MIN (
        DateTable[DATEKEY] ) )
    )
)

Current No. of Sales Filter =
COUNTROWS (
    FILTER (
        Winesales,
        FILTER ( ALL ( DateTable ), DateTable[DATEKEY] < MIN (
        DateTable[DATEKEY] ) )
    )
)
```

The measure using CALCULATETABLE would be the correct measure because if you attempt to use the measure using FILTER, you get an error, as shown in Figure 19-3.

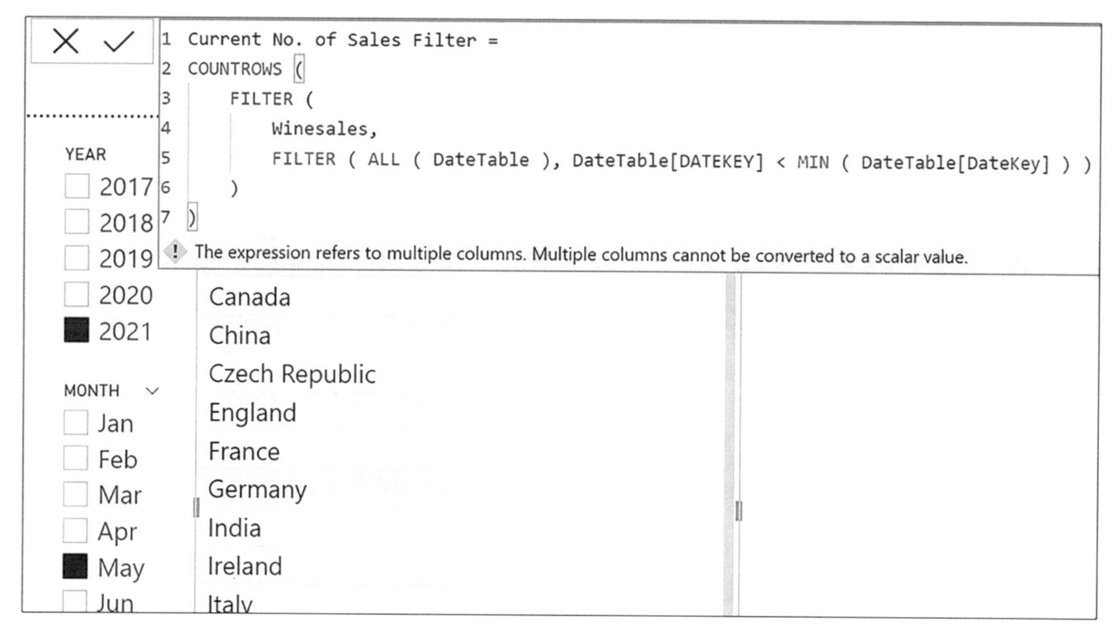

Figure 19-3. *Using FILTER to filter the DateTable returns an error*

If we consider that these functions are interchangeable, why does CALCULATETABLE work, but FILTER does not? To understand the error when using FILTER, we must look more closely at what the inner FILTER expression is generating in memory.

1. The inner FILTER iterates over the DateTable to find all dates up to the end of the prior month.

2. The inner FILTER creates a new virtual DateTable containing just these dates.

3. The outer FILTER then uses the virtual DateTable to filter the rows of the Winesales table, iterating each row in the Winesales table accordingly.

What is the criterion by which each row in the Winesales table will be filtered in step 3? You can't filter a row by values in an entire table, and so we get this error:

"The expression refers to multiple columns. Multiple columns cannot be converted to a scalar value."

The table generated by FILTER is the "multiple columns" alluded to in the error message, and it tells us that if using FILTER, we can only return scalar values for the criterion to filter rows. It's not possible to use a table expression in the filter expression of FILTER, only predicates.

How does the CALCULATETABLE measure differ? In the correct measure:

1. The inner FILTER iterates over the DateTable to find all dates up to the end of the prior month.
2. The inner FILTER then generates a virtual DateTable containing just these dates.
3. CALCULATETABLE generates a virtual Winesales table that can be filtered by the virtual DateTable generated by FILTER. This is simply a table filter and therefore is used in the same way as any table filter that would normally be placed inside CALCULATE.

In other words, the virtual table generated by FILTER provides the *new filter context* for CALCULATETABLE by which the virtual Winesales table can be filtered. The rows of the virtual Winesales table can then be counted.

At this stage of exploring CALCULATETABLE, hopefully, you have worked out that if you want to calculate the "Current No. of Sales", the following measure, using CALCULATE, would be much simpler to write and not return an error:

```
Current No. of Sales =
CALCULATE (
    COUNTROWS ( Winesales ),
    FILTER ( ALL ( DateTable ), DateTable[DATEKEY]
    < MIN ( DateTable[DATEKEY] ) ) )
```

However, we are exploring the difference between CALCULATETABLE and FILTER, and this measure does not illustrate this. But more than this, make a mental note of the expression using CALCULATETABLE as it will be a "building block" in more complex expressions that follow later in this chapter. Here is the expression again that calculates how many sales in each region there have been up to the end of the prior month:

```
Current No. of Sales CalculateTable =
COUNTROWS (
    CALCULATETABLE (
```

```
        Winesales,
        FILTER ( ALL ( DateTable ), DateTable[DATEKEY] < MIN (
        DateTable[DATEKEY] ) )
    )
)
```

Specifically, we will be using this expression as a constituent part of the calculations for "New Regions" and "Returning Regions" later.

We've established that CALCULATETABLE will modify the filter context when generating a virtual table, but we haven't yet found a useful application for this function. Where CALCULATETABLE really comes into its own is when you reference an *expanded* table in this function's filter argument so that it can then be used as a table filter.

CALCULATETABLE and Table Expansion

Just as with CALCULATE, you can use expanded tables as filter expressions to modify the filter context inside CALCULATETABLE. So, for instance, the following table expression

=CALCULATETABLE (Regions, Winesales)

will return a virtual Regions table containing only the rows of this table that are in the current filter in the expanded Winesales table. If we count the rows of the Regions table generated by CALCULATETABLE, this would be an alternative way of finding how many distinct regions we have sales within the current filter context. So these two measures, both using the expanded Winesales table, return the same values:

```
Distinct Regions #1 =
CALCULATE (
    COUNTROWS ( Regions ), Winesales )

Distinct Regions #2=
 COUNTROWS (
    CALCULATETABLE ( Regions, Winesales )
```

By understanding that a table filter inside CALCULATETABLE will use an expanded table where applicable, we can now use this knowledge to resolve more challenging calculations. One of these more challenging calculations is finding "new" and/or "returning" entities, such as new and returning customers or new and returning sales regions.

Calculating "New" Entities

Typically, this would involve discovering how many new customers or new sales regions there are within a specific month, quarter, or year, perhaps further refined by considering only sales for a specific salesperson.

For example, you have been asked to show in how many *new* regions your salespeople have made sales in any given month. You do this by using the following "New Regions" measure that uses CALCULATETABLE:

```
New Regions =
VAR CurrentRegions =
    CALCULATETABLE ( Regions, Winesales )

VAR PreviousRegions =
    CALCULATETABLE (
        Regions,
        CALCULATETABLE ( Winesales,
      FILTER ( ALL ( DateTable ), DateTable[DATEKEY]
    < MIN ( DateTable[DATEKEY] ) ) ) )

RETURN
    COUNTROWS ( EXCEPT ( CurrentRegions, PreviousRegions ) )
```

You can see the result of this measure in the Table visual in Figure 19-4.

Note You could substitute "Customers" for "Regions" if you want to find new customers.

YEAR	MONTH	SALESPERSON	New Regions
2017	Jan	Abel	5
2017	Jan	Blanchet	4
2017	Jan	Charron	2
2017	Jan	Denis	4
2017	Jan	Leblanc	3
2017	Jan	Reyer	3
2017	Feb	Abel	2
2017	Feb	Blanchet	4
2017	Feb	Charron	6
2017	Feb	Denis	3
2017	Feb	Leblanc	2
Total			**20**

Figure 19-4. *Calculating the number of new regions for each salesperson in each month*

We can appreciate that the "New Regions" measure is quite a challenge to understand, so let's separate the three component expressions within the measure as follows:

1. The "CurrentRegions" variable
2. The "PreviousRegions variable
3. The Return statement

By taking the measure apart, piece by piece like this, we can now explain each component.

1. **The "CurrentRegions" variable**

   This variable uses this expression:

   CALCULATETABLE (Regions, Winesales)

   Here, CALCULATETABLE uses the expanded Winesales table as the filter for Regions, therefore generating a Regions table that contains only the regions in which the salesperson (in the current filter context) has made sales *in the current month* (the month in the current filter context). Let's take the evaluation for salesperson "Abel" in "February 2017", which returns **2**, as our example; see Figure 19-5.

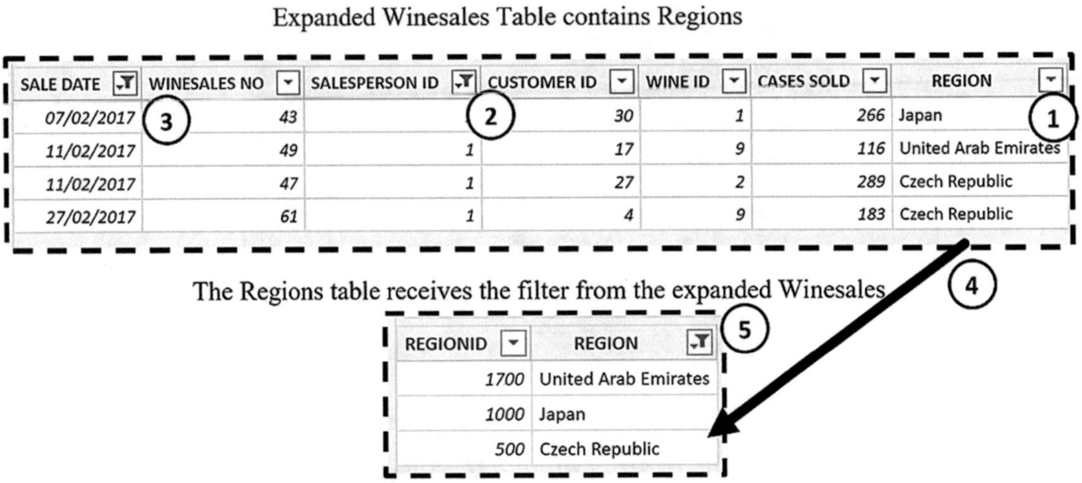

Expanded Winesales Table contains Regions

SALE DATE	WINESALES NO	SALESPERSON ID	CUSTOMER ID	WINE ID	CASES SOLD	REGION
07/02/2017	43		30	1	266	Japan
11/02/2017	49	1	17	9	116	United Arab Emirates
11/02/2017	47	1	27	2	289	Czech Republic
27/02/2017	61	1	4	9	183	Czech Republic

The Regions table receives the filter from the expanded Winesales

REGIONID	REGION
1700	United Arab Emirates
1000	Japan
500	Czech Republic

***Figure 19-5.** The evaluation of the "CurrentRegions" variable for "Abel" in "February 2017". 1. The expanded Winesales table contains columns from the Regions table. 2. The Winesales table is filtered for "Abel". 3. The Winesales table is filtered for "February 2017". 4. The expanded Winesales table is used to filter the Regions table that now only contains regions where "Abel" has made sales in "February 2017." 5. CALCULATETABLE generates a virtual table from the filtered Regions table*

Let's now move on to look at the second component.

2. **The "PreviousRegions" variable**

   This variable uses this expression:

```
CALCULATETABLE ( Regions,
        CALCULATETABLE ( Winesales,
                FILTER ( ALL ( DateTable ), DateTable[DATEKEY]
                        < MIN ( DateTable[DATEKEY] ) ) ) )
```

   Remember that we used the nested CALCULATETABLE expression (highlighted) when we calculated the "Current No. of Sales" measure (see Figure 19-2).

   Here, CALCULATETABLE also uses the expanded Winesales table as the filter for Regions, but this time the expanded Winesales table has been filtered (using FILTER) to contain only sales up to the last date of the prior month. This filter is applied on top of the filters from the SalesPeople table. CALCULATETABLE uses the filtered expanded Winesales table to generate a Regions table that contains the regions in which the salesperson has made sales *up to the end of the prior month*; see Figure 19-6.

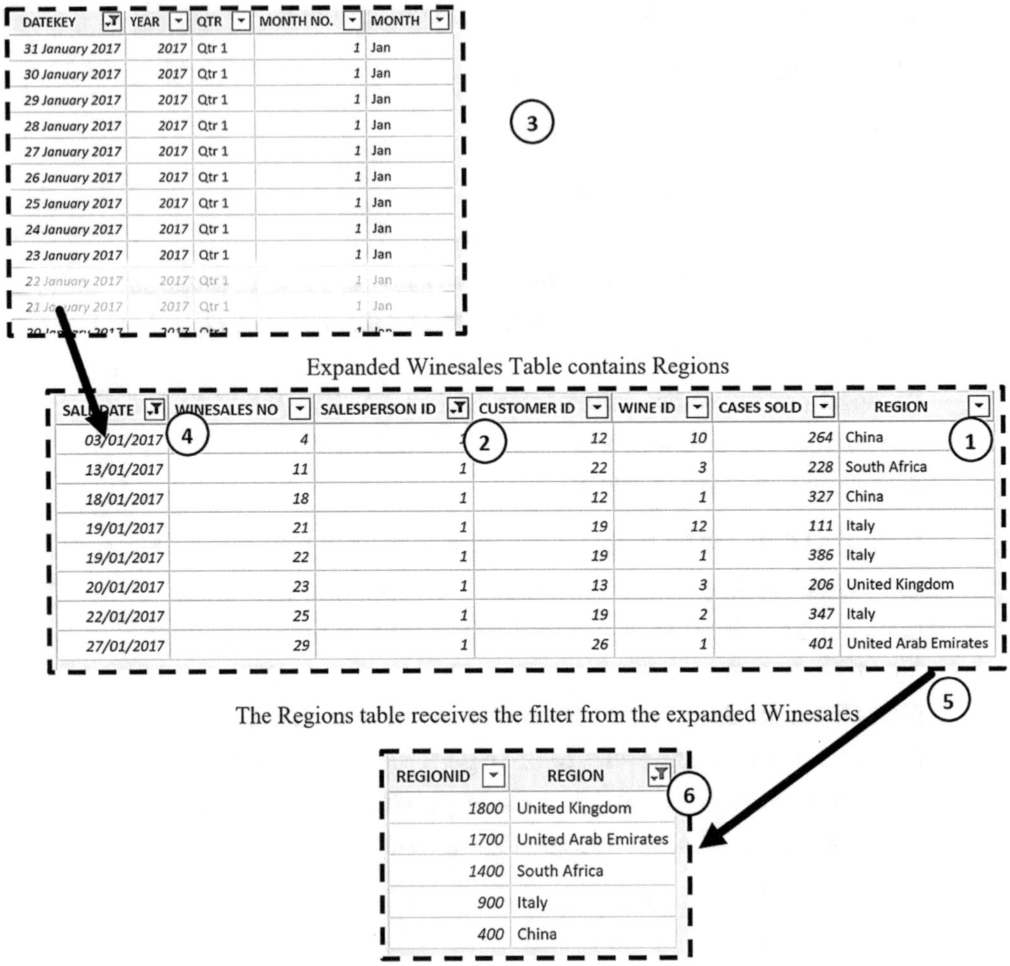

DATEKEY	YEAR	QTR	MONTH NO.	MONTH
31 January 2017	2017	Qtr 1	1	Jan
30 January 2017	2017	Qtr 1	1	Jan
29 January 2017	2017	Qtr 1	1	Jan
28 January 2017	2017	Qtr 1	1	Jan
27 January 2017	2017	Qtr 1	1	Jan
26 January 2017	2017	Qtr 1	1	Jan
25 January 2017	2017	Qtr 1	1	Jan
24 January 2017	2017	Qtr 1	1	Jan
23 January 2017	2017	Qtr 1	1	Jan
22 January 2017	2017	Qtr 1	1	Jan
21 January 2017	2017	Qtr 1	1	Jan

3

Expanded Winesales Table contains Regions

SALEDATE	WINESALES NO	SALESPERSON ID	CUSTOMER ID	WINE ID	CASES SOLD	REGION
03/01/2017	4	1	12	10	264	China
13/01/2017	11	1	22	3	228	South Africa
18/01/2017	18	1	12	1	327	China
19/01/2017	21	1	19	12	111	Italy
19/01/2017	22	1	19	1	386	Italy
20/01/2017	23	1	13	3	206	United Kingdom
22/01/2017	25	1	19	2	347	Italy
27/01/2017	29	1	26	1	401	United Arab Emirates

4 2 1

The Regions table receives the filter from the expanded Winesales 5

REGIONID	REGION
1800	United Kingdom
1700	United Arab Emirates
1400	South Africa
900	Italy
400	China

6

***Figure 19-6.** The evaluation of the "PreviousRegions" variable for "Abel" in "February 2017". 1. The expanded Winesales table contains columns from the Regions table. 2. The Winesales table is filtered for "Abel". 3. FILTER inside CALCULATETABLE generates a filtered DateTable containing dates up to and including "31 January 2017". 4. The DateTable table generated by FILTER is used to filter the expanded Winesales table. 5. The expanded Winesales table is used to filter the Regions table that now only contains regions where "Abel" has made sales up to "31 January 2017". 6. CALCULATETABLE generates a virtual table from the filtered Regions table*

So the variables have generated two in-memory tables using CALCULATETABLE as follows:

CurrentRegions – Holds the regions in which the salesperson has made sales in the month in the current filter context

PreviousRegions – Holds the regions in which the salesperson has made sales up to the last date of the prior month in the current filter context

3. **The RETURN statement**

   The RETURN statement uses the EXCEPT function to return a table that contains only the rows of the table in the first argument that are not in the table of the second argument; see Figure 19-7.

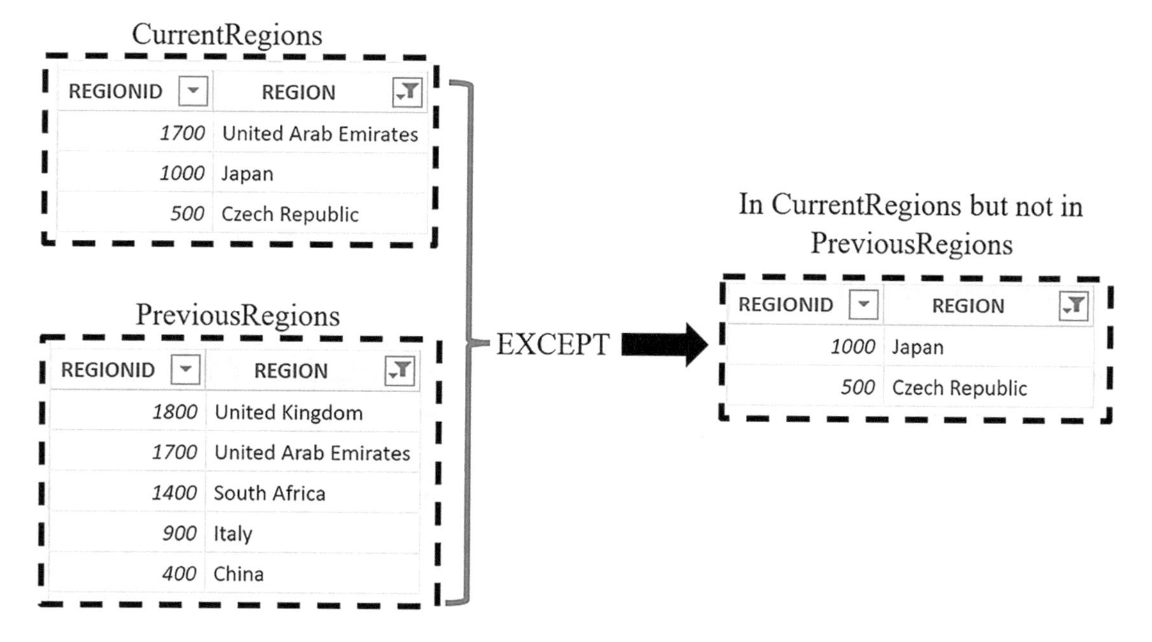

Figure 19-7. *The EXCEPT function returns all the rows in the first table that are not in the second table*

COUNTROWS then counts the rows in the virtual table generated by EXCEPT and returns **2** rows for "Abel" in "February 2017".

Calculating "Returning" Entities

To find "Returning Regions", which is the regions where salespeople have previously made sales in any month, we can use the function INTERSECT in place of EXCEPT as follows:

```
Returning Regions =
VAR CurrentRegions =
    CALCULATETABLE ( Regions, Winesales )

VAR PreviousRegions =
    CALCULATETABLE (
        Regions,
        CALCULATETABLE ( Winesales,
      FILTER ( ALL ( DateTable ), DateTable[DATEKEY]
    < MIN ( DateTable[DATEKEY] ) ) ) )

RETURN
    COUNTROWS ( INTERSECT ( CurrentRegions, PreviousRegions ) )
```

The table generated by INTERSECT contains all the rows in the first table that are also in the second table.

You can see the output of the two measures in Figure 19-8. Note how the Total row has been removed from the Table visual. The Total values calculated (or the absence of a value) are correct, but ambiguous. Remember that on the evaluation of the Total row, filters on YEAR, MONTH, and SALESPERSON will have been removed so the "New Regions" measure, for example, would calculate how many new regions there were for all salespeople in all months of all years, which is the same as the total number of regions.

YEAR	MONTH	SALESPERSON	New Regions	Returning Regions
2017	Jan	Abel	5	
2017	Jan	Blanchet	4	
2017	Jan	Charron	2	
2017	Jan	Denis	4	
2017	Jan	Leblanc	3	
2017	Jan	Reyer	3	
2017	Feb	Abel	2	1
2017	Feb	Blanchet	4	
2017	Feb	Charron	6	
2017	Feb	Denis	3	2
2017	Feb	Leblanc	2	1
2017	Feb	Reyer	2	1
2017	Mar	Abel	1	

Figure 19-8. *The "New Regions" and "Returning Regions" measures in a Table visual. Note the absence of the Total row*

What lies at the root of these expressions is using CALCULATETABLE to create two sets of data for comparisons. You can then use EXCEPT and INTERSECT to find either values that are the same or values that are different respectively; see Figure 19-9.

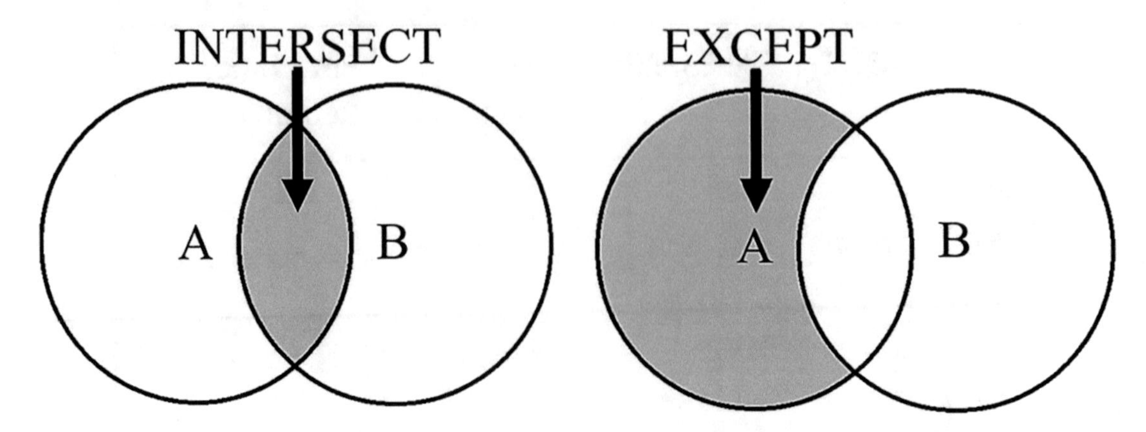

***Figure 19-9.** You can use the INTERSECT and EXCEPT functions to return sets of values*

The benefit of understanding these expressions using CALCULATETABLE is that they can be repurposed for many different scenarios. For example, rather than finding new regions in the current month for each salesperson, you could analyze the number of new customers there are for each wine compared to the previous month, as follows:

```
New Customers from Previous Month =
VAR CurrentCustomers =
    CALCULATETABLE ( Customers, Winesales )
VAR PreviousMthsCustomers =
    CALCULATETABLE (
        Customers,
        CALCULATETABLE ( Winesales,
      PREVIOUSMONTH(DateTable[DATEKEY])) )

RETURN
  COUNTROWS ( EXCEPT ( CurrentCustomers, PreviousMthsCustomers ) )
```

This measure tells us that for "Bordeaux" wine, in "December 2021", there were **4** new customers compared to the customers in "November 2021"; see Figure 19-10. Again, note the evaluation of the Total row, which, although not summing the values for each wine, is correct because it tells us that there were 13 new customers for *all wines* in "December 2021" compared to the previous month. Renaming the Total may make this value less ambiguous.

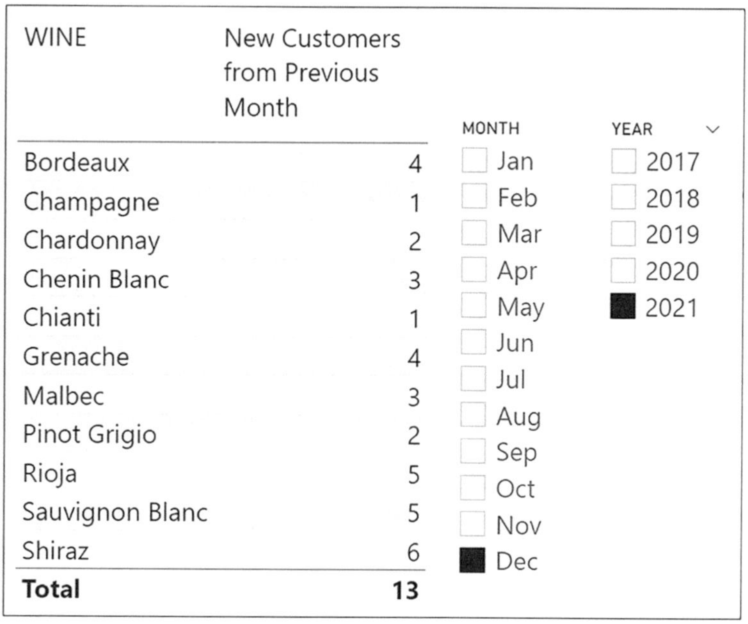

WINE	New Customers from Previous Month
Bordeaux	4
Champagne	1
Chardonnay	2
Chenin Blanc	3
Chianti	1
Grenache	4
Malbec	3
Pinot Grigio	2
Rioja	5
Sauvignon Blanc	5
Shiraz	6
Total	**13**

MONTH	YEAR
☐ Jan	☐ 2017
☐ Feb	☐ 2018
☐ Mar	☐ 2019
☐ Apr	☐ 2020
☐ May	■ 2021
☐ Jun	
☐ Jul	
☐ Aug	
☐ Sep	
☐ Oct	
☐ Nov	
■ Dec	

***Figure 19-10.** The evaluation of the "New Customers from Previous Month" measure, repurposing the "New Regions" measure*

I think you'll agree that the DAX expressions you've authored in this chapter using CALCULATETABLE and expanded tables bear no comparison in their complexity to the simple measures using SUM and AVERAGE with which you started out. Throughout this book, we've paid particular attention to how the DAX expressions work, understanding the detail beneath and getting to grips with the difficult concepts that underpin the DAX language.

Now all that remains is for you to put your newfound knowledge to good use. Spend your day finding data to analyze using DAX. Don't give up if at first things don't go your way. Persevere and keep with it. There is no silver bullet; everyone who has ever mastered DAX has worked hard to get where they are.

But nothing replaces that glowing feeling of finding a solution to a calculation that you initially thought was impossible to solve.

Happy DAXing!

Index

A

Active/inactive relationships
  comparison dimension
    data table, 224
    edit relationship dialog, 222, 223
    fact table, 222
    measures, 225
    slicers, 221, 223
    USERELATIONSHIP function, 226
    wines dimension, 224
  DateTable filters, 219
  measures, 218
  table relationship, 218
  USERELATIONSHIP function, 219
ALLEXCEPT function, 127, 128
ALL function
  aggregating totals, 259–265
  table expansion, 318
ALL function/variations
  ALLEXCEPT function, 127, 128
  ALLSELECTED function, 129–131
  CALCULATE, 115, 133–143
  calculating percentages, 123
  data modifier, 110
  description, 109
  dimensions, 113
  dimension table, 117–119
  fact tables, 111–117
  grand total cases, 116
  multiple columns, 126
  SUPPLIER column, 120–126
  syntaxes, 110
  tables/remove filters, 111
  TYPE column, 125
ALLSELECTED function, 129–131

B

Bidirectional relationships, 214
BLANK() function, 165–167

C

Calculated columns
  AND (&), 25
  context transition, 293–296
  Excel formulas, 23
  expression, 24
  fields list, 24
  OR (||), 25
  RELATED function (*see* RELATED function)
CALCULATE function, 71, 133
  AND/OR filters, 79–81
  complex filters, 81–84
  DateTable dimension, 71–73
  details, 136
  DIVIDE function, 77
  error message, 82
  evaluation, 137, 138, 140
  expressions, 134
  FILTER function, 139
  filters, 74, 75
  in-memory virtual tables, 84
  measures, 80

© Alison Box 2022
A. Box, *Up and Running with DAX for Power BI*, https://doi.org/10.1007/978-1-4842-8188-8

CALCULATE function (*cont.*)
  modification, 136
  multiple filter, 78
  OR expression, 83
  return different results, 132
  similar expressions, 135
  single filters, 77, 78
  square bracket, 76
  SUPPLIER column, 131
  syntax, 75
  table function, 132
  total cases, 73, 74
  virtual table, 84
  WINE column, 133
  Wines dimension, 132
CALCULATETABLE function
  FILTER function
    calculation, 346
    DateTable returns, 347
    error message, 348
    expression, 347
    filter context, 344
    modification, 344
    virtual table, 345
  syntax, 343
  table expansion
    component expressions, 351, 353
    entities, 350–355
    evaluation, 352, 354, 355
    EXCEPT function, 355
    expression, 349
    INTERSECT/EXCEPT expressions, 356–359
COALESCE function, 171, 172
CONCATENATEX function
  arguments, 190
  COUNTROWS function, 193
  multiple selections, 192
  parameter tables, 195–198
  problem scenarios, 192, 193
  SELECTEDVALUE function, 195
  TOPN expression, 194
  value returns, 190
  VALUES function, 194
  WINE column, 191
Constants, 178–181
Context transition
  aggregated values, 227
  aggregating totals
    ALL function, 259–265
    dimensions, 252–258
    matrix visuals, 262
    row-level calculation, 253
    SUMMARIZE function, 265–270
    virtual tables, 259
  attributes, 229, 230
  calculated columns, 231, 234, 277, 278, 293
    CUMULATIVE TOTAL, 294
    cumulative totals, 293
    DAYS DIFFERENCE, 295
    FILTER function, 293
    RETURN statement, 295
    row context, 294
  CALCULATE function, 232
  data analysis, 271
  definition, 229
  description, 227
  expressions, 228, 230, 234
  filter context, 229
  numeric ranges, 275–279
  parameter table, 277
  RANKX function, 272–275
  row context, 228
  SUMMARIZE, 285–292
  surprising results

AVERAGE function, 238–242
calculated columns, 237
cumulative totals, 243
MAX function, 242–247
measures, 247–251
SUMX expression, 249
total cases, 251
TopN percent
analysis, 280
dynamic ranking, 285
parameter tables, 281
requirements, 279
row selection, 283, 284
slicers, 280
top/bottom percent selection, 280, 281
Wines dimension, 235, 236
COUNTROWS function, 278
CROSSFILTER function, 212, 213, 216, 311
table expansion, 333–336
Current filter context, 183

D

Data analysis expression (DAX)
concepts, 1
data model, 3
expression, 1
formula bar, 16
functions, 86–88
many-to-one relationships, 3
table tools tab, 15
Denormalization, 28

E

Empty values *vs.* Zero, 165
BLANK() function, 165–167
COALESCE function, 171, 172
ISBLANK function, 168
measures, 169, 170
testing, 169, 170
Excel formulas, calculated columns, 23

F, G

Filter context
evaluation, 47
factors, 48
fact table, 50
in-memory dimension, 49
measures, 47
multiple filters, 53–56
fact table, 55
in-memory tables, 54
propagation, 54
roles, 56
slicer/page-level filter, 53
propagation, 50
total row, 52
WINE column, 51
FILTER function, 278
CALCULATE
AVERAGE/MAX, 99
calculation, 93
error messages, 97
incorrect results, 92
iterators, 94
measure, 92
profit wines, 97
propagate filters, 94
requirement, 91
requirements, 97
scenario, 96
source code, 98
step-by-step guide, 93, 95

FILTER function (*cont.*)
    COUNTROWS, 90
    row expression, 90, 91
    syntax, 89
Filter propagation
    bidirectional filters, 210–213
    cross filter, 214–216
    CROSSFILTER function, 210, 212
    customers table, 210
    fact tables, 209
    model view, 209
Formatting/unformatting expression, 19, 20

H

HASONEVALUE function, 274

I, J

Implicit measures, 32–34
ISBLANK function, 168
Iterators, 59
    aggregating iterators, 59
    SUMX function (*see* SUMX function)
    total row
        constituent expressions, 68
        evaluation, 68
        incorrect result, 66, 68
        SUM function, 69

K

KEEPFILTERS function, 108

L

LOOKUPVALUE function
    approaches, 301
    calculated column, 300, 301
    definition, 302
    many-to-one relationship, 299
    syntax, 300
    table records, 298
    transaction records, 299

M

Measures, 32
    benefits, 38
    CALCULATE function, 80
    COUNTROWS function, 38
    definition, 41, 42
    dimensions, 38, 39
    DISTINCTCOUNT function, 39, 40
    editor, 36
    error message, 45
    Excel pivot table, 44
    explicit measure, 34
    filtered data, 41
    formatting group, 37
    implicit measure, 32–34
    reporting tools, 42
    return scalar values, 42–46
    table creation, 35
    visuals, 42
    visual table, 36, 37

N, O

Non matching values
    blank entry, 12, 13
    linking columns, 8, 9
    scenarios, 7
    visualisations pane, 9, 10
    Wines dimension, 11

P, Q

Power BI Desktop, 2

R

RANKX function, 272–275
RELATED function, 26
  advantages, 28
  CUSTOMER ID, 26
  denormalization, 28
  description, 28
  hiding tables, 29
  regions table, 27, 28
  Sales revenue values, 30–33
  VLOOKUP function, 27
  Winesales table, 26
Row context, 56, 57

S

SELECTEDVALUE function, 276, 281
  current filter, 188, 189
  evaluation, 185
  expression, 187
  multiple items, 187
  parameter table, 184
  results, 188
  slicer filters, 186
  syntax, 184
  user selections, 186
  VALUES function, 204
Start/snowflakes
  data model, 4
  dimensions, 5
  fact tables, 4
  inactive relationship, 7
  star schema, 5
SUMMARIZE function
  calculation steps, 287
  context transition
    ALLEXEPT function, 269
    analytics, 268
    calculated column, 269
    calculated tables, 266
    DateTable, 266
    description, 265
    measures, 270
    syntax, 265
    table creation, 267
  like/like sales calculation, 287
  matrix visual, 289, 292
  measure returns, 290
  multiselection, 286
  virtual table, 288
SUMX function
  average price, 66
  calculated column, 61, 62
  implicit measure, 62
  maximum/average sales, 64
  RELATED function, 63
  row-level calculation, 60
  SUMX, AVERAGEX, and
    MAXX, 65
  syntax, 61
  X aggregators, 61
Syntax
  AND/OR expression, 19
  column name, 17
  Excel formulas, 16
  Excel formulas/DAX expressions, 18
  formula bar, 16
  IntelliSense list, 17

T

Table expansion
  ALLEXCEPT/SUMMARIZE
    functions, 323
  ALL function, 318, 319
  ALLSELECTED, 341

Table expansion (*cont.*)
  base tables, 319
  CROSSFILTER function, 333–336
  data model, 312, 320, 321
  description, 311
  expanded tables, 319, 320
  expansion results, 318–323
  filters
    code generation, 316
    column information, 313–317
    description, 313
    real evaluation, 315
    WINECOUNTRY column, 316
  leverage tables
    approaches, 323
    base/expanded tables, 327, 329, 332
    CALCULATE, 325, 326
    CROSSFILTER function, 333
    customers table, 324
    distinct regions, 326
    measures, 331
    RELATED function, 325
    snowflake dimensions, 324
    table visuals, 328
  RELATED function, 322
  removes filters, 339
  snowflake schemas, 337–342
Table functions
  column filters
    Bordeaux, 107
    CALCULATE, 106
    description, 99
    difference, 100
    difference results, 106
    less efficient, 100–104
    measures return, 101
    table filters, 104–108
    technical terms, 104
  FILTER (*see* FILTER function)
  KEEPFILTERS, 108
  scalar functions, 87
  table expressions
    CALCULATE, 89
    calculated tables, 87, 88
    filter arguments, 89
    VALUES, 88
  types, 86–88
Tables, *see* Active/inactive relationships
Time intelligence functions
  annual totals and averages, 156–158
  base date, 149, 150
  consecutive transactions, 160
  cumulative totals, 155, 156
  DATEADD, 154
  date dimension
    built-in option, 145
    column option, 149
    DATEKEY column, 147
    hierarchies, 144–146
    options pane, 145
    table creation, 146–149
    table tools tab, 147
  DATESINPERIOD function, 157
  DATESYTD function, 154
  description, 143
  differences (dates), 162, 163
  FIRSTNONBLANK/LASTNONBLANK
      functions, 158–161
  LASTDATE option, 152
  LASTNONBLANK/
      LASTNONBLANKVALUE
      expressions, 161
  PREVIOUSMONTH/YEAR, 153
  return value, 151
  SAMEPERIODLASTYEAR, 153
  scalar value, 152

unique dates, 150
VALUES function, 158
TREATAS function
DateTable/Targets table, 305
evaluation, 309
fact table, 306
many-to-many relationship, 307
reporting targets, 303
results, 310
SalesPeople table, 304
syntax, 307
targets table, 302, 303
VALUES function, 308

U

USERELATIONSHIP function, 219

V, W, X, Y, Z

VALUES function
converting columns, 207, 208
edit interactions, 205
error message, 202
lost filters, 205–207
references, 199
requirements, 199
scalar function, 201
table function/
virtual table, 199
table/scalar function, 200–204
total row selection, 202
Variables
advantages, 173
calculated columns, 174
constants, 178–181
declarations, 174
nested measures/
expressions, 176
performance, 174–176
readability, 176, 177
reduce complexity, 177, 178
VAR/RETURN keywords, 173
Virtual relationships, 297
VLOOKUP function, 298

Printed in the United States
by Baker & Taylor Publisher Services